Advances in
ATOMIC, MOLECULAR, AND OPTICAL PHYSICS

VOLUME 40

Editors

BENJAMIN BEDERSON
New York University
New York, New York

HERBERT WALTHER
Max-Planck-Institut für Quantenoptik
Garching bei Munchen
Germany

Editorial Board

P. R. BERMAN
University of Michigan
Ann Arbor, Michigan

M. GAVRILA
F.O.M. Instituut voor Atoom-en Molecuulfysica
Amsterdam
The Netherlands

M. INOKUTI
Argonne National Laboratory
Argonne, Illinois

W. D. PHILLIPS
National Institute for Standards and Technology
Gaithersburg, Maryland

Founding Editor

SIR DAVID R. BATES

Supplements

1. *Atoms in Intense Laser Fields,* Mihai Gavrila, Ed.
2. *Cavity Quantum Electrodynamics,* Paul R. Berman, Ed.
3. *Cross Section Data,* Mitio Inokuti, Ed.

ADVANCES IN

ATOMIC, MOLECULAR, AND OPTICAL PHYSICS

Edited by

Benjamin Bederson
DEPARTMENT OF PHYSICS
NEW YORK UNIVERSITY
NEW YORK, NEW YORK

Herbert Walther
UNIVERSITY OF MUNICH AND
MAX-PLANK-INSTITUT FÜR QUANTENOPTIK
MUNICH, GERMANY

Volume 40

ACADEMIC PRESS

San Diego   London   Boston
New York   Sydney   Tokyo   Toronto

This book is printed on acid-free paper.

Copyright © 1999 by Academic Press

All rights reserved.
No part of this publication may be reproduced or transmitted in any form or by any means, electronic or mechanical, including photocopy, recording, or any information storage and retrieval system, without permission in writing from the publisher.

The appearance of code at the bottom of the first page of a chapter in this book indicates the Publisher's consent that copies of the chapter may be made for personal or internal use of specific clients. This consent is given on the condition, however, that the copier pay the stated per-copy fee through the Copyright Clearance Center, Inc. (222 Rosewood Drive, Danvers, Massachusetts 01923), for copying beyond that permitted by Sections 107 or 108 of the U.S. Copyright Law. This consent does not extend to other kinds of copying, such as copying for general distribution, for advertising or promotional purposes, for creating new collective works, or for resale. Copy fees for chapters are as shown on the title pages; if no fee code appears on the chapter title page, the copy fee is the same as for current chapters. 1049-250X/99 $30.00

Academic Press
525 B Street, Suite 1900, San Diego, CA 92101-4495, USA
http://www.apnet.com

Academic Press
24–28 Oval Road, London NW1 7DX, UK
http://www.hbuk.co.uk./ap/

International Standard Serial Number: 1049-250X
International Standard Book Number: 0-12-003840-4

Printed in the United States of America
98 99 00 01 02 MV 9 8 7 6 5 4 3 2 1

Contents

CONTRIBUTORS vii

Electric Dipole Moments of Leptons
Eugene D. Commins

I. Introduction	1
II. Theoretical Models of Lepton EDMs	6
III. The Electron EDM	14
IV. The Muon EDM	36
V. Weak Dipole Moment and EDM of the Tau Lepton	39
VI. Magnetic and Electric Dipole Moments of Neutrinos	44
VII. Acknowledgments	48
VIII. References	49

High-Precision Calculations for the Ground and Excited States of the Lithium Atom
Frederick W. King

I. Introduction	57
II. Computational Approaches	58
III. Some Mathematical Issues	63
IV. Nonrelativistic Energies	67
V. Relativistic Corrections to the Energies	80
VI. Specific Mass Shift Correction to the Energy Levels	83
VII. Quantum Electrodynamic Corrections	88
VIII. The First Ionization Potential	90
IX. Hyperfine Coupling Constants	93
X. Other Properties	100
XI. Outlook	103
XII. Acknowledgments	105
XIII. References	105

Storage Ring Laser Spectroscopy
Thomas U. Kühl

I. Introduction	114
II. Properties of Existing Heavy-Ion Storage Rings	115
III. Kinematic Effects in Storage Rings	117
IV. Laser Experiments in the Electron Cooler	122
V. Laser Cooling	131

VI. A Test of Special Relativity in the Storage Ring	142
VII. Quantum Electrodynamics in Strong Fields Probed by Laser Spectroscopy in Highly Charged Ions	146
VIII. Conclusion and Outlook	155
IX. Acknowledgments	156
X. References	157

Laser Cooling of Solids

Carl E. Mungan and Timothy R. Gosnell

I. Introduction	161
II. Historical Review of the Thermodynamics of Fluorescence Cooling	165
III. Working Substances for Fluorescence Cooling	175
IV. Laser Cooling of Ytterbium-Doped ZBLANP Glass	188
V. Fundamental Limits	204
VI. Conclusions and Prospects	221
VII. Acknowledgments	223
VIII. Notes	224
IX. References	224

Optical Pattern Formation

L. A. Lugiato, M. Brambilla, and A. Gatti

I. Introduction	229
II. General Features About OPF	233
III. OPF and Solitary Structures in Cavities	269
IV. Quantum Fluctuations and Optical Pattern Formation	278
V. Acknowledgments	295
VI. References	295
SUBJECT INDEX	307
CONTENTS OF VOLUMES IN THIS SERIES	313

Contributors

Numbers in parentheses indicate pages on which the author's contributions begin.

EUGENE D. COMMINS (1), Physics Department, University of California at Berkeley, 366 Le Conte Hall # 7300, Berkeley, California 94720-7300

FREDERICK W. KING (57), Department of Chemistry, University of Wisconsin–Eau Claire, Eau Claire, Wisconsin 54702-4004

THOMAS U. KÜHL (113), Gesellschaft für Schwerionenforschung, Planckstr, 1, D-64291 Darmstadt, Germany and Johannes-Gutenberg-Universität, Mainz, Germany. Presently at University of California, Lawrence Livermore National Laboratory, Livermore, California

CARL E. MUNGAN (161), Department of Physics, The University of West Florida, Pensacola, Florida 32514-5751

TIMOTHY R. GOSNELL (161), Condensed Matter and Thermal Physics Group, Mail Stop E543, Los Alamos National Laboratory, Los Alamos, New Mexico 87545

L. A. LUGIATO (229), Istituto Nazionale di Fisica per la Materia, Università di Milano, Via Celoria 16, 20133 Milano, Italy

M. BRAMBILLA (229), Istituto Nazionale di Fisica per la Materia, Dipartimento Interateneo di Fisica del Politecnico di Bari, Via E. Oraboua 4, 7016 Bari, Italy

A. GATTI (229), Istituto Nazionale di Fisica per la Materia, Università di Milano, Via Celoria 16, 20133 Milano, Italy

Advances in
ATOMIC, MOLECULAR, AND OPTICAL PHYSICS

VOLUME 40

ELECTRIC DIPOLE MOMENTS OF LEPTONS

EUGENE D. COMMINS
Physics Department, University of California at Berkeley,
Berkeley, California

I. Introduction	1
II. Theoretical Models of Lepton EDMs	6
A. Proper Lorentz-Invariant EDM Interaction	6
B. The Standard Model	7
C. Supersymmetric Models	10
D. Multi-Higgs Models	12
E. Left-Right Symmetric Models	12
III. The Electron EDM	14
A. Schiff's Theorem	14
B. Calculation of the Enhancement Factor for Paramagnetic Atoms	15
C. P,T-Odd Electron–Nucleon Interaction	21
D. Atomic EDM Effects Caused by the Nuclear Magnetic Moment	24
E. Experimental Searches for the Electron EDM	25
F. EDMs of Paramagnetic Molecules	32
IV. The Muon EDM	36
A. Muonic Atoms Are Not Practical for an EDM Search	36
B. Precession of Relativistic Muons in a Storage Ring	37
V. Weak Dipole Moment and EDM of the Tau Lepton	39
VI. Magnetic and Electric Dipole Moments of Neutrinos	44
VII. Acknowledgments	48
VIII. References	49

I. Introduction

There is no experimental evidence that any elementary particle, nucleus, or atom possesses a permanent electric dipole moment (EDM). Nevertheless, experimental searches for EDMs and theoretical investigations of their possible magnitudes have attracted great interest for many years, and continue to do so. The observation of a nonzero EDM would directly reveal the violation of time reversal (T) invariance as well as of parity (P); the possible existence of EDMs is intimately connected to these fundamental symmetries. Experimental search for the neutron EDM began almost 50 years ago (Purcell and Ramsey, 1950; Smith *et al.*, 1957), and continues with refined methods in the present era (Smith *et al.*, 1990; Ramsey, 1990; Altarev *et al.*, 1996). In this review, we concentrate on the electric dipole moments of the electron, muon, tau lepton, and their associated neutrinos. Needless to say, among these leptons the electron is most accessible to experimental observation. For this reason, and also because the methods employed for it are those of low-energy physics—atomic and molecular beams, magnetic resonance, optical pumping, laser spectroscopy, and so on—it is appropriate that in this review we devote most of our attention to the electron.

To begin our discussion, let us show that a particle cannot possess an EDM unless both P and T are violated (Landau, 1957). Consider a particle of spin 1/2, and assume that it has an EDM **d**, as well as a spin magnetic dipole moment $\boldsymbol{\mu}$. Both moments must lie along the spin direction because the spin is the only vector available to orient the particle. We write the Hamiltonians H_M, H_E that describe the interaction of $\boldsymbol{\mu}$ with a magnetic field **B**, and of **d** with an electric field **E**, in the nonrelativistic limit:

$$H_M = -\boldsymbol{\mu} \cdot \mathbf{B} = -\mu\boldsymbol{\sigma} \cdot \mathbf{B} \tag{1}$$

$$H_E = -\mathbf{d} \cdot \mathbf{E} = -d\boldsymbol{\sigma} \cdot \mathbf{E} \tag{2}$$

where $\boldsymbol{\sigma}$ is the Pauli spin operator. Under a space inversion or parity (P) transformation, the axial vectors $\boldsymbol{\sigma}$ and **B** remain unchanged, but the polar vector **E** changes sign. Hence H_M is invariant under P, but H_E is not. Furthermore, under a time reversal (T) transformation, **E** remains invariant, but the angular momentum-like $\boldsymbol{\sigma}$ changes sign, as does **B**, which is generated by electric currents. Hence, whereas H_M is invariant under parity and time reversal transformations, H_E changes sign under each.

Now, P is violated maximally in charged weak interactions, and so is charge conjugation (C) symmetry, whereas P and C are also violated in neutral weak interactions. Moreover the combined symmetry CP is violated in the decays of neutral K mesons, and it has been shown that this CP violation is equivalent to T-violation (as is demanded by CPT invariance). Hence the existence of a P,T-violating EDM appears quite possible, because CP violation and the weak interaction could induce an EDM by means of radiative corrections to the P,C,T-conserving electromagnetic interaction.

Although the parameters that describe CP violation in kaon decay have been measured ever more precisely in the last three decades (Steinberger, 1990; Kleinknecht, 1990; Wolfenstein, 1986, 1991, 1996), the fundamental explanation for this phenomenon remains very much in doubt, and a wide variety of plausible theoretical models of CP violation have been presented. According to the standard model, the electron EDM d_e and the neutron EDM d_n are both too small by many orders of magnitude to be detected by any experiment now or in the foreseeable future. However, in various well-motivated extensions of the standard model, d_e and/or d_n are sufficiently large to be detected by practical experiments. In some models, even the muon EDM d_μ and the tau EDM d_τ are large enough that one might hope to observe them, although less sensitive experimental methods are available for them than for d_n or d_e.

To detect an electric dipole moment it is usually necessary to place the particle of interest in an external electric field E and observe the change in energy that is proportional to E (a linear Stark effect). The *quadratic* Stark effect, which arises because of an *induced* electric dipole moment and gives an energy shift proportional to E^2, violates neither P nor T and is of no significance in the present discussion. (Various atoms and the neutron itself have nonzero electric polarizability, but this does not violate P or T).

It is often said that certain polar molecules possess "permanent" electric dipole moments, and exhibit a linear Stark effect. However, in this phenomenon the Stark effect is *not* really linear for sufficiently small E, and involves no violation of P or T. In fact, the electric dipole moment of a polar diatomic molecule lies along the internuclear axis, which is fixed in the molecule but not fixed in space, and precesses about the total angular momentum **J**. In a $^1\Sigma$ molecule **J** is perpendicular to the internuclear axis, and here it can be shown that there is no first-order Stark effect. A molecule in a $^2\Sigma$ state has nonzero projection of total electron spin on the internuclear axis, and the interaction of electron spin with nuclear rotation gives rise to γ doubling of the rotational levels. However, the resultant splittings are very small compared to rotational splittings, and the Stark effect is quadratic for small applied electric fields. In other cases, where the projection of electron orbital angular momentum and/or spin on the internuclear axis is nonzero, one has Λ or Ω doubling of rotational levels. Once again, if the external electric field applied to the molecule is sufficiently weak that the Stark energy is substantially smaller than the splitting from Λ or Ω doubling, the Stark effect is not linear, but quadratic (Penney, 1931; van Vleck, 1932). The Stark shift becomes linear in E only for much larger electric fields (but ones that may still be extremely small on the scale employed for many experiments in chemistry and molecular physics).

Obviously it is impractical to observe the electron EDM by placing a free electron in an external E field, because the electron would quickly be accelerated out of the region of observation. However, one can observe the spin precession of a relativistic free electron in a magnetic field by means of a "g-2" experiment. If the electron possesses an EDM, the precession angular velocity is slightly modified because in the electron's rest frame there exists not only a magnetic field, but also a motional E field to which the EDM is coupled (Bargmann *et al.*, 1959). This was actually the basis for one of the earliest attempts to observe the electron EDM (Nelson *et al.*, 1959), and is still the only practical method available for the muon (Bailey *et al.*, 1978, 1979).

However, a far more sensitive method exists for the electron, in which one searches for the EDM d_a of a neutral paramagnetic atom containing an unpaired electron. At first sight, this approach appears useless, because even if the unpaired electron possessed an EDM d_e, the atom itself would not exhibit a linear Stark effect to first order in d_e in the limit of nonrelativistic quantum mechanics. For neither the atom nor any of its constituents is accelerated in the external electric field E, and in the nonrelativistic limit where atomic forces are electrostatic the average E field at the nucleus or at any electron must be zero. (The electronic and nuclear charges rearrange themselves to cancel the external E field.) This argument can easily be cast in quantum mechanical form (Schiff, 1963; Khriplovich, 1991a) and is known as "Schiff's theorem." However, as was first pointed out by Schiff, the theorem can be evaded when one takes into account magnetic hyperfine structure, and more importantly, for a nucleus of finite size if the nucleon electric dipole moment distribution is not the same as the nuclear charge distribution. In this case the nucleus may possess a "Schiff moment" as the result of the EDM of an unpaired nucleon, and thus the atom may possess an EDM.

As for the valence electron, it was first demonstrated by P. G. H. Sandars (Sandars, 1965, 1966) that Schiff's theorem is evaded here also if relativistic effects are taken into account. (Qualitatively, Schiff's theorem fails when relativity is included because the force on an electron is no longer purely electrostatic, but now also includes a "motional" contribution.) Sandars showed that in paramagnetic atoms such as the alkalis and thallium the ratio $R = d_a/d_e$ is of order $10\, Z^3\alpha^2$, where Z is the atomic number and α is the fine structure constant. Thus $|R|$ can be much larger than unity, and in fact one calculates $R = 120$ for cesium where $Z = 55$, whereas $R = -585$ for thallium where $Z = 81$. Enhancement factors R for certain paramagnetic molecules are orders of magnitude larger. Sandars' important discovery provides the basis for all modern electron EDM searches. One attempts to observe the linear Stark effect of the appropriate neutral paramagnetic atom or molecule in an external electric field, and interprets the result in terms of d_e by means of a calculated enhancement factor R.

It is important to note, however, that an atomic or molecular EDM could arise from other contributions besides a nucleon EDM and/or an electron EDM. P,T-odd nucleon–nucleon (NN) interactions might generate an EDM of the nucleus (Sushkov et al., 1984; Flambaum et al., 1985, 1986; Khatsymovsky et al., 1988; Haxton and Henley, 1983; He and McKellar, 1997). Furthermore, a nucleus with nuclear spin $I \geq 1$ could possess a magnetic quadrupole moment M originating from P,T odd nucleon–nucleon couplings and/or from nucleonic EDMs (Khriplovich, 1976, 1991a; Dmitriev et al., 1994). In a paramagnetic atom, M would couple to the magnetic field arising from the spin and spatial distribution of the unpaired electron, resulting in an atomic or molecular EDM. Because this interaction is magnetic, it would not be constrained by Schiff's theorem. P,T-odd electron–nucleon (eN) interactions might exist (Bouchiat, 1975; Hinds et al., 1976; Dzuba et al., 1985; Martensson-Pendrill, 1985; Flambaum et al., 1985; Barr, 1992a,b; Barr, 1993a). These, as well as the P,T-odd NN interactions, could appear in one or several nonderivative coupling forms: "scalar," "tensor," and "pseudoscalar." (P,T-odd electron-electron interactions are also possible, but these are likely to yield an extremely small contribution). C,T-odd (P-even) e–N and N-N interactions, and possible T-odd beta decay couplings, could cause a P,T-odd atomic or molecular EDM through radiative corrections involving the usual weak interactions of the standard model (Khriplovich, 1991b; Conti and Khriplovich, 1992). And finally, magnetic monopole–antimonopole pairs in the nucleus could also give rise to an atomic EDM (Flambaum and Murray, 1997).

A very precise experiment carried out by E. N. Fortson and co-workers (Jabobs et al., 1995) on the ground $^1S_0$ state of the diamagnetic atom ^{199}Hg is sensitive primarily to the Schiff moment and through it the P,T-odd N–N interaction, as well as the tensor form of the P,T-odd eN interaction. Meanwhile an experiment on the $6^2P_{1/2}$ ground state of the paramagnetic atom ^{205}Tl (Commins et al., 1994) is mainly sensitive to d_e, to the scalar form of the eN interaction, and to possible magnetic monopole–antimonopole pairs in the nucleus.

Having briefly mentioned the muon, we turn next to the tau lepton, which must of course be investigated by the methods of high-energy physics. Taus are produced in e^+e^- collisions at colliding beam accelerators:

$$e^+ + e^- \to \tau^+ + \tau^- \qquad (3)$$

In lowest order of perturbation theory two distinct amplitudes contribute to this reaction: single photon exchange (electromagnetic interaction, see Fig. 1a) and single Z^0 exchange (neutral weak interaction, see Fig. 1b). We may then consider P,T-odd radiative corrections to each amplitude, in which case we not only introduce the possibility of an electric dipole moment d_τ but also a "weak" dipole

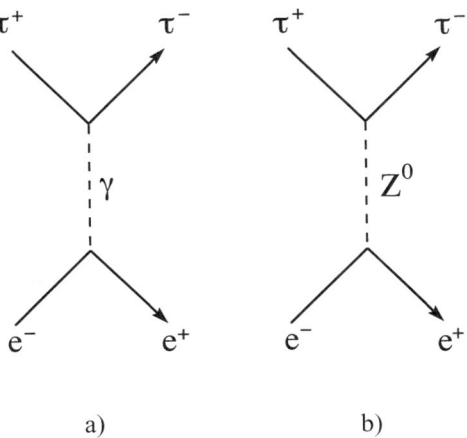

FIG. 1. Feynman diagrams corresponding to lowest-order amplitudes for $e^+e^- \to \tau^+\tau^-$ by: (a) photon exchange, (b) Z^0 exchange.

moment $\tilde{d}_\tau$. When the CM energy $\sqrt{s}$ in reaction (3) is greater than the tau pair production threshold of $2\,m_\tau = 2 \times 1.777$ GeV, but much less than the Z mass $m_Z = 91$ GeV, the electromagnetic amplitude of Fig. 1a is far more significant than the neutral weak amplitude of Fig. 1b. However, at $\sqrt{s} \approx m_Z$ (where production of a real Z at the Z resonance or Z "pole" occurs in experiments conducted at LEP), the neutral weak amplitude greatly dominates. By searching for certain P,T-odd correlations at the Z pole between the momenta of the initial electron and the decay products of the τ^+, τ^-, one places an upper limit on $\tilde{d}_\tau$. Although no explicit relationship exists between d_τ and $\tilde{d}_\tau$, these are expected to be roughly the same in many models of CP violation. Thus from a limit on $\tilde{d}_\tau$ we may obtain a model-dependent limit on d_τ. Furthermore, comparison of the observed partial width of the Z resonance for decay to $\tau^+\tau^-$ with that predicted by the standard model yields an upper limit on d_τ.

II. Theoretical Models of Lepton EDMs

A. Proper Lorentz-Invariant EDM Interaction

Let us write a gauge-invariant, proper Lorentz-invariant formulation of the interaction of the EDM of a spin-1/2 fermion with an electromagnetic field. We recall that the analogous formulation for an anomalous magnetic moment ("Pauli moment") is given by the well-known Lagrangian density:

$$L_{\text{Pauli}} = -\kappa \frac{\mu_B}{2} \overline{\Psi}\, \sigma^{\mu\nu} F_{\mu\nu} \qquad (4)$$

Here Ψ is the Dirac field for the fermion, $\overline{\Psi}$ is the Dirac conjugate field, $\sigma^{\mu\nu} = (i/2)(\gamma^\mu \gamma^\nu - \gamma^\nu \gamma^\mu)$ where γ^μ, γ^ν are the usual 4×4 Dirac matrices, $F_{\mu\nu}$ is the electromagnetic field tensor, μ_B is the Bohr magneton, and κ is an appropriate constant. (Here and throughout this review we use the conventions in (Bjorken and Drell, 1964).) The Lagrangian density of Eq. (4) is invariant under P and T. We now render it P,T-odd by making the replacement: $\sigma^{\mu\nu} \to \gamma^5 \sigma^{\mu\nu}$, where $\gamma^5 = i\gamma^0\gamma^1\gamma^2\gamma^3$. Also, we replace $\kappa \mu_B$ by id, where d is the real electric dipole moment (and $i = \sqrt{-1}$ is included so the Hamiltonian that results is Hermitian). Thus we obtain (Salpeter, 1958):

$$L_{\text{EDM}} = -i\frac{d}{2} \overline{\Psi} \gamma^5 \sigma^{\mu\nu} \Psi F_{\mu\nu} \qquad (5)$$

A simple manipulation of Eq. (5) yields:

$$L_{\text{EDM}} = d\overline{\Psi}[\mathbf{\Sigma} \cdot \mathbf{E} - i\boldsymbol{\alpha} \cdot \mathbf{B}]\Psi \qquad (6)$$

TABLE 1
RANGES OF PREDICTED VALUES FOR d_e
ACCORDING TO VARIOUS THEORETICAL MODELS.

Model	d_e, e cm
Standard model	$< 10^{-38}$
Supersymmetric	10^{-26}–10^{-28}
Multi-Higgs	10^{-26}–10^{-28}
Left-right symmetric	10^{-26}–10^{-28}

where

$$\Sigma = \begin{pmatrix} \sigma & 0 \\ 0 & \sigma \end{pmatrix}, \quad \alpha = \begin{pmatrix} 0 & \sigma \\ \sigma & 0 \end{pmatrix}$$

and **E,B** are the electric and magnetic fields, respectively. One can easily see that Eq. (6) yields the following single-particle Dirac EDM Hamiltonian:

$$H_{\text{EDM}} = -d\gamma^0 \Sigma \cdot \mathbf{E} + id\boldsymbol{\gamma} \cdot \mathbf{B} \tag{7}$$

In the nonrelativistic limit the first term on the right-hand side of Eq. (7) reduces to the right-hand side of (2), whereas the second term on the right-hand side gives no contribution.

The Lagrangian density L_{EDM} appearing in Eq. (5) is not renormalizable, and it can only contribute to the EDM of a lepton by virtue of loop corrections to the lepton–photon interaction of quantum electrodynamics. We next consider briefly what types of loops can occur in various theories of *CP* violation. Theoretical models of the electron EDM have been reviewed in detail by Bernreuther and Suzuki (Bernreuther, 1991b) and by Barr (Barr, 1993a). The predictions are briefly summarized in Table 1.

B. THE STANDARD MODEL

It is known from many experiments in particle physics that the quark mass eigenstates *d, s, b* are not identical with the corresponding weak interaction eigenstates. In the standard model, this is accounted for by writing the (Hermitian conjugate) charged weak current of the quarks as:

$$J_\lambda^\dagger = \bar{P}_L \gamma_\lambda (1 - \gamma^5) U N_L \tag{8}$$

where P_L, N_L are separate column vectors of left-handed quark fields with electric charges $+2/3\ e$, $-1/3\ e$, respectively:

$$P_L = \begin{pmatrix} u \\ c \\ t \end{pmatrix}_L \quad N_L = \begin{pmatrix} d \\ s \\ b \end{pmatrix}_L \tag{9}$$

and U is a 3×3 unitary matrix (Kobayashi and Maskawa, 1973) called the "Cabbibo–Kobayashi–Maskawa" (CKM) matrix. Most generally, U contains $3 \times 3 = 9$ complex numbers or 18 real parameters. The unitarity condition $U^\dagger U = I$ imposes 9 constraints, and one overall phase is arbitrary, so the number of independent real parameters would seem to be 8. However, the relative phases of u, c, t and the relative phases of d, s, b are completely arbitrary. When this is taken into account, 4 degrees of freedom remain in U. Thus, a real orthogonal 3×3 matrix, which is characterized by only 3 real parameters, is inadequate. Instead, we need three "Cabibbo-like" angles and an additional real parameter δ, which may be interpreted as a CP-violating phase. In a standard notation (Gilman et al., 1996), U can be written as:

$$U = \begin{pmatrix} V_{ud} & V_{us} & V_{ub} \\ V_{cd} & V_{cs} & V_{cb} \\ V_{td} & V_{ts} & V_{tb} \end{pmatrix} \quad (10)$$

$$= \begin{pmatrix} c_{12}c_{13} & s_{12}c_{13} & s_{13}e^{-i\delta} \\ -s_{12}c_{23} - c_{12}s_{23}s_{13}e^{i\delta} & c_{12}c_{23} - s_{12}s_{23}s_{13}e^{i\delta} & s_{23}c_{13} \\ s_{12}s_{23} - c_{12}c_{23}s_{13}e^{i\delta} & -c_{12}s_{23} - -s_{12}c_{23}s_{13}e^{i\delta} & c_{23}c_{13} \end{pmatrix}$$

where $c_{ij} = \cos\theta_{ij}$, $s_{ij} = \sin\theta_{ij}$, and $i,j = 1,2,3$ are generation labels. The angles $\theta_{12}, \theta_{23}, \theta_{13}$ can all be placed in the first quadrant with appropriate definitions of quark phases. Thus, all the s_{ij}, c_{ij} are positive. It can be shown (Jarlskog, 1985a, 1985b) that all CP-violating amplitudes are proportional to

$$J = s_{12}s_{13}s_{23}c_{12}c_{13}^2 c_{23} \sin(\delta) \quad (11)$$

The standard model does not really explain CP violation, because it does not provide us with the tools for calculating J from first principles. It merely reveals how CP violation can be accommodated in the six-quark model (CKM matrix) in terms of J.

In the standard model the neutron EDM could arise from valence quark EDMs, but this cannot occur at the one-loop level. At the two-loop level, individual diagrams do have complex phases that contribute to the neutron EDM. However, it has been shown that the sum of these diagrams over all quark flavors yields zero (Shabalin, 1978, 1983; Donoghue, 1978). Thus, according to the standard model the neutron EDM appears only at the three-loop level. Even here there are suppressions and cancellations, and one finds (Czarnecki and Krause, 1997) that:

$$d_n \text{ (std. model)} \approx 10^{-34} \, e \text{ cm} \quad (12)$$

where $e = 4.8 \cdot 10^{-10}$ esu is the electronic charge. (However, it is possible that certain "interquark" diagrams may yield somewhat larger contributions to the

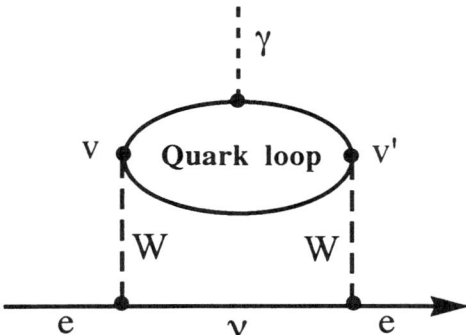

FIG. 2. This two-loop diagram cannot contribute to the electron EDM, because although a factor V_{ij} from the CKM matrix appears at vertex v, a factor V_{ij}^* appears at vertex v'. Thus there is no net CP-violating phase.

neutron EDM, perhaps of order 10^{-31} e cm to 10^{-32} e cm; see for example He, McKellar, and Pakvasa, 1989).

In the standard model with massless neutrinos, there is no analog to the CKM matrix in the lepton sector, and thus no analogous way to generate CP violation. For the electron EDM to arise here we would require coupling to virtual quarks via virtual $W^{\pm}$. Naively one might expect a contribution from the two-loop diagram of Fig. 2. However, for each contribution V_{ij} from the CKM matrix at one vertex v, there is a contribution $V_{ij}*$ at the other vertex v'; hence, the overall amplitude cannot contain a CP-violating phase. (This is also the reason why there is no contribution to the neutron EDM at the one-loop level). Next one can consider contributions to the electron EDM at the three-loop level. This situation was analyzed in some detail (Hoogeveen, 1990), but it was subsequently shown (Pospelov and Khriplovich, 1991) that the various three-loop diagrams cancel, yielding a net contribution of zero in the absence of gluonic corrections to the quark lines. (See Fig. 3.) Thus, in the standard model the electron EDM is predicted to be extremely small:

$$d_e(\text{std. model}) \leq 10^{-38} \; e \; \text{cm} \tag{13}$$

even when gluonic corrections are included. Results (12) and (13) are 6 to 10 orders of magnitude smaller than the ultimate sensitivities one might expect in practical experiments on the neutron and electron. (At present the experimental limit for the neutron is $|d_n| < 8 \cdot 10^{-26}$ e cm, (Altarev et al., 1996; Smith et al., 1990)).

One can incorporate finite and distinct neutrino masses into the standard model, and construct a CKM-like matrix for the lepton sector. However, given the present experimental limits on neutrino masses, this results in possible values for

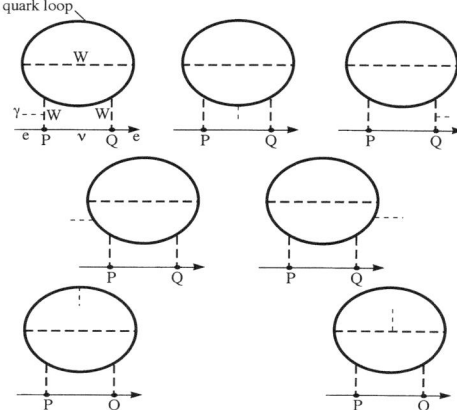

FIG. 3. Pospelov and Khriplovich (1991) proved that the sum of the contributions to d_e from these three-loop diagrams is zero, according to the standard model. If each diagram is disconnected from the lepton line at P,Q one has the two-loop contributions to the EDM of an (on-mass-shell) W boson. Thus they also proved that the EDM of a W boson is zero in the two-loop approximation.

d_e that are much smaller than the limit given in Eq. (13); (Bernreuther and Suzuki, 1991). Although the standard model prediction (13) for the lepton EDM is proportional to the lepton mass, and therefore 2 or 3 orders of magnitude larger for μ or τ than for the electron, the experimental sensitivities for μ, τ at present are poorer by 7 to 9 orders of magnitude than for the electron. Obviously, then, all the standard model predictions are hopelessly small. Hence, if d_n, d_e, d_μ, and/or d_τ were to be observed, it would clearly signify the existence of physics beyond the standard model. In the following paragraphs we summarize very briefly and superficially some of the main points of three popular alternatives to the standard model: supersymmetric, multi-Higgs, and left-right symmetric models. Other possibilities have also been considered.

C. SUPERSYMMETRIC MODELS

Supersymmetric (SUSY) models are motivated by the desire to give a natural explanation for the "gauge hierarchy problem"; for reviews see (Nilles, 1984; Haber and Kane, 1985). In the standard model, electroweak symmetry breaking is induced by the Higgs mechanism, which imparts masses to $W^\pm$ and Z of the order of 100 GeV (the "weak scale"). The mass of the Higgs boson itself is still unknown, but it must be less than ≈ 1 TeV if unitarity is to be preserved in standard model perturbation calculations. However, radiative corrections to the Higgs mass are quadratically divergent, and cannot be controlled within the weak scale unless "fine tuning" is imposed that appears artificial and contrived to many authors.

Supersymmetric models attempt to avoid this problem in a natural way by linking physics at the weak scale to physics at the Planck scale.

In the minimal supersymmetric standard model (MSSM), which is an extension of the standard model with $N = 1$ supergravity (and indeed in all SUSY models), many new hypothetical particles appear. For each fermion (lepton or quark) one introduces a supersymmetric bosonic partner (slepton, squark); for each standard model boson (gluons, Z^0, $W^{\pm}$, photon) a supersymmetric fermionic partner called a "gaugino" is invoked (gluinos, zino, winos, photino). In addition even the simplest SUSY models require at least two Higgs supermultiplets. The wealth of hypothetical new particles and their couplings (none of the new particles has been observed) yields new phases in addition to δ, and hence new opportunities for *CP* violation. It now becomes possible to generate an electron EDM at the one-loop level (Fig. 4). Many variations on this theme have been considered in the last few years (Fischler *et al.*, 1992; Kizukuri and Oshimo, 1992; Barr and Segre, 1993; Babu and Barr, 1994a,b; Falk *et al.*, 1995; Falk and Olive, 1996; Matsuda and Tanimoto, 1995). Grand unified SUSY models have attracted special interest recently (Barbieri *et al.*, 1995; Dimopoulos and Hall, 1995; Barbieri *et al.*, 1996, 1997). In some of the latter models the large mass of the top quark plays an essential role. It induces a splitting between sfermion masses of the third generation with respect to the first two generations, which, together with CKM-like mixing angles and phases appearing in gaugino–matter interactions, result in values of d_e and d_n that are very close to the present experimental upper limits. In such theories one also expects lepton flavor violation: the "forbidden" transition

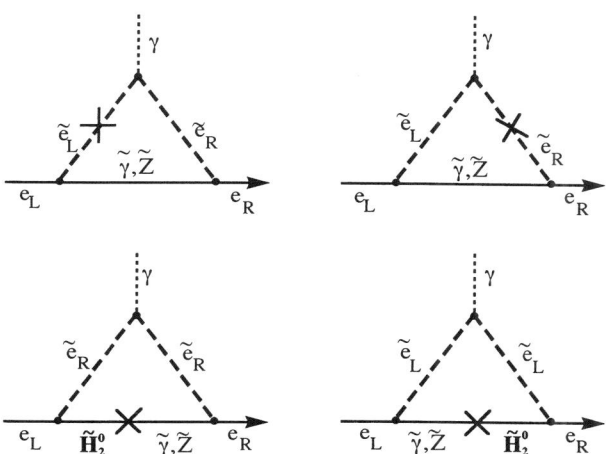

FIG. 4. Typical one-loop contributions to d_e in supersymmetric models. $\tilde{e}_L$, $\tilde{e}_R$: selectrons; $\tilde{\gamma}$: photino; $\tilde{Z}$: zino; $\tilde{H}_2^0$: Higgsino. The **X** indicates schematically where *CP* violation occurs.

$\mu \to e\gamma$ should actually occur with a transition probability closely related to the predicted value of d_e.

D. MULTI-HIGGS MODELS

In the simplest form of the standard model there is only one Higgs boson. However, in various extensions two or more Higgs bosons could appear. *CP* violation could then arise in a variety of new ways (Barr, 1992a,b; 1993a,b; Kuo and Xu, 1992; Mahanta, 1992; Deshpande and He, 1994; Hayashi *et al.*, 1995; Matsuda and Tanimoto, 1995). In particular, *CP* violation could appear directly in the coupling of one Higgs field to another. An electron EDM as large as 10^{-27} *e* cm (close to the present experimental limit) could be generated from two-loop diagrams, as in (Fig. 5). Here the lepton chirality change occurs at the $\phi\ell\ell$ vertex, and the lepton EDM is proportional to the lepton mass. Models of this type might also give rise to an appreciable scalar *P,T*-odd *eN* interaction (see Fig. 6 and Section III.C), which would contribute to the EDM of a paramagnetic atom (Barr, 1992a,b; 1993a). In one possible class of multi-Higgs models, the lepton EDM at the one-loop level is proportional to the cube of the lepton mass, and might be quite substantial for the tau lepton. (See section V.)

E. LEFT-RIGHT SYMMETRIC MODELS

Left-right symmetric models based on the gauge group $SU(2)_L \otimes SU(2)_R \otimes U(1)$ are motivated by the desire to find a natural explanation for the very striking phenomenon of parity violation in weak interactions (Pati and Salam, 1974; Mohapatra and Pati, 1975; Mohapatra and Sidhu, 1977; Mohapatra and Senjanovic, 1980, 1981). Here, space inversion symmetry (*P*) is valid before spontaneous

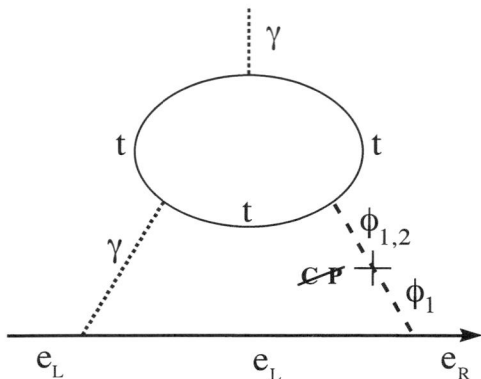

FIG. 5. Two-loop contribution to d_e in a multi-Higgs model. ϕ_1, ϕ_2 are two Higgs fields, with a *CP*-violating coupling between them; *t* is a top quark.

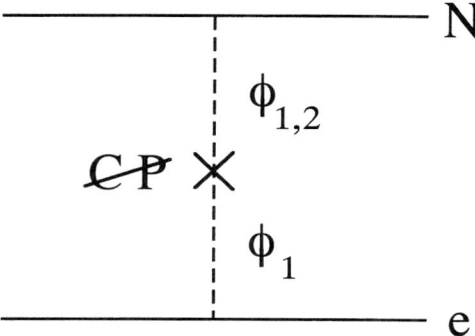

FIG. 6. Two-Higgs model of *CP* violation could contribute at tree-level to a scalar P,T odd e–N interaction, as illustrated in this diagram.

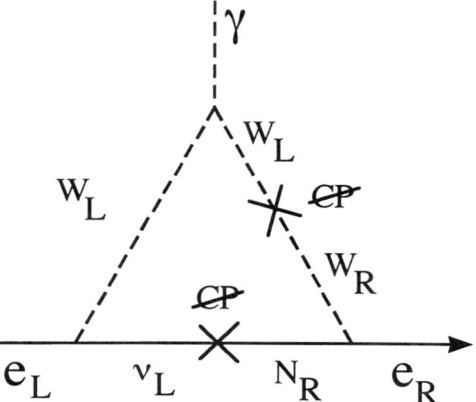

FIG. 7. One-loop contribution to d_e in a left-right-symmetric model. $W_{L,R}$ are left- and right-handed W bosons respectively and N_R is a massive right-handed neutrino.

symmetry breaking. In the simplest left-right symmetric model two Higgs multiplets appear, the first of which is a triplet χ_R, which transforms like (1, 3) under $SU(2)_L \otimes SU(2)_R$. It gives rise to a very large mass of the right-handed intermediate vector boson W_R and thus breaks parity symmetry. A complex doublet ϕ transforming like (2, 2) under $SU(2)_L \otimes SU(2)_R$ contributes to the mass of both W_R and W_L and causes mixing between them. These models also contain a right-handed neutrino N_R for each left-handed neutrino ν_L. The N_R acquires a large Majorana mass from χ_R and mixes with the light neutrino ν_L by means of ϕ. *CP* violation can occur at the one-loop level from the phases associated with W_L–W_R and N_R–ν_L mixing (Liu, 1986; Nieves *et al.*, 1986) (see Fig. 7). Once again, some predictions for d_e are close to the present experimental upper limit.

III. The Electron EDM

A. Schiff's Theorem

We now derive Schiff's theorem for a paramagnetic atom and show how it fails for the valence electron when relativistic motion is taken into account. For purposes of this discussion we assume the one-electron central field approximation, and begin by writing the one-electron Dirac Hamiltonian for an atom in an external electric field $\mathbf{E}_e$ in the absence of an electron EDM:

$$H = c\boldsymbol{\alpha} \cdot \mathbf{p} + mc^2\gamma^0 - e(\Phi_i + \Phi_e) \tag{14}$$

Here $e > 0$ (the charge of the electron is $-e$), while Φ_i is the atomic electrostatic potential, assumed to be central, and $\Phi_e = -\mathbf{E}_e \cdot \mathbf{r}$ is the external electrostatic potential. An eigenstate of H will be denoted by $|\psi\rangle$.

We next introduce the EDM Hamiltonian as a perturbation. From Eq. (7) with $\mathbf{B} = 0$, we have:

$$H_{\text{EDM}} = -d_e \gamma^0 \boldsymbol{\Sigma} \cdot \mathbf{E} \tag{15}$$

where $\mathbf{E}$ is the total electric field: $\mathbf{E} = -\boldsymbol{\nabla}(\Phi_i + \Phi_e)$. It is convenient to separate the right-hand side of Eq. (15) into two parts as follows:

$$H_{\text{EDM}} = -d_e \boldsymbol{\Sigma} \cdot \mathbf{E} - d_e(\gamma^0 - 1)\boldsymbol{\Sigma} \cdot \mathbf{E} \tag{16}$$

The first term on the right-hand side of Eq. (16) is the only portion that survives in the nonrelativistic limit, but as we shall now demonstrate, it contributes nothing to the first-order energy shift arising from H_{EDM}. (This is Schiff's theorem.) We write:

$$-d_e \boldsymbol{\Sigma} \cdot \mathbf{E} = \frac{d_e}{e} \boldsymbol{\Sigma} \cdot \boldsymbol{\nabla} e\Phi = \frac{id_e}{e}[\boldsymbol{\Sigma} \cdot \mathbf{p}, e\Phi]$$

Making use of Eq. (14) this becomes:

$$-d_e \boldsymbol{\Sigma} \cdot \mathbf{E} = -\frac{id_e}{e}[\boldsymbol{\Sigma} \cdot \mathbf{p}, (H - c\boldsymbol{\alpha} \cdot \mathbf{p} - mc^2\gamma^0)]$$

$$= -\frac{id_e}{e}[\boldsymbol{\Sigma} \cdot \mathbf{p}, H]$$

where the last step follows because $[\Sigma_i, \alpha_j] = [\Sigma_i, \gamma^0] = 0$. Therefore, we obtain:

$$\langle \psi | -d_e \boldsymbol{\Sigma} \cdot \mathbf{E} | \psi \rangle = -\left\langle \psi \left| \frac{id_e}{e}[\boldsymbol{\Sigma} \cdot \mathbf{p}, H] \right| \psi \right\rangle = 0$$

which holds because $|\psi\rangle$ is an eigenstate of H.

We thus conclude that only the second term on the right-hand side of Eq. (16) can contribute to a first-order energy shift:

$$\Delta E = \langle \psi | -d_e(\gamma^0 - 1)\mathbf{\Sigma} \cdot \mathbf{E} | \psi \rangle \qquad (17)$$

Although the right-hand side of Eq. (15) was separated into "nonrelativistic" and "relativistic" parts in (16), it is sometimes convenient to deal directly with Eq. (15) by writing:

$$\langle -d_e\gamma^0\mathbf{\Sigma} \cdot \mathbf{E} \rangle = -\frac{id_e}{e}\langle [\gamma^0\mathbf{\Sigma} \cdot \mathbf{p}, (H - c\boldsymbol{\alpha} \cdot \mathbf{p} - mc^2\gamma^0)] \rangle$$

$$= \frac{icd_e}{e}\langle [\gamma^0\mathbf{\Sigma} \cdot \mathbf{p}, \boldsymbol{\alpha} \cdot \mathbf{p}] \rangle$$

Now taking into account the identities $\boldsymbol{\alpha} = \gamma^5\mathbf{\Sigma} = \mathbf{\Sigma}\gamma^5$, $\gamma^0\gamma^5 = -\gamma^0\gamma^5$, and $\mathbf{\Sigma} \cdot \mathbf{p}\mathbf{\Sigma} \cdot \mathbf{p} = \mathbf{p}^2$, we obtain:

$$\Delta E = \langle -d_e\gamma^0\mathbf{\Sigma} \cdot \mathbf{E} \rangle = \frac{2icd_e}{e}\langle \gamma^0\gamma^5\mathbf{p}^2 \rangle \qquad (18)$$

which is a form frequently employed in numerical calculations.

B. Calculation of the Enhancement Factor for Paramagnetic Atoms

Recalling that $|\psi\rangle$ is an eigenstate of Hamiltonian H, which includes the term $-e\Phi_e = e\mathbf{E}_e \cdot \mathbf{r}$, we treat the latter term as a perturbation on the atomic Hamiltonian H_0 with no external field:

$$H_0 = c\boldsymbol{\alpha} \cdot \mathbf{p} + mc^2\gamma^0 - e\Phi_i \qquad (19)$$

and express $|\psi\rangle$ in terms of the eigenstates $|\psi_n\rangle$ of H_0 to first order in E_e:

$$|\psi\rangle = |\psi_0\rangle + eE_e\sum_n{}'\frac{|\psi_n\rangle\langle\psi_n|z|\psi_0\rangle}{E_0 - E_n} = |\psi_0\rangle + eE_e|\eta\rangle \qquad (20)$$

Here $|\psi_0\rangle$ is that state to which $|\psi\rangle$ reduces when $\mathbf{E}_e = 0$, we have assumed that $\mathbf{E}_e$ is in the z direction, the E_n are energy eigenvalues of H_0 corresponding to the $|\psi_n\rangle$, the prime on the summation indicates that $n = 0$ is omitted, and:

$$|\eta\rangle = \sum_n{}'\frac{|\psi_n\rangle\langle\psi_n|z|\psi_0\rangle}{E_0 - E_n} \qquad (21)$$

Now substituting Eq. (20) in Eq. (17), and retaining only terms of first order in E_e, we obtain:

$$\Delta E = -d_eE_e\langle\psi_0|(\gamma^0 - 1)\Sigma_z|\psi_0\rangle \\ - ed_eE_e[\langle\eta|(\gamma^0-1)\mathbf{\Sigma}\cdot\mathbf{E}_i|\psi_0\rangle + \langle\psi_0|(\gamma^0-1)\mathbf{\Sigma}\cdot\mathbf{E}_i|\eta\rangle] \qquad (22)$$

The atomic electric dipole moment is defined by $d_a = -(\partial \Delta E)/(\partial E_e)$. Thus the enhancement factor $R = d_a/d_e$ is given by:

$$R = \langle \psi_0 | (\gamma^0 - 1) \Sigma_z | \psi_0 \rangle \\ + e[\langle \eta | (\gamma^0 - 1) \Sigma \cdot \mathbf{E}_i | \psi_0 \rangle + \langle \psi_0 | (\gamma^0 - 1) \Sigma \cdot \mathbf{E}_i | \eta \rangle] \quad (23)$$

In the case of the $1\,^2s_{1/2}$ state of a hydrogenic atom, one can show (Sandars, 1968) that the first term on the right-hand side of Eq. (23) is $\approx \frac{1}{6} Z^2 \alpha^2$, while the second and third terms together give $\approx -\frac{13}{6} Z^2 \alpha^2$ (where $Z\alpha \ll 1$ is assumed). For the $2\,^2s_{1/2}$ state of hydrogen the second and third terms are much larger because of the close proximity of the $2\,^2p_{1/2}$ state; here one obtains $R(H, 2\,^2S_{1/2}) \approx 120$. Unfortunately, however, the $2\,^2s_{1/2}$ state is not suitable for a sensitive electron EDM experiment because its lifetime is very short in a strong external electric field.

For neutral paramagnetic atoms of large Z, the second and third terms of Eq. (23) together give the contribution:

$$2e \sum \frac{\langle \psi_0 | (\gamma^0 - 1) \Sigma \cdot \mathbf{E}_i | \psi_n \rangle \langle \psi_n | z | \psi_0 \rangle}{E_0 - E_n}$$

Assuming that a single term dominates in this sum, we make a crude estimate of the various factors as follows: $\langle \psi_n | z | \psi_0 \rangle \approx a_0$; $E_0 - E_n \approx .05\, e^2/a_0$; $\langle \psi_0 | (\gamma^0 - 1) \Sigma \cdot \mathbf{E}_i | \psi_n \rangle \approx Z^2\alpha^2 \cdot Ze/a_0^2$. Thus the second and third terms of Eq. (23) together yield a contribution $\approx 10\, Z^3 \alpha^2$; this would give ≈ 100 for cesium and ≈ 300 for thallium. Meanwhile the contribution of the first term of Eq. (23) is much smaller, of order $Z^2\alpha^2$. Henceforth we ignore this term and write:

$$R = 2e \langle \psi_0 | (\gamma^0 - 1) \Sigma \cdot \mathbf{E}_i | \eta \rangle \quad (24)$$

We now sketch the calculation of R, restricting ourselves to the case where $|\psi_0\rangle$ is a state with $J = 1/2$, $m_J = 1/2$, which includes the alkali atoms and thallium. Then the four-component Dirac wave-function corresponding to $|\psi_0\rangle$ can be written:

$$\psi^\ell_{J=1/2, m=1/2} = \begin{pmatrix} \dfrac{iG_{\ell, J=1/2}(r)}{r} \phi^\ell_{1/2, 1/2} \\ \dfrac{F_{\ell, 1/2}(r)}{r} \sigma \cdot \hat{\mathbf{r}} \phi^\ell_{1/2, 1/2} \end{pmatrix} \quad (25)$$

where we employ the notation of Bjorken and Drell (19), in which:

$$\phi^{\ell=0}_{1/2, 1/2} = \begin{pmatrix} Y_0^0 \\ 0 \end{pmatrix}, \qquad \phi^{\ell=1}_{1/2, 1/2} = \begin{pmatrix} \sqrt{\tfrac{1}{3}} Y_1^0 \\ -\sqrt{\tfrac{2}{3}} Y_1^1 \end{pmatrix} \quad (26)$$

and where

$$\sigma \cdot \hat{\mathbf{r}} \phi^0_{1/2,1/2} = \phi^1_{1/2,1/2}, \qquad \sigma \cdot \hat{\mathbf{r}} \phi^1_{1/2,1/2} = \phi^0_{1/2,1/2}$$

Inserting Eq. (25) in Dirac's equation $(H_0 - E_0)|\psi_0\rangle = 0$, defining $W_0 = E_0 - mc^2$, and choosing atomic units where $\hbar = e = m_e = 1, c = \alpha^{-1} = 137.036$, we obtain the well-known coupled radial equations:

$$\frac{\partial G_{\ell,1/2}}{\partial r} + \frac{\kappa}{r} G_{\ell,1/2} = \alpha \left(W_0 + \frac{2}{\alpha^2} + \Phi_i \right) F_{\ell,1/2} \qquad (27)$$

$$\frac{\partial F_{\ell,1/2}}{\partial r} - \frac{\kappa}{r} F_{\ell,1/2} = -\alpha (W_0 + \Phi_i) G_{\ell,1/2} \qquad (28)$$

where $\kappa = +1$ for $\ell = 1$, while $\kappa = -1$ for $\ell = 0$. If Φ_i is specified, these equations can be solved analytically or numerically subject to the condition that $\psi^\ell_{1/2,1/2}$ is normalized to unity.

To find a useful expression for $|\eta\rangle$ we apply the operator $H_0 - E_0$ to both sides of Eq. (21) and make use of the completeness relation:

$$\sum |\psi_n\rangle\langle\psi_n| = 1$$

to obtain:

$$(H_0 - E_0)|\eta\rangle = -z|\psi_0\rangle \qquad (29)$$

which is known as the Sternheimer equation (Sandars, 1966, 1968; Sandars and Sternheimer, 1975). It can be seen from Eq. (21) that $|\eta\rangle$ and $|\psi_0\rangle$ must be of opposite parity, and that, a priori, $|\eta\rangle$ can contain $J = 1/2$ and $J = 3/2$ components. However, because $\mathbf{\Sigma} \cdot \mathbf{E}_i$ is a pseudoscalar operator, only the $J = 1/2$ component of $|\eta\rangle$ can contribute to Eq. (24). Writing the wave-function of this component as:

$$\eta^L_{1/2,1/2} = \begin{pmatrix} \dfrac{iG^S_{L,1/2}}{r} \phi^L_{1/2,1/2} \\ \dfrac{F^S_{L,1/2}}{r} \sigma \cdot \hat{\mathbf{r}} \phi^L_{1/2,1/2} \end{pmatrix} \qquad (30)$$

with $L = 0$ or 1, we carry out standard manipulations on the Sternheimer equation (29) to arrive at the coupled radial equations:

$$r \frac{\partial G^S_{1,1/2}}{\partial r} + G^S_{1,1/2} - \alpha r \left(W_0 + \Phi_i + \frac{2}{\alpha^2} \right) F^S_{1,1/2} = -\frac{\alpha r^2}{3} F_{0,1/2} \qquad (31)$$

$$r \frac{\partial F^S_{1,1/2}}{\partial r} - F^S_{1,1/2} + \alpha r (W_0 + \Phi_i) G^S_{1,1/2} = \frac{\alpha r^2}{3} G_{0,1/2} \qquad (32)$$

for $^2S_{1/2}$ enhancement factors (as in the ground states of alkali atoms), or:

$$r\frac{\partial G^S_{0,1/2}}{\partial r} - G^S_{0,1/2} - \alpha r\left(W_0 + \Phi_i + \frac{2}{\alpha^2}\right)F^S_{0,1/2} = -\frac{\alpha r^2}{3}F_{1,1/2} \quad (33)$$

$$r\frac{\partial F^S_{0,1/2}}{\partial r} + F^S_{0,1/2} + \alpha r(W_0 + \Phi_i)G^S_{0,1/2} = \frac{\alpha r^2}{3}G_{1,1/2} \quad (34)$$

for the $6\,^2P_{1/2}$ ground state of thallium.

Two features of η are worthy of note. First, unlike $\psi^\ell_{1/2,1/2}$, $\eta^\ell_{1/2,1/2}$ is not normalized to unity. Instead the magnitudes of $F^S_{L,1/2}$, $G^S_{L,1/2}$ are determined as solutions to the inhomogeneous differential equations (31, 32), or (33, 34) and the conditions $F^S_{L,1/2} \to 0$, $G^S_{L,1/2} \to 0$ as $r \to \infty$. Second, η includes contributions from orbitals $|\psi_n\rangle$ that one might naively think were prohibited by the Pauli principle (e.g., the $6s^2 6s$ orbital in thallium). However, it can be shown that these must be included and that they account for the contribution of some core-excited orbitals (Bouchiat and Bouchiat, 1975; Neuffer and Commins, 1977).

From the solutions to (27, 28) and (31, 32) or (33, 34) we may write R as follows (where all quantities are in atomic units):

$$R = 2\langle \psi_0 | (\gamma^0 - 1)\Sigma \cdot \mathbf{E}_i | \eta \rangle$$

$$= 4\int \begin{pmatrix} \frac{iG_{\ell,J=1/2}(r)}{r}\phi^\ell_{1/2,1/2} \\ \frac{F_{\ell,1/2}(r)}{r}\sigma\cdot\hat{\mathbf{r}}\phi^\ell_{1/2,1/2} \end{pmatrix}^\dagger \begin{pmatrix} 0 & 0 \\ 0 & 1 \end{pmatrix}\frac{\partial\Phi_i}{\partial r}\sigma\cdot\hat{\mathbf{r}} \begin{pmatrix} \frac{iG^S_{L,1/2}}{r}\phi^L_{1/2,1/2} \\ \frac{F^S_{L,1/2}}{r}\sigma\cdot\hat{\mathbf{r}}\phi^L_{1/2,1/2} \end{pmatrix} d^3r$$

which yields

$$R = 4\int_0^\infty F_{\ell,1/2} F^S_{L,1/2} \frac{\partial\Phi_i}{\partial r} dr \quad (35)$$

Numerical calculations of R based on (35) and using a variety of semi-empirical methods have been carried out by a number of authors: (Sandars, 1965, 1966; Sandars and Sternheimer, 1975; Flambaum, 1976; Johnson et al., 1985). Also, ab initio calculations based on the Hartree-Fock method have been done (Johnson et al., 1986; Kraftmakher, 1988; Hartley et al., 1990, Liu and Kelly, 1992). For illustration we here employ the following semi-empirical "modified Tietz" potential:

$$\Phi_i = \frac{(Z-1)}{r(1+br)^2}e^{-ar} + \frac{1}{r} \qquad r > R_{nuc} \quad (36)$$

$$\Phi_i = \left[\frac{(Z-1)}{(1+br)^2}e^{-ar} + 1\right]\left(3 - \frac{r^2}{R^2_{nuc}}\right)\frac{1}{2R_{nuc}} \qquad r < R_{nuc} \quad (37)$$

to repeat one of the calculations of R for the $6\,^2P_{1/2}$ state of thallium published by Johnson et al. (1985). In Eqs. (36, 37) a uniform nuclear charge density is chosen inside the radius $R_{\text{nuc}} = 1.2 \cdot 10^{-13} A^{1/3}$ cm $= 1.34 \cdot 10^{-4}$ in atomic units for $A = 205$. (As is well known, $J = 1/2$ solutions (27,28) to the Dirac equation for a point nucleus are singular at the origin, but the singularity disappears if the nucleus is of finite size. Here we have chosen a uniform nuclear charge density for simplicity, but the result for R is not sensitive to the details of the nuclear charge distribution.) Without the exponential shielding factor e^{-ar}, Eqs. (36, 37) define the "Tietz" potential, which yields a good approximate solution to the Thomas-Fermi equation. The factor e^{-ar} is inserted to account for the exponential decrease in electron density for large r. Parameters a = .2579 and b = 2.5937 are chosen so that the calculated and observed $6\,^2P_{1/2}$ and $7\,^2P_{1/2}$ energies agree. Solutions (27, 28) for the potential given in Eqs. (36, 37) were used by Neuffer and Commins (1977) to calculate energy levels, hyperfine structure splittings, allowed $E1$ and forbidden $M1$ transition amplitudes, parity nonconserving (PNC) $E1$ amplitudes, and other observables; and the results are in reasonably good agreement with experimental data.

The result of this calculation is $R = -679$. Figure 8 shows the integrand of (35): $I = 4F_{\ell,1/2} F^S_{L,1/2} (\partial \Phi_i/\partial r)$ plotted as a function of r. It can be seen that the magnitude of I rises from zero to a maximum at the nuclear radius and then drops rapidly to zero at a distance $r \approx 0.015$ atomic units. In other words, essentially the entire contribution to R comes from $r \leq a_0/Z$.

Such calculations must be corrected to account for screening of the external electric field by the electron core. (It can be seen from the simplest classical arguments that screening must occur. For example, if an external electric field E_0 is applied to a singly charged ion of atomic number Z, the magnitude of the field at the nucleus must be reduced to E_0/Z by core polarization, because the acceleration of the ion (and the nucleus) in field E_0 is proportional to the ion's total charge.) On the other hand, one expects that the screening correction should be relatively minor, because from Eq. (20) it is evident that the external electric field has its most significant effect at distances $r \approx a_0$.

In early calculations of R, (Sandars, 1966; Sandars and Sternheimer, 1975) a simple phenomenological screening factor:

$$f(r) = \frac{\alpha_C + (Z-1)r^3}{\alpha_C Z + (Z-1)r^3} \tag{38}$$

was included on the right-hand side of Eq. (29). Here α_C is the core polarizability, an empirically determined parameter, and the function $f(r)$ varies smoothly from Z^{-1} at $r = 0$ to unity at large values of r. However, a Hartree-Fock calculation (Dzuba et al., 1986) shows that the radial dependence of the applied electric field inside the atom is actually quite complicated. In recent calculations of R,

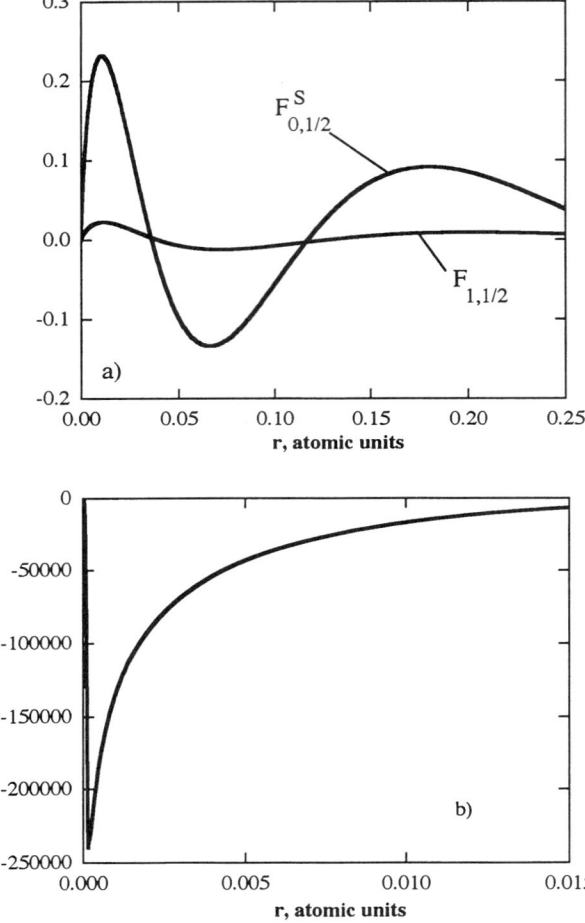

FIG. 8. Results of calculation of enhancement factor $R(^{205}Tl)$ using modified Tietz potential, described in the text. (a) The "small component" functions $F_{1,1/2}(r)$, $F^S_{0,1/2}(r)$ are plotted versus r; (b) the integrand of Eq. 35 is plotted versus r.

many-body perturbation theory has been used to include core polarization effects; see for example (Johnson, 1986).

The slow convergence of ab initio Hartree-Fock calculations for thallium caused some difficulties for a time (Kraftmakher, 1988; Hartley et al., 1990) but these were resolved by Liu and Kelly (1992), whose result $R(^{205}Tl) = -585$ appears to be reliable to 5–10% accuracy, and is reasonably close to the values obtained by semi-empirical methods. Various calculations of R are summarized in Table 2.

TABLE 2
CALCULATED ENHANCEMENT FACTORS $R = d_a/d_e$ FOR PARAMAGNETIC ATOMS.

Atom	State	Enhancement factor	
		Semi-empirical calculation	Ab initio calculation
Li	$2\,^2S_{1/2}$	.0043[a]	
Na	$3\,^2S_{1/2}$	.32[a]	
K	$4\,^2S_{1/2}$	2.42[a]	
Rb	$5\,^2S_{1/2}$	24[a]	
"	"	16 to 22[b]	**24.6**[b]
Cs	$6\,^2S_{1/2}$	119[a]	
"	"	80.3 to 106[b]	**114.9**[b]
Fr	$7\,^2S_{1/2}$	**1150**[a]	
Tl	$6\,^2P_{1/2}$	−700[c]	
"	"	−502 to −607[b]	−1041.0[b]
"	"	−500[d]	
"	"		−301[e]
"	"		−179[f]
"	"		**−585**[g]

Note: The preferred value is shown in boldface.
References: (a) (Sandars, 1966), (b) (Johnson et al., 1986), (c) (Sandars and Sternheimer, 1975), (d) (Flambaum, 1976), (e) (Kraftmakher, 1988), (f) (Hartley et al., 1990), (g) (Liu and Kelly, 1992).

C. P,T-Odd Electron–Nucleon Interaction

As previously mentioned, the EDM d_a of a paramagnetic atom can arise from a P,T-odd electron–nucleon interaction. If we limit ourselves to nonderivative terms, the various possibilities for P-odd electron–nucleon couplings are easily written by analogy from the theory of nuclear beta decay as follows:

$$\begin{aligned} \bar{N}N \cdot \bar{e}\gamma^5 e & \quad \text{S–PS (scalar–pseudoscalar)} \\ \bar{N}\gamma^\mu N \cdot \bar{e}\gamma_\mu \gamma^5 e & \quad \text{V–A (vector–pseudovector)} \\ \bar{N}\sigma^{\mu\nu} N \cdot \bar{e}\sigma_{\mu\nu}\gamma^5 e & \quad \text{T–PT (tensor–pseudotensor)} \\ \bar{N}\gamma^\mu \gamma^5 N \cdot \bar{e}\gamma_\mu e & \quad \text{A–V (pseudovector–vector)} \\ \bar{N}\gamma^5 N \cdot \bar{e}e & \quad \text{PS–S (pseudoscalar–scalar)} \end{aligned} \qquad (39)$$

In Table 3 we summarize the transformation properties under time reversal of bilinear forms $\bar{u}' F u$ (where u, u' are Dirac spinors and F is a 4×4 matrix). From Table 3 we see that among the five possibilities listed in Eq. (39), only the couplings S–PS ("scalar"), T–PT ("tensor"), and PS–S ("pseudoscalar") are odd under time reversal. We may then write an effective P,T-odd Hamiltonian density as follows:

TABLE 3
TRANSFORMATION PROPERTIES OF BILINEAR FORMS
UNDER TIME REVERSAL AND COMPLEX CONJUGATION.

Original form	After time reversal	After complex conjugation
$\bar{u}'u$	$\bar{u}u'$	$\bar{u}u'$
$\bar{u}'\gamma^\mu u$	$\bar{u}\gamma_\mu u'$	$\bar{u}\gamma^\mu u'$
$\bar{u}'\sigma^{\mu\nu}u$	$-\bar{u}\sigma_{\mu\nu}u'$	$\bar{u}\sigma^{\mu\nu}u'$
$\bar{u}'\sigma^{\mu\nu}\gamma^5 u$	$\bar{u}'\sigma_{\mu\nu}\gamma^5 u'$	$-\bar{u}\sigma^{\mu\nu}\gamma^5 u'$
$\bar{u}'\gamma^\mu\gamma^5 u$	$\bar{u}\gamma_\mu\gamma^5 u'$	$\bar{u}\gamma^\mu\gamma^5 u'$
$\bar{u}'\gamma^5 u$	$-\bar{u}\gamma^5 u'$	$-\bar{u}\gamma^5 u'$

$$H_{e-N} = i\frac{G_F}{\sqrt{2}}[C_S \sum_{i=1}^{A} \bar{N}_i N_i \cdot \bar{e}\gamma^5 e$$

$$+ C_T \sum_{i=1}^{A} \bar{N}_i \sigma^{\mu\nu} N_i \cdot \bar{e}\sigma_{\mu\nu}\gamma^5 e + C_P \sum_{i=1}^{A} \bar{N}_i \gamma^5 N_i \cdot \bar{e}e] \quad (40)$$

Here it is convenient to express the coupling strengths in terms of Fermi's constant G_F. The sums are taken over all A nucleons in the nucleus (for simplicity, we do not differentiate between neutrons and protons, although this could be done), C_S, C_T, and C_P are real coupling constants, and, taking into account the complex conjugation properties of the various bilinear forms in Table 3, we include a factor of i so that H is Hermitian. In the nonrelativistic limit for the nucleons, the term in C_P vanishes; we neglect it henceforth. In that same limit, the scalar and tensor terms yield the following effective one-electron Hamiltonian:

$$H_{e-N} = \frac{iG_F}{\sqrt{2}}[AC_S\gamma^0_e\gamma^5_e + 2C_T\boldsymbol{\gamma}_e \cdot \boldsymbol{\sigma}_N]n(\mathbf{r}) \quad (41)$$

where $\boldsymbol{\sigma}_n$ is the Pauli spin operator of the last unpaired nucleon and $n(\mathbf{r})$ is the nucleon density. The factor A in the C_S term reflects the fact that here the nucleons add coherently. In addition the matrix element of each term in Eq. (41) receives a factor $\approx Z\alpha$ from the Dirac matrices $\gamma^0\gamma^5$ or $\boldsymbol{\gamma}$, (which couple large and small components) and another factor of Z because of the zero-range nature of the interaction (the nucleon density is $n(\mathbf{r}) \approx \delta^3(\mathbf{r})$). Thus, matrix elements of the scalar term vary roughly as $AZ^2 \approx Z^3$, (as does the enhancement factor R). Consequently, if nonzero atomic EDMs were to be observed in atoms of various atomic numbers Z, it would be difficult (although not impossible) to determine from the Z dependence alone whether d_a arose from d_e or the P,T-odd scalar coupling, or both. The scalar term in Eq. (41) is analogous to the dominant contribution to ordinary atomic parity nonconservation (PNC), which arises from the coupling of the axial electronic neutral weak current to the vector nucleonic neutral weak current via Z^0 exchange. Matrix elements of the tensor term vary roughly as Z^2, and this term is analogous to the much smaller nuclear spin-dependent *PNC* contribution arising from vector electronic–axial nucleonic coupling.

For purposes of illustration we once again assume a uniform nucleon density within the nuclear volume $V = (4\pi/3)R_{nuc}^3$ and thus obtain from Eq. (41) the effective short-range scalar electronic Hamiltonian:

$$H^S_{e-N} = i\frac{G_F}{\sqrt{2}}\frac{3A}{4\pi R_{nuc}^3}C_S\gamma^0\gamma^5 \qquad r \le R_{nuc} \tag{42}$$

The first-order shift in the energy of a paramagnetic atom in the presence of external electric field E_e is then:

$$\begin{aligned}\Delta E_S &= \langle\psi|H^S_{e-N}|\psi\rangle = 2eE_e\langle\psi_0|H^S_{e-N}|\eta\rangle \\ &= 2i\frac{G_F}{\sqrt{2}}\frac{3AC_S}{4\pi R_{nuc}^3}eE_e\int_{\text{nuc. volume}}\psi_0^\dagger\gamma^0\gamma^5\eta d^3\mathbf{r}\end{aligned} \tag{43}$$

Hence, employing Eqs. (25), (26), and (30) we obtain the following contribution to the atomic EDM:

$$\begin{aligned}d_a^{e-N,\text{ scalar}} &= -\frac{\Delta E_S}{eE_e} \\ &= -\frac{6}{4\pi\sqrt{2}}\frac{G_F AC_S}{R_{nuc}^3}\int_0^{R_{nuc}}[G_{\ell,1/2}F^S_{L,1/2} + G^S_{L,1/2}F_{\ell,1/2}]\,dr\end{aligned} \tag{44}$$

It is easy to show that the integrand can be expressed as a power series $ar^2 + br^4 + \ldots$ where a, b are constants, and that $d_a^{e-N,\text{ scalar}}$ is in fact very insensitive to R_{nuc}, as it is to the shape of the nucleon density distribution. Employing the same numerical values in Eq. (44) of $F_{\ell,1/2}$, $G_{\ell,1/2}$, $F^S_{L,1/2}$, and $G^S_{L,1/2}$ that were used in the Tietz-potential calculation of $R(^{205}Tl)$, we obtain:

$$\begin{aligned}d_a^{e-N,\text{ scalar}} &= 6.8\cdot 10^4\, G_F C_S \\ &= 1.5\cdot 10^{-9}\, C_S\end{aligned} \tag{45}$$

in atomic units. At present the upper limit on $d_a(^{205}Tl)$ is $4.4 \cdot 10^{-16}$ in atomic units. Hence, if we assume that there are no other contributions to d_a, (45) provides the following upper limit on C_S:

$$|C_S| \le 2.9 \cdot 10^{-7} \tag{46}$$

In the calculation just described, electron screening was neglected. Assuming that this gives a correction of no more than 30%, we finally obtain:

$$|C_S| \le 4 \cdot 10^{-7} \tag{47}$$

More sophisticated calculations arrive at essentially the same result (Martensson-Pendrill and Lindroth, 1991).

D. Atomic EDM Effects Caused by the Nuclear Magnetic Moment

If an atomic nucleus has spin and a nuclear magnetic moment, several related effects can occur to generate a nonzero d_a from d_e and/or the scalar P,T-odd $e-N$ interaction, even if the atom has closed shells and is thus diamagnetic. These effects have been calculated (Flambaum and Khriplovich, 1985; Martensson-Pendrill and Oster, 1987) following a suggestion by Fortson (1983). First, Eq. (7) implies that an electron EDM d_e should interact with a magnetic field according to the formula: $H_{\text{mag}} = id_e \gamma \cdot \mathbf{B}$, a contribution we have hitherto ignored. In an atom containing a point nucleus with nonzero spin and a nuclear magnetic moment $\boldsymbol{\mu}$, the magnetic field due to $\boldsymbol{\mu}$ is:

$$\mathbf{B} = \left[\frac{3\boldsymbol{\mu} \cdot \mathbf{rr} - \boldsymbol{\mu} \mathbf{r} \cdot \mathbf{r}}{r^5}\right] + \frac{8\pi}{3} \boldsymbol{\mu} \delta^3(\mathbf{r}) \tag{48}$$

The quantity in brackets on the right-hand side of Eq. (48) is frequently called the "dipole" term, whereas the second term is the "contact" term. The field $\mathbf{B}$ is employed in the following expression for d_a, assuming as usual that the external electric field is in the z direction:

$$d_a = \sum_n \frac{\langle \psi_0 | H_{\text{mag}} | \psi_n \rangle \langle \psi_n | ez | \psi_0 \rangle + \langle \psi_0 | ez | \psi_n \rangle \langle \psi_n | H_{\text{mag}} | \psi_0 \rangle}{E_0 - E_n} \tag{49}$$

Among the matrix elements in Eq. (49), those of the form $(8\pi i/3)d_e \langle s_{1/2} | \boldsymbol{\mu} \cdot \gamma \delta^3(\mathbf{r}) | p_{1/2} \rangle$ arising from the contact term are infinite for a point nucleus and large ("relativistically enhanced") for finite nuclear size. Analogous matrix elements between $s_{1/2}$ and $p_{3/2}$ orbitals are obtained from the dipole part of Eq. (48) in the limit of a point nucleus, and do not have the same relativistic enhancement. Thus, when one sums over all electrons in a diamagnetic atom, the contributions of $p_{1/2}$ and $p_{3/2}$ subshells do not cancel (as they would in the nonrelativistic limit), and a nonzero result remains. It is interesting to note that the tensor P,T-odd $e-N$ interaction generates matrix elements of similar form to those in Eq. (49), and its effects are calculated for a diamagnetic atom in similar fashion. The main difference is that the tensor interaction is purely "contact."

In a second effect, we consider simultaneously the perturbation due to d_e: $H_{\text{EDM}} = -d_e \gamma^0 \boldsymbol{\Sigma} \cdot \mathbf{E}$, and that due to the hyperfine interaction, $H_{\text{hfs}} = (e/c)\mathbf{r} \times \boldsymbol{\alpha} \cdot (\boldsymbol{\mu}/r^3)$. This gives rise to an atomic EDM d_a in third-order of perturbation:

$$d_a = \sum_{n,k} \frac{\langle \psi_0 | -d_e \gamma^0 \boldsymbol{\Sigma} \cdot \mathbf{E}_i | \psi_n \rangle \langle \psi_n | \frac{e}{c}\mathbf{r} \times \boldsymbol{\alpha} \cdot \frac{\boldsymbol{\mu}}{r^3} | \psi_k \rangle \langle \psi_k | ez | \psi_0 \rangle}{(E_0 - E_n)(E_0 - E_k)} + \cdots \tag{50}$$

where the ellipses refer to permutations of the various operators. Here, because electron spin appears in the first *and* second matrix elements, the sum over all paired electron spins does not vanish. Calculations reveal that the contribution of Eq. (50) is much larger than that of Eq. (49) (Flambaum and Khriplovich, 1985; Martensson-Pendrill and Oster, 1987). One finds that for the two diamagnetic atoms ^{129}Xe, ^{199}Hg on which EDM experiments have been performed, $R(^{129}Xe) = [d_a/d_e]_{Xe} = -.0008$, whereas $R(^{199}Hg) = [d_a/d_e]_{Hg} = -.014$. Thus, the enhancement factors for these diamagnetic atoms are small, but not zero. Similar calculations have been performed for the diamagnetic molecule *TlF*.

Finally, we note that an expression similar to (50) is obtained if H_{EDM} is replaced by the scalar *P,T*-odd interaction. Here, in spite of the fact that one has a third-order expression, the very precise experimental result obtained for $d_a(^{199}Hg)$ by Fortson and co-workers (Jacobs *et al.*, 1995) yields a limit on C_S that is nearly comparable to that achieved with ^{205}Tl (see Eq. (47)).

E. EXPERIMENTAL SEARCHES FOR THE ELECTRON EDM

The first useful limits on d_e were established by Salpeter (1958), who analyzed the consequences of various experimental results known at the time concerning the Lamb shift in hydrogen, the metastability of the 2s state in hydrogen, the absence of $K \rightarrow L1$ x-ray transitions in heavy atoms, and the hyperfine splitting in the ground state of positronium. These gave $d_e \leq 1.5 \cdot 10^{-13}$ *e* cm. Shortly thereafter, using a *g*-2 experiment with relativistic electrons, Crane and collaborators obtained the limit $|d_e| \leq 3 \cdot 10^{-15}$ *e* cm (Nelson *et al.*, 1959). In 1968, Lipworth and co-workers (Weisskopf *et al.*, 1968) completed an atomic beam magnetic resonance experiment on *Cs* and found $|d(Cs)| \leq 3.7 \cdot 10^{-22}$ *e* cm, which implies $|d_e| \leq 3 \cdot 10^{-24}$ *e* cm. A similar result was obtained for thallium: d(*Tl*) = $(1.3 \pm 2.4) \cdot 10^{-21}$ *e* cm, eventually yielding $d_e = (2 \pm 4) \cdot 10^{-24}$ *e* cm (Gould, 1970). At about the same time, Player and Sandars (1970), working with the $^3P_2$ metastable state of xenon, obtained: $d_e = (0.7 \pm 2.2) \cdot 10^{-24}$ *e* cm. No further substantial progress on d_e was made until 1989, when L. Hunter and co-workers (Murthy *et al.*, 1989) performed an optical pumping experiment on cesium, and obtained the result $d(Cs) = (-1.8 \pm 6.7 \pm 1.8) \cdot 10^{-24}$ *e* cm, which implies $d_e \leq 9 \cdot 10^{-26}$ *e* cm. Hunter's ingenious experiment, carried out with modest resources at a small institution, represented a major advance. In a molecular beam experiment on the molecule *TlF* designed primarily to place a limit on the proton EDM and on the Schiff moment of the thallium nucleus, various other useful limits were obtained by Hinds and co-workers (Cho *et al.*, 1991).

Electron EDM experiments have been suggested that would utilize the methods of atom trapping and cooling (Gould, 1995; Chu, 1997). Heinzen and co-workers (Bijlsma *et al.*, 1994) have analyzed the limitations imposed by collisions on the precision that can be achieved in such an experiment. The trapping of cesium atoms in cold solid helium has been studied experimentally, and it has been

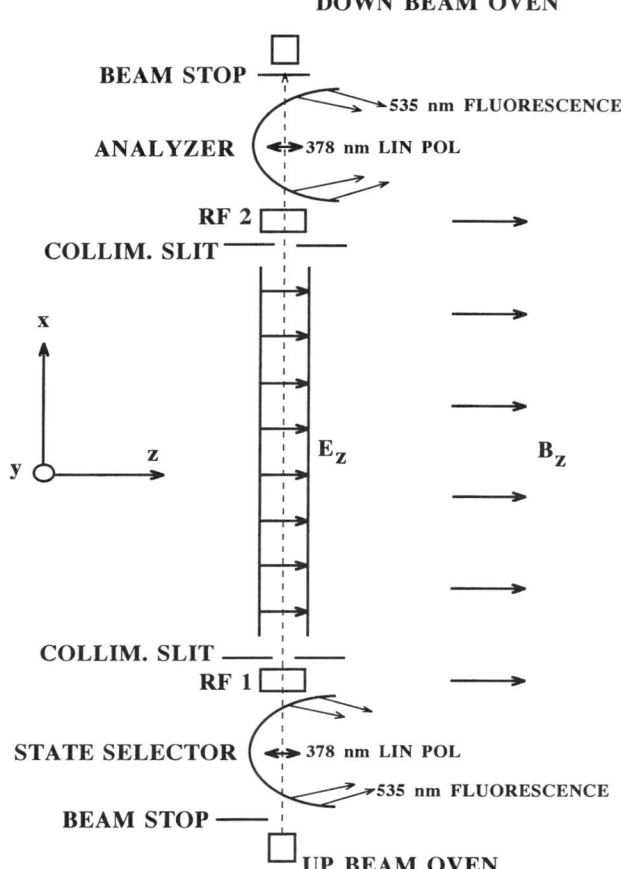

FIG. 9. Schematic diagram of apparatus used in Berkeley experiment to search for electron EDM d_e in ^{205}Tl, (Commins et al., 1994).

proposed that this offers interesting possibilities for an electron EDM search (Kanorsky et al., 1996).

At present the best limits on d_e and C_S are derived from the Berkeley experiment on ^{205}Tl (Abdullah et al., 1990; Commins et al., 1994), the principal features of which will now be summarized. The atomic beam magnetic resonance method with separated oscillating fields was employed (see Fig. 9). The experiment was performed in a weak uniform magnetic field that defined the axis of quantization z (typically $B_Z = 0.4$ G). A strong electric field **E** (typically 107 kV/cm) was placed between two oscillating rf field regions and was nominally parallel to **B**. In order to minimize an important possible systematic effect (the "$\mathbf{E} \times \mathbf{v}$" effect) two counterpropagating beams of atomic Tl were utilized, which travelled in the $\pm x$ directions (vertical to minimize the effects of gravity).

We follow the up-going beam in order to explain the main points. As the beam emerged from the oven it consisted almost entirely of atoms in an incoherent mixture of the ground state components $F = 1$, $m_F = +1, 0$, and -1; and $F = 0$, $m_F = 0$, with essentially equal populations. (See Fig. 10 for the energy levels of ^{205}Tl). In the state selector the atomic beam was intersected by a laser beam propagating in the y direction, linearly polarized in the z direction, and tuned to the $E1$ transition $6P_{1/2}$, $F = 1 \rightarrow 7S$, $F = 1$ at 378 nm (see Fig. 11). Here the selection rule $\Delta m_F = 0$ applies; moreover there can be no transition $F = 1$, $m_F = 0 \rightarrow F = 1$, $m_F = 0$, because the corresponding Clebsch-Gordan coefficient is zero. Atoms excited to the $7S$, $F = 1$, $m_F = \pm 1$ states decayed spontaneously in the following ways: (a) to $6P_{1/2}$, $F = 1$, $m_F = \pm 1$ states from which they were repumped; (b) to $6P_{1/2}$, $F = 1$, $m_F = 0$ where they remained; (c) to $6P_{1/2}$, $F = 0$, $m_F = 0$ where they remained to play no further role; or (d) to the metastable state $6P_{3/2}$, accompanied by fluorescence at 535 nm, which was detected. Because the mean life of $6P_{3/2}$ (≈ 0.2 s) is long compared to the transit time of the beam through the apparatus, atoms arriving in this state remained and played no further role.

Consequently, as the beam emerged from the state selector region, the $6P_{1/2}$, $F = 1$ level contained only atoms in the $m_F = 0$ sublevel, the $m_F = \pm 1$ components having been depopulated. We may represent the $F = 1$ state by the three-component spinor:

$$\psi = \begin{pmatrix} 0 \\ 1 \\ 0 \end{pmatrix} \tag{51}$$

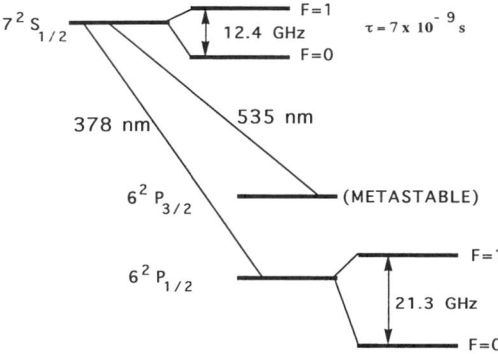

FIG. 10. Low-lying energy levels of ^{205}Tl, (not to scale).

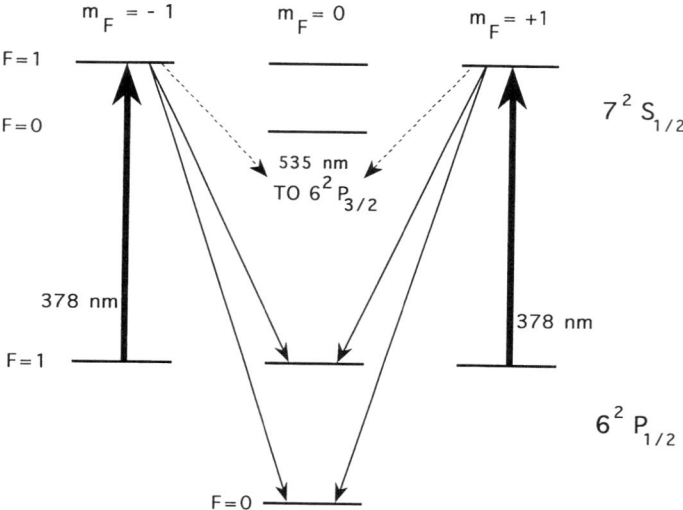

FIG. 11. Schematic diagram showing state selection by laser optical pumping in ^{205}Tl experiment (Commins et al., 1994).

in the rotating frame defined by the static magnetic field $\mathbf{B} = B_0 \hat{\mathbf{z}}$. The beam next traversed the first rf region (RF1 in Fig. 9) where a magnetic field $B_1 \cos(\omega t) \hat{\mathbf{x}}$ was applied at frequency ω tuned to the $M1$ transitions $6P_{1/2}, F = 1: m_F = 0 \to m_F = \pm 1$. On resonance, for a monoenergetic beam, and for the appropriate field magnitude B_1, ψ was transformed as follows in RF1:

$$\psi \to \psi' = \frac{-i}{\sqrt{2}} \begin{pmatrix} 1 \\ 0 \\ 1 \end{pmatrix} \quad (52)$$

Thus as the $F = 1$ atoms emerged from RF1 they were in a coherent superposition of $m_F = \pm 1$ components. We may regard everything described so far as initial preparation of the atomic state. The remainder of the apparatus may be thought of as an interferometer, in which a phase shift between the $m_F = \pm 1$ components proportional to d_a and to an applied electric field was introduced and detected. The beam next passed through the electric field E of length $L = 100$ cm. If $d_a \neq 0$, ψ' underwent the following transformation:

$$\psi' \to \psi'' = \frac{-i}{\sqrt{2}} \begin{pmatrix} \exp(-i\eta_E \epsilon) \\ 0 \\ \exp(+i\eta_E \epsilon) \end{pmatrix} \approx \frac{-i}{\sqrt{2}} \begin{pmatrix} (1 - i\eta_E \epsilon) \\ 0 \\ (1 + i\eta_E \epsilon) \end{pmatrix} \quad (53)$$

where $\eta_E = \pm 1$ for $E > 0$, $E < 0$, respectively, and:

$$\epsilon = \frac{-d_a |E|}{\hbar} \frac{L}{v} \tag{54}$$

where v was the beam velocity (typically $v \approx 4 \cdot 10^4 =$ cm/s). Next the beam passed through the second rf region RF2, containing an rf field oscillating coherently with that of RF1 in the same direction, with the same frequency and similar amplitude, but shifted by a phase $\alpha = \pm\pi/4$ or $\pm 3\pi/4$. On resonance and for $\alpha = \pm\pi/4$, ψ'' was transformed as follows:

$$\psi'' \to \psi''' = \begin{bmatrix} \pm \eta_B e^{-i\eta_B \alpha} \dfrac{(1 \mp \eta_E \eta_B \epsilon)}{2} \\ -\dfrac{1}{\sqrt{2}}(1 \pm \eta_E \eta_B \epsilon) \\ \mp \eta_B e^{i\eta_B \alpha} \dfrac{(1 \mp \eta_E \eta_B \epsilon)}{2} \end{bmatrix} \tag{55}$$

and where $\eta_B = \pm 1$ for $B_0 > 0$, $B_0 < 0$ respectively. Finally the atoms entered the analyzer region, where a second laser beam, directed along the y axis, with z linear polarization, and tuned to the transition $6P_{1/2}, F = 1 \to 7S, F = 1$ at 378 nm as before, intersected the atomic beam. Once again only atoms in the $6P_{1/2}, F = 1$, $m_F = \pm 1$ states were excited to the $7S$ state. The fluorescence at 535 nm accompanying decay of $7S$ atoms in the analyzer region was detected. Its intensity was proportional to the populations of $6P_{1/2}, F = 1, m_F = \pm 1$ levels just prior to laser excitation; hence from Eq. (55), it was proportional to:

$$S = |1 \mp \eta_E \eta_B \epsilon|^2 \approx 1 \mp 2\eta_E \eta_B \epsilon \tag{56}$$

By observing the change in S when E or B was reversed, it was possible to measure ϵ and thus d_a. In Eq. (56), the term in ϵ is proportional to the P,T-odd pseudoscalar $\mathbf{E} \cdot \mathbf{B}$. By means of automatically controlled beam stops it was possible to switch back and forth periodically from the up-beam to the down-beam. The state selector, RF1, RF2 and the analyzer for the up-beam became the analyzer, RF2, RF1, and the state selector, respectively, for the down-beam. For the beam velocities and electric fields employed in this experiment, $d_e = 1 \cdot 10^{-27}$ e cm would correspond to $|\epsilon| \approx 3 \cdot 10^{-7}$.

In the idealized description just given, we have ignored a number of important features, including the beam velocity distributions, the deviation of applied radio frequency from resonance, the quadratic Stark effect, and various important systematic effects. However, in the actual experiment all of these features and in particular the systematics required very careful attention and an elaborate set of auxiliary measurements. The final result was:

$$d_a = [-1.05 \pm .70 \pm .59] \cdot 10^{-24} \, e \, \text{cm} \qquad (57)$$

where the first uncertainty in Eq. (57) is statistical, and the second is systematic. Assuming an enhancement factor of $R = -585$, and ignoring all possible contributions to d_a except for d_e, one obtains from Eq. (57) the result:

$$d_e = [1.8 \pm 1.2 \pm 1.0] \cdot 10^{-27} \, e \, \text{cm} \qquad (58)$$

As noted in Section III.C, result (57) can also be used to place a limit on C_S given by Eq. (47), assuming that the sole contribution to d_a is the scalar P,T-odd $e-N$ interaction. Finally, as noted earlier, result (57) can be used to place limits on P-even, T-odd $e-e$ and $e-N$ interactions, on certain T-odd beta decay couplings (Commins et al., 1994) and on possible magnetic monopole–antimonopole pairs in the nucleus (Flambaum and Murray, 1997).

In approximate order of importance, the main sources of noise in the experiment just described were (a) temporal magnetic fluctuations arising from small variations in the current passing through magnetic field coils; (b) atomic beam intensity fluctuations; (c) laser power and frequency fluctuations; and (d) shot noise. The main effects contributing to the systematic uncertainty were, in order of importance, the **Exv** effect, a related geometric phase effect, and possible fluctuations in **B** correlated with the sign of **E** due to charging and leakage currents. A careful analysis of these effects led to the conclusion that the combined uncertainty in result (57) could be reduced by at least an order of magnitude with several essential improvements, which have been implemented since 1994 and will now be described.

The new apparatus (see Fig. 12) functions according to the same general plan as the previous one, but utilizes two up-beams issuing from a common oven with two source slits, as well as two down-beams. These beams are separated by 2.5 cm and pass through separate state selectors, collimating slits, rf regions, and analyzer-detector regions, and *opposite* electric fields. However, the magnetic field is essentially the same for each beam. Thus, the EDM asymmetry is of opposite sign for the two beams but many sources of noise are highly correlated ("common-mode") for the two beams. In particular this includes all the sources of noise just mentioned except for shot noise. We have demonstrated conclusively by experiment that when the difference between the signals (Eq. 56) is taken for the two beams, the terms in ϵ add but the noise is reduced by about a factor of 10.

In another basic improvement, each up-beam and each down-beam actually consists of atomic sodium as well as atomic thallium issuing simultaneously from each source slit. (This requires a rather sophisticated oven design because thallium and sodium have very different vapor pressures.) Because the atomic number of sodium is $Z = 11$, the enhancement factor is only $R(Na) \approx 0.3$, whereas the contribution from the scalar P,T-odd $e-N$ interaction, proportional to Z^3, is also negligible. Thus, sodium cannot exhibit an observable EDM or P,T-odd effect from

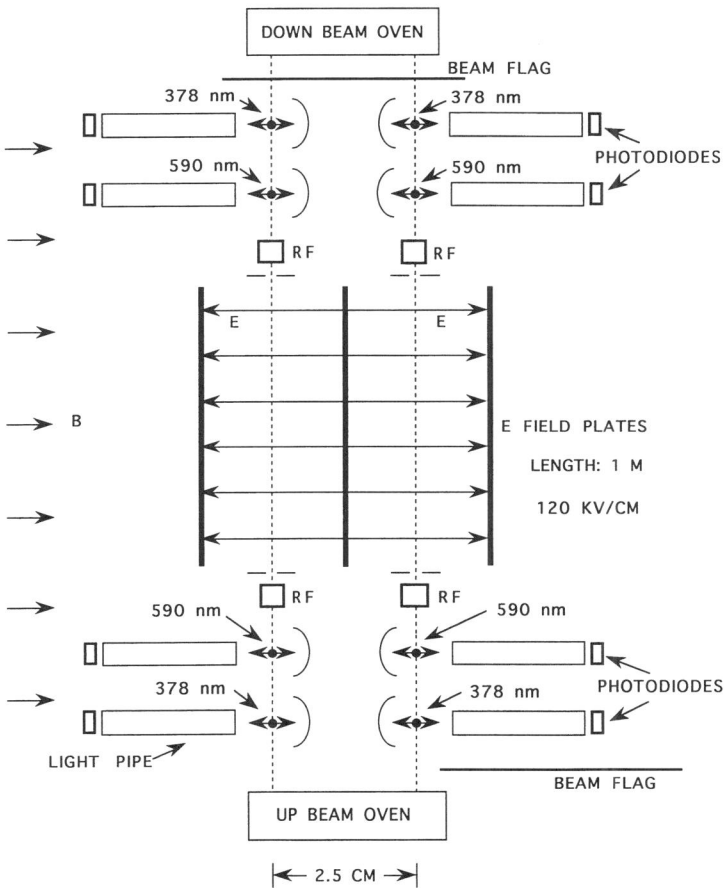

FIG. 12. Schematic diagram showing major modifications in Berkeley ^{205}Tl experiment, completed between 1995 and 1997. There are now two parallel up-beams separated by 2.5 cm, as well as two parallel down-beams. Each beam consists of ^{23}Na and ^{205}Tl emitted simultaneously from each source. Optical pumping of thallium (378 nm) and sodium (590 nm) is achieved with laser beams perpendicular to page, with linear polarizations indicated by short double-arrows.

the scalar $e-N$ interaction, but is even more sensitive to the Ev effect, the geometric phase effect, and leakage and charging currents than thallium. Thus, sodium functions as an ideal "null comparison." Additional improvements include more efficient fluorescence detection and higher atomic beam intensities. Detailed experimental investigations of this system show that it has the potential to reduce the systematic uncertainties in Eq. (57) by at least an order of magnitude.

F. EDMs of Paramagnetic Molecules

Certain paramagnetic polar diatomic molecules are very attractive candidates for experimental electron EDM searches, because their enhancement factors are orders of magnitude larger than that of atomic thallium (Sushkov and Flambaum, 1978; Gorshkov et al., 1979; Sushkov et al., 1984; Flambaum and Khriplovich, 1985; Kozlov, 1985, 1988; Kozlov et al., 1988; Kozlov and Ezhov, 1994; Kozlov and Labzowsky, 1995). The list of proposed molecules includes *BaF, CdF, YbF, HgF, PbF, PbO,* and several others as well. Most have $^2\Sigma_{1/2}$ or $^2\Pi_{1/2}$ electronic ground states. Here, as usual, the left superscript $2\Sigma + 1$ refers to the projection Σ of electron spin on the internuclear axis, the letters Σ, Π refer to the projection Λ (= 0, 1 respectively) of electronic orbital angular momentum on the internuclear axis, while the right subscript denotes the projection Ω of total electronic angular momentum on the same axis. The energy level spectrum of any of these molecules is typically quite complicated because of electron and nuclear spin. For $^2\Sigma_{1/2}$ molecules (which conform to Hund's case (b)), each rotational level above the lowest is split into two components ("γ doubling") by the interaction between electron spin and nuclear rotation. Generally speaking this interaction is weak and the splittings thus generated are small compared to energy differences between adjacent rotational levels. The hyperfine coupling of electron spin to fluorine nuclear spin results in splittings roughly comparable to or larger than γ doubling. However, in a molecule *MF,* where M = heavy metal atom and F = fluorine, the hyperfine splittings arising from interaction between electron spin and M nuclear spin (when the latter is nonzero) can be much larger; they are comparable to rotational splittings. For a $^2\Pi_{1/2}$ molecule, (the description of which is usually intermediate between Hund's cases (a) and (c)) "γ" doubling is replaced by the much larger "Ω" doubling, and once again there are very substantial effects due to hyperfine interaction between electron spin and M nuclear spin.

Typically the M atom in its normal state has two 6s valence electrons, while the ground fluorine configuration is $1s^2 \ldots 2p^5$. In the *MF* molecule, one of the 6s electrons is transferred to the fluorine, thereby completing its p shell and creating an ionic bond, with a corresponding molecular dipole moment that is typically 3–5 Debye units. The remaining 6s electron moves in a highly polarized orbit in the strong internal electric field E_i directed along the internuclear axis between M^+ and F^-. This unpaired electron is the analog of the valence electron in cesium or thallium.

As is the case for atoms, the basic idea here is to search for d_e by observing an energy shift that reverses with an external electric field E_e. From previous discussion we know that in the atomic case the shift can be expressed schematically as follows:

$$\Delta E \approx \langle \psi_0 | -d_e(\gamma^0 - 1)\Sigma \cdot \mathbf{E}_i | \chi \rangle \frac{\langle \chi | e\mathbf{E}_e \cdot \mathbf{r} | \psi_0 \rangle}{E_0 - E_\chi} \tag{59}$$

where $|\psi_0\rangle$ is the state of interest and $|\chi\rangle$ is an appropriate nearby state of opposite parity. For a typical atom $E_0 - E_\chi \gg \langle\chi|e\mathbf{E}_e \cdot \mathbf{r}|\psi_0\rangle$; in other words, the atom can be only weakly polarized by the most intense external electric fields realizable in the laboratory. However, in a polar diatomic molecule, $E_0 - E_\chi$ is the splitting between two adjacent spin-rotational levels of opposite parity and is 10^{-3}–10^{-4} of the corresponding atomic energy difference. It is then possible to polarize some types of molecules almost completely with practical applied electric fields, so that the internuclear axis lies along the direction of $\mathbf{E}_e$. In this circumstance the molecule can have a very large enhancement factor.

The molecule YbF is of special interest, because of efforts by E. Hinds and co-workers to create a practical electron EDM experiment by the molecular beam method (Hinds and Sauer, 1997), in the course of which accurate laser-rf double resonance measurements have been performed on the spin-rotation structure of low-lying molecular levels (Sauer et al., 1995, 1996). Experiments on YbF are also planned by the St. Petersburg group (Ezhov, 1997). Ytterbium has seven stable isotopes, of which five have zero nuclear spin. Ignoring P,T-odd effects for the moment, one finds that the spin-rotation structure of YbF for a spin-zero Yb isotope is described by the following Hamiltonian:

$$H_{\text{spin-rot}} = B\mathbf{N}^2 + \gamma\mathbf{S} \cdot \mathbf{N} + b\mathbf{I} \cdot \mathbf{S} + c\mathbf{I} \cdot \hat{\mathbf{n}}\mathbf{S} \cdot \hat{\mathbf{n}} + C\mathbf{I} \cdot \mathbf{N} \quad (60)$$

Here $\mathbf{N}$, $\mathbf{S}$, and $\mathbf{I}$ are the rotational angular momentum, electron spin, and fluorine nuclear spin ($I = 1/2$), respectively, and $\hat{\mathbf{n}}$ is a unit vector along the internuclear axis from M^+ to F^-. The quantities b and γ exhibit appreciable centrifugal distortion, described by the relations:

$$b = b_0 + b_1 N(N+1) \quad (61)$$
$$\gamma = \gamma_0 + \gamma_1 N(N+1) + \gamma_2 [N(N+1)]^2$$

The constants B, b_0, b_1, γ_0, γ_1, γ_2, c, C, and the molecular dipole moment μ_e were measured by Sauer et al. (1996) for the $X\,^2\Sigma_{1/2}$, $v = 0$ rotational manifold, and are displayed in Table 4. The hyperfine structure of the rotational levels can be understood as follows: we first couple $\mathbf{N}$ and $\mathbf{S}$ to form $\mathbf{J}$, which then couples to $\mathbf{I}$ to form $\mathbf{F}$. The state thus constructed can be expressed as $|(NS)J, I; F, M\rangle$ or simply $|(J), F, M\rangle$. Because $S = 1/2$, one has $J = 1/2$ for $N = 0$, and $J = N \pm 1/2$ for $N > 0$. Because $I = 1/2$ also, we have $F = N + 1$ and $F = N$ for $J = N + 1/2$, while $F = N$ and $N - 1$ for $J = N - 1/2$. From Eq. (60) it can then be shown that, apart from a common term $BN(N+1)$, the hyperfine levels have the following energies:

$$E_{N+1} = \frac{\gamma}{2}N + \frac{b}{4} + \frac{c}{4(2N+3)} + \frac{CN}{2} \quad (62)$$

TABLE 4
MEASURED CONSTANTS FOR THE MOLECULE ^{174}YbF, $X^2\Sigma^+$, $v = 0$ MANIFOLD.
SYMBOLS ARE DEFINED IN TEXT. (SAUER ET AL., 1996)

Molecular parameters	Measured values
γ_0 (MHz)	-13.42400 (16)
γ_1 (kHz)	3.9823 (11)
γ_2 (mHz)	-25 (1)
b_0 (MHz)	141.7956 (5)
b_1 (kHz)	$-.510$ (11)
c (MHz)	85.4026 (14)
C (kHz)	20.38 (13)
μ_e (D) [dipole moment, Debye units]	3.91
B (cm^{-1})	0.2412927 (7) = 7234 MHz

$$E_{N-1} = -\frac{\gamma}{2}(N + 1) + \frac{b}{4} - \frac{c}{4(2N - 1)} - \frac{C(N + 1)}{2} \quad (63)$$

$$E_{N^\pm} = -\frac{(\gamma + b + C)}{4}$$

$$\pm \frac{1}{4}\sqrt{(\gamma - C)^2(2N + 1)^2 + (2b + c - 2C)(2b + c - 2\gamma)} \quad (64)$$

with corresponding eigenfunctions:

$$|N^\pm, M\rangle = x^\pm |(N + 1/2), N, M\rangle + y^\pm (N - 1/2), N, M\rangle \quad (65)$$

where

$$\frac{x^\pm}{y^\pm} = -\frac{\langle(N + 1/2), N, M|H|(N - 1/2), N, M\rangle}{\langle(N + 1/2), N, M|H|(N + 1/2), N, M\rangle - E_{N^\pm}} \quad (66)$$

For the important special case of the ground rotational level $N = 0$, one obtains:

$$E(F = 1) = \frac{b}{4} + \frac{c}{12}$$

$$E(F = 0) = -\frac{3b}{4} - \frac{c}{4}$$

which yields the ground $N = 0$ state hyperfine splitting:

$$E(F = 1) - E(F = 0) = b + \frac{c}{3} = 170 \text{ MHz} \quad (67)$$

We now consider the effect of d_e and the scalar P,T-odd $e-N$ interaction on the aforementioned spin-rotation structure. Kozlov and Labzowsky (1995) have re-

TABLE 5
CALCULATED VALUES OF W^d AND W_1^{PT} FOR SELECTED PARAMAGNETIC DIATOMIC MOLECULES.

Molecule	$W^d[10^{25}\,\text{Hz}\,e^{-1}\,\text{cm}^{-1}]$	$W_1^{PT}[10^3\,\text{Hz}]$	Reference
BaF	−0.35 to −0.41	−11 to −13	Kozlov and Labzowsky, 1995
YbF	−1.5	−48	,, ,, ,, ,,
HgF	−4.7	−203	,, ,, ,, ,,
PbF	1.4	55	,, ,, ,, ,,
PbO	0.13		DeMille, 1997
CsXe	0.017		Kozlov and Yashchuk, 1996

viewed various methods for calculating these effects. As in the case of atoms, one can take a semi-empirical or an ab initio approach. It was shown by Kozlov (1985) that an effective semi-empirical method is based on the close connection between the matrix elements of the P,T-odd operators, and magnetic hyperfine structure operators for coupling of electron spin to nonzero M nuclear spin. Because for several molecules of interest (and in particular for YbF) the hyperfine structure constants are known from experiment, these data can be used to calculate quite accurately the electron spin density near the M nucleus, without direct knowledge of the complicated electronic wave-function of the molecule (Kozlov and Ezhov, 1994). The results of these calculations are conveniently expressed by an additional P,T-odd effective Hamiltonian H' that must be added to $H_{\text{spin-rot}}$:

$$H' = (W_1^{P,T}C_S + W^d d_e)\mathbf{S} \cdot \hat{\mathbf{n}} \quad (68)$$

The calculated numerical coefficients $W_1^{P,T}$ and W^d are summarized in Table 5. We recall that for a sufficiently strong external electric field E_e ($\approx 3 \cdot 10^4$ V/cm for YbF) $\hat{\mathbf{n}}$ would be polarized along E_e and the energy shift would be given directly by coefficients $W_1^{P,T}$ and/or $W^d d_e$.

In the experiment of Hinds and co-workers (Hinds, 1997), a molecular beam of YbF is created, and the $F = 1$, $N = 0$ level is depopulated by optical pumping. Then an adiabatic Raman transition transfers the $F = 0$ population to a coherent superposition of $F = 1$, $m_F = \pm 1$ levels. The beam then passes through a region of electric field E_Z in which there is a magnetic field B_Z nominally parallel (or antiparallel) to E_Z, and a phase shift is developed between the $m_F = \pm 1$ components that is analogous to that described in previous paragraphs for the Berkeley thallium experiment. For analysis and detection a second adiabatic Raman transition transfers the $F = 1$ population back to $F = 0$, and a probe laser excites $F = 0$ to the $A^2\Pi_{1/2}$ level, from which fluorescence is detected. One searches for a change in the signal that is proportional to the relative sign of E_Z and B_Z.

A serious practical difficulty for alkaline earth-fluorides and YbF in particular is that they are radicals and thus chemically unstable. Hence they can only be used in molecular beam experiments, and there it is difficult to prepare sources to

TABLE 6
SUMMARY OF RESULTS FROM NEUTRON, ATOMIC, AND MOLECULAR EDM EXPERIMENTS.

P,T viol. param.	System	Upper limit	Reference
d_n	n	$8 \cdot 10^{-26}$ e cm	Smith et al., 1990, Altarev et al., 1996
$d_{mol}(TlF)$	TlF	$4.6 \cdot 10^{-23}$ e cm	Cho et al., 1991
d_{proton}	TlF	$1 \cdot 10^{-23}$ e cm	Cho et al., 1991
$d_a(^{199}Hg)$	^{199}Hg	$8.7 \cdot 10^{-28}$ e cm	Jacobs et al., 1995
$d_a(^{205}Tl)$	^{205}Tl	$2.3 \cdot 10^{-24}$ e cm	Commins et al., 1994
Schiff moment Q_S	^{199}Hg	$2.2 \cdot 10^{-11}$ e cm^3	Jacobs et al., 1995
" " "	^{205}Tl	$1 \cdot 10^{-9}$ e cm^3	Cho et al., 1991
$\eta^{(a)}$	^{199}Hg	$1.6 \cdot 10^{-3}$	Jacobs et al., 1995
$\eta_q^{(b)}$	^{199}Hg	$3.4 \cdot 10^{-6}$	Jacobs et al., 1995
C_T	^{199}Hg	$1.3 \cdot 10^{-8}$	Jacobs et al., 1995
C_S	^{205}Tl	$4 \cdot 10^{-7}$	Commins et al., 1994
d_e	^{205}Tl	$4 \cdot 10^{-27}$ e cm	Commins et al., 1994
QCD phase $\bar{\Theta}_{QCD}$	n	$4 \cdot 10^{-10}$	Smith et al., 1990
Supersym. ϵ_q^{SUSY}	^{199}Hg	$7 \cdot 10^{-3}$	Jacobs et al., 1995 (Fischler et al., 1992)
Supersym. ϵ_e^{SUSY}	^{205}Tl	$4 \cdot 10^{-2}$	Commins et al., 1994 (Fischler et al., 1992)
Mag. monopole const B_0	^{205}Tl	$2.5 \cdot 10^{-5} e$ (TeV)$^{-2}$	Flambaum and Murray, 1997

(a) Appears in: $i\eta(G_F/\sqrt{2})\bar{n}n \cdot \bar{n}\gamma^5 n$
(b) Appears in $i\eta(G_F/\sqrt{2})\bar{q}q \cdot \bar{q}\gamma^5 q$

yield sufficiently intense signals. However, a chemically stable molecule such as *PbO* might be employed in a cell experiment, which could have many practical advantages (DeMille, 1997). The ground state of *PbO* is $^1\Sigma_0$, but there is an excited electronic state $^3\Sigma_1$ at 16025 cm^{-1} with two unpaired electrons. It has been suggested that one might observe Faraday rotation in the optical transition ($^1\Sigma_0 \to {}^3\Sigma_1$) in an external electric field (Sushkov and Flambaum, 1978; Flambaum, 1987; Barkov et al., 1988). The enhancement factor for *PbO* is estimated to be ≈ 100 times that for atomic thallium, but considerably smaller than for *YbF*. Possible experiments with van der Waals molecules have also been suggested (Kozlov and Yashchuk, 1996).

To conclude this section, we summarize in Table 6 the limits obtained by various atomic and molecular experiments, and also include results of experimental searches for the electric dipole moment of the neutron.

IV. The Muon EDM

A. Muonic Atoms Are Not Practical for an EDM Search

Until now the only way to search for the muon EDM d_μ is to observe precession of free relativistic muons in a magnetic field (g-2 experiment). Before discussing this method let us answer a frequently asked question: Why not use a muonic atom and take advantage of the enhancement factor as one does for the electron EDM? To see why this is not practical we note that when a μ^- is captured by an atom, it quickly cascades down from excited levels to the $1s$ shell, and remains there until it is captured by the nucleus, or else decays (with mean life $\approx 2 \times 10^{-6}$ s). The capture rate on the nucleus is approximately proportional to Z^4, and is about equal to the decay rate for $Z = 6$. The enhancement factor R for a $1s$ orbital in the Coulomb field of a point nucleus with atomic number Z is easily shown to be independent of the lepton mass, and it has the value (Sandars, 1968):

$$R = -2 \frac{Z^2 \alpha^2}{2\gamma - 1} \tag{69}$$

where $\gamma = \sqrt{1 - Z^2\alpha^2}$. Of course, in a multi-electron muonic atom R must be corrected for screening; because the muon is very close to the nucleus this is now a major correction and could reduce R by a factor $\approx Z^{-1}$: $R_{\text{eff}} \approx R/Z$. In an external electric field E_e, the frequency shift of a suitable muonic atom ground-state magnetic sublevel arising from the EDM d_μ would be $\Delta v = (|R_{\text{eff}}|d_\mu E_e)/h$. The uncertainty principle requires $\tau \, \Delta v \geq 1$ for observation of a single muonic atom, and $\tau \, \Delta v \geq N^{-1/2}$ for observation of N muonic atoms. Thus we require:

$$N \geq \left(\frac{h}{|R_{\text{eff}}|\tau E_e d_\mu}\right)^2$$

Because $|R_{\text{eff}}|$ is approximately proportional to $Z^2 \cdot Z^{-1} = Z$, and τ is approximately proportional to Z^{-4} for $Z \gg 6$, N increases in proportion to Z^6 for $Z \gg 6$. Hence we choose $Z = 6$, for which $\tau = 10^{-6}$ s and $|R_{\text{eff}}| \approx .0007$. Also choosing $E_e = 100{,}000$ V/cm $= 330$ esu/cm and $d_\mu = 7 \cdot 10^{-19}$ e cm $= 3.4 \cdot 10^{-28}$ esu-cm (the present experimental upper limit), we find $N \geq 10^{15}$, a number that is many orders of magnitude beyond the realm of possibility. Thus muonic atoms cannot be used because their numbers are too few, their enhancement factors are too small, and their lifetimes are too short.

B. Precession of Relativistic Muons in a Storage Ring

In an important experiment carried out at the CERN muon storage ring by Bailey and co-workers (Bailey et al., 1978, 1979) the limit:

$$d_\mu \leq 7 \cdot 10^{-19} \, e \text{ cm} \tag{70}$$

was established simultaneously with a precise measurement of the muon g-factor anomaly $a = \frac{1}{2}(g - 2)$. Muons from pion decay with an initial longitudinal polarization $P \geq 95\%$ travelled around the 14 m diameter storage ring in a horizontal plane. A homogeneous vertical magnetic field $B = 1.47$ T was applied, and weak vertical focussing was provided by an electrostatic quadrupole field. In such circumstances it can be shown (Bargmann et al., 1959; Hagedorn, 1963) that the precession angular velocity ω of the muon spin relative to its velocity $\boldsymbol{\beta}$ (which is perpendicular to $\mathbf{B}$ and $\mathbf{E}$) is given by:

$$\omega = -\frac{e}{m_\mu c}\left[a\mathbf{B} + \left(\frac{1}{\gamma^2 - 1} - a\right)\boldsymbol{\beta} \times \mathbf{E} + \frac{1}{2}f(\mathbf{E} + \boldsymbol{\beta} \times \mathbf{B})\right] \quad (71)$$

where $\boldsymbol{\beta}$ is in units where $c = 1$, $\gamma = (1 - \beta^2)^{-1/2}$, and

$$f = 2d_\mu \left(\frac{e\hbar}{2m_\mu c}\right)^{-1} \quad (72)$$

Actually the muon momentum (3.094 GeV/c) was chosen so that the second term in Eq. (71) vanishes:

$$\gamma = [1 + a^{-1}]^{1/2} = 29.3$$

and furthermore, $\mathbf{E}$ was negligible. Thus Eq. (71) can be written:

$$\omega = \omega_a + \omega_{\text{EDM}} \quad (73)$$

where

$$\omega_a = -\frac{e}{m_\mu c} a\mathbf{B} \quad (74)$$

and

$$\omega_{\text{EDM}} = -\frac{e}{2m_\mu c} f(\boldsymbol{\beta} \times \mathbf{B}) \quad (75)$$

Figure 13 shows the orientation of the vectors $\boldsymbol{\beta}$, ω_a, ω_{EDM}, $\mathbf{B}$, and a unit vector $\hat{\mathbf{s}}$ in the direction of the expectation value of muon spin. The angle δ between ω_a and ω_{EDM} is given by:

$$\delta = \arctan\frac{\omega_{\text{EDM}}}{\omega_a} \approx \frac{\omega_{\text{EDM}}}{\omega_a} = \frac{f\beta}{2a} \quad (76)$$

Consequently, one obtains a vertically oscillating component of the muon polarization $\hat{\mathbf{s}}$. This could be measured by recording separately the electrons emitted from muon decay into the upper and lower half-spaces. (Here, as in the measurement of g-2 itself, advantage was taken of the parity-violating beta decay asymmetry of polarized muons). It was also possible to place a limit almost as precise on δ by comparing the observed frequency $\omega \approx \omega_a(1 + \delta^2)^{1/2}$ with ω_a.

A dedicated experiment of the same type has been proposed with a much larger muon storage ring at Brookhaven National Laboratory, where it appears possible in principle to improve the limit given in Eq. (70) by approximately 4 orders of

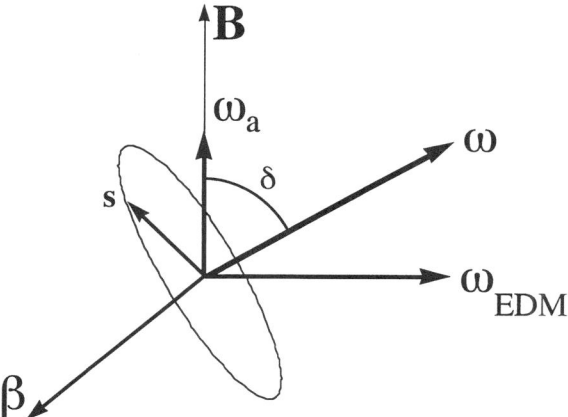

FIG. 13. Schematic diagram (not to scale) of relevant vectors in muon spin-precession experiment (Bailey *et al.*, 1978, 1979). Symbols defined in text.

magnitude (Semertzidis, 1997). According to a 2-Higgs theoretical model of *CP* violation (Bernreuther, 1992a; Barger *et al.*, 1997) the muon EDM could be sufficiently large to be detected in such an experiment.

V. Weak Dipole Moment and EDM of the Tau Lepton

We now consider questions related to the EDM of the tau lepton. Although taus can be generated in a variety of ways, they are most often produced in e^+e^- collisions at colliding beam facilities. As we have remarked previously, the reaction:

$$e^+ + e^- \to (\gamma, Z) \to \begin{array}{c} \tau^+ \\ \downarrow \\ A^+ \bar{\nu}_\tau \end{array} + \begin{array}{c} \tau^- \\ \downarrow \\ B^- \nu_\tau \end{array} \qquad (77)$$

proceeds in lowest order by two amplitudes: photon exchange (Fig. 1a); and Z^0 exchange (Fig. 1b). The mean life of the tau is only $2.9 \cdot 10^{-13}$ s, and it decays in a variety of modes, represented schematically in reaction (77) by $A^+ \bar{\nu}_\tau$ and $B^- \nu_\tau$:

$$\tau^- \to e^- \bar{\nu}_e \nu_\tau$$
$$\mu^- \bar{\nu}_\mu \nu_\tau$$
$$\pi^- \nu_\tau$$
$$\cdots$$

When the e^+e^- CM energy is in the vicinity of the Z resonance (91 GeV), the Z amplitude greatly dominates over the photon amplitude, and the latter can be neglected.

In principle, CP violation effects could occur at the eeZ vertex, at the $Z\tau\tau$ vertex, and/or in the decays of τ^+ and τ^-. However, if e^+ and e^- are unpolarized it can be shown (Bernreuther et al., 1989) that there are no observable CP violating effects at the eeZ vertex. Furthermore, it seems most likely that new CP-violation effects would be associated with the vertex at which there is the largest momentum transfer (Bernreuther, 1991); hence we concentrate our attention on the $Z\tau\tau$ vertex. In order to account for CP violation, we must then construct a generalization of the CP-violating Lagrangian density (Eq. (5)). Including the electromagnetic coupling, it is written as follows:

$$L_{CP} = -\frac{i}{2}\bar{\psi}\gamma^5\sigma^{\mu\nu}\psi[d_\tau F_{\mu\nu} + \tilde{d}_\tau(\partial_\mu Z_\nu - \partial_\nu Z_\mu)] \tag{78}$$

Here Z_ν is the Z^0 boson vector potential (analogous to the electromagnetic 4-potential A_ν), and the quantities d_τ, $\tilde{d}_\tau$ are the EDM and the "weak dipole moment" (WDM), respectively. (Strictly speaking, these quantities can depend on q^2 where q is the 4-momentum transfer; $d_\tau(q^2 = 0)$ is defined to be the EDM, while $\tilde{d}_\tau(q^2 = m_Z^2)$ is defined as the WDM. Although there is no explicit relationship between d_τ and $\tilde{d}_\tau$, they are expected to be roughly the same in most models of CP violation. Near the Z resonance only the $\tilde{d}_\tau$ term is important in Eq. (78).

One motivation to search for $\tilde{d}_\tau$ is that in certain 2-Higgs models of CP violation, lepton flavor-diagonal amplitudes at the one-loop level can generate a lepton EDM proportional to the cube of the lepton mass (Bernreuther and Nachtmann, 1989; Geng and Ng, 1989; Bernreuther et al., 1992b):

$$d \approx \frac{eG_F m^3}{4\pi^2 m_\phi^2} \tag{79}$$

where m_ϕ is the Higgs mass. This might yield $d_\tau \approx 2 \cdot 10^{-19}$ e cm, and essentially the same argument would apply for $\tilde{d}_\tau$.

The transition amplitude T for reaction (77) to a specific final state can be written quite generally as:

$$T = T_{SM} + T_{CP} \tag{80}$$

where T_{SM} is the CP-conserving (standard model) part, and T_{CP} is the CP-violating part. Then the differential cross section is proportional to:

$$|T_{SM} + T_{CP}|^2 = (|T_{SM}|^2 + |T_{CP}|^2) + (T_{SM}^*T_{CP} + cc) \tag{81}$$

The first quantity in parentheses on the right-hand side of Eq. (81) is CP-even and is proportional to the total cross section, or in other words, to the partial width

$\Gamma_{\tau\tau}$ for Z decay to $\tau\tau$. New physics beyond the standard model would be required for $T_{CP} \neq 0$, because according to the standard model, CP-violation effects in the $Z\tau\tau$ vertex are negligibly small. If we assume that the "new physics" causes no change in the CP-conserving amplitude T_{SM} (and this is a very questionable assumption), then we can compare the observed partial width $\Gamma_{\tau\tau}$ to that predicted by the standard model, and thus obtain a limit on $\tilde{d}_\tau$. (With these assumptions it can be shown that $\Delta\Gamma \approx (|\tilde{d}_\tau|^2 m_Z^3)/24\pi$). This procedure yields the limit:

$$|\tilde{d}_\tau| < 2.2 \cdot 10^{-17} \, e \, \text{cm} \tag{82}$$

However, to improve the limit and to avoid the questionable assumption just mentioned, we must exploit the second term in parentheses on the right-hand side of Eq. (81). This "interference" term manifests itself in certain CP-odd observables (correlations) in the differential cross section. (Here we do not get into difficulty by assuming that the standard model applies to the CP-conserving factor T_{SM}^*). The correlations are related to the spins of the outgoing $\tau^+\tau^-$, but these spins are not directly observable. Instead, because of their short lifetime, the taus decay very close to the point where they were created, and transfer information about their spins to the energies and momenta of the decay products A^+, B^-. Thus one must observe the latter quantities in order to measure $\tilde{d}_\tau$.

In principle, the transition amplitude T can have real and imaginary parts. The latter would arise from final state interactions and/or from absorptive parts in the propagators of unstable particles among the decay products. For example, ρ mesons, which have an extremely short lifetime, can occur as intermediate states in the products of tau decay. In order to allow for the possibility that T_{CP} has an imaginary part, we must assume also that $\tilde{d}_\tau$ may have an imaginary part.

In a series of papers, Bernreuther and co-workers (Bernreuther and Nachtmann, 1989; Bernreuther et al., 1989; Bernreuther et al., 1991, Bernreuther et al., 1997) have analyzed the various observable correlations in reaction (77), assuming that e^+ and e^- are unpolarized. In this case the most general CP-odd correlation may be expressed in terms of the contractions of certain irreducible tensors A_i of rank n written in terms of the $\tau^\pm$ spin polarizations $\mathbf{s}_\pm$ and the direction of the τ^+ momentum $\hat{\mathbf{k}}_+$, with n factors of a unit vector $\hat{\mathbf{q}}_e$ in the direction of e^- momentum. For $n \leq 2$, there are 14 such tensors A_i, which are displayed in Table 7.

Under a CPT ($\equiv \Theta$) transformation, $\hat{\mathbf{k}}_+ \to -\hat{\mathbf{k}}_+$, $\mathbf{s}_\pm \to -\mathbf{s}_\mp$, and $\hat{\mathbf{q}}_e \to -\hat{\mathbf{q}}_e$. Let us define the Θ parity of tensor A_i by:

$$(-1)^n A^\Theta = \eta_\Theta A \tag{83}$$

where $\eta_\Theta = \pm 1$. Then it follows that A_i is CPT-even ($\eta_\Theta = +1$) for $i = 2, 6, 7, 8, 12, 13,$ and 14; while A_i is CPT-odd ($\eta_\Theta = -1$) for $i = 1, 3, 4, 5, 9, 10,$ and 11. It can be shown that if CPT invariance is assumed, and we ignore the imaginary parts of all transition amplitudes, then all of the CPT-odd correlations give

TABLE 7
IRREDUCIBLE CP-ODD TENSORS A_i CONSTRUCTED FROM $\tau^\pm$ SPIN POLARIZATIONS $\mathbf{s}_\pm$ AND DIRECTION OF τ^+ MOMENTUM $\hat{\mathbf{k}}_+$. (BERNREUTHER AND NACHTMANN, 1989; BERNREUTHER ET AL., 1989; BERNREUTHER ET AL., 1991; BERNREUTHER ET AL., 1997)

	i	A_i	η_Θ
$n = 0$	1	$\hat{\mathbf{k}}_+ \cdot (\mathbf{s}_+ - \mathbf{s}_-)$	−
	2	$\hat{\mathbf{k}}_+ \cdot (\mathbf{s}_+ \times \mathbf{s}_-)$	+
$n = 1$	3	$\mathbf{s}_+ - \mathbf{s}_-$	−
	4	$[\hat{\mathbf{k}}_+ \cdot (\mathbf{s}_+ - \mathbf{s}_-)]\hat{\mathbf{k}}_+$	−
	5	$(\mathbf{s}_+ \times \mathbf{s}_-) \times \hat{\mathbf{k}}_+$	−
	6	$(\mathbf{s}_+ - \mathbf{s}_-) \times \hat{\mathbf{k}}_+$	+
	7	$\mathbf{s}_+ \times \mathbf{s}_-$	+
	8	$[\hat{\mathbf{k}}_+ \cdot (\mathbf{s}_+ \times \mathbf{s}_-)]\hat{\mathbf{k}}_+$	+
$n = 2$	9	$\hat{k}_{+i}(s_+ - s_-)_j + (i \leftrightarrow j)$	−
	10	$[\hat{\mathbf{k}}_+ \cdot (\mathbf{s}_+ - \mathbf{s}_-)](\hat{k}_{+i}\hat{k}_{+j} - \frac{1}{3}\delta_{ij})$	−
	11	$\hat{k}_{+i}(\hat{\mathbf{k}}_+ \times (\mathbf{s}_+ \times \mathbf{s}_-))_j + (i \leftrightarrow j)$	−
	12	$\hat{k}_{+i}(\hat{\mathbf{k}}_+ \times (\mathbf{s}_+ - \mathbf{s}_-))_j + (i \leftrightarrow j)$	+
	13	$[\hat{\mathbf{k}}_+ \cdot (\mathbf{s}_+ \times \mathbf{s}_-)](\hat{k}_{+i}\hat{k}_{+j} - \frac{1}{3}\delta_{ij})$	+
	14	$\hat{k}_{+i}(\mathbf{s}_+ \times \mathbf{s}_-)_j + (i \leftrightarrow j) - \frac{2}{3}\delta_{ij}$	+

zero after integration over phase space. However, if we allow for the possibility that T (and thus $\tilde{d}_\tau$) have imaginary parts, this is no longer true.

If we assume Eq. (78), ignore the electromagnetic contribution to Eq. (78) at the Z pole, and use the standard model for T_{SM} in the second term in parentheses in Eq. (81), we find (Bernreuther et al., 1991; Akers et al., 1995) that there are only two significant correlations, given by tensors A_{11} (CPT odd) and A_{12} (CPT even). The observables O corresponding to these tensors can be expressed in terms of the WDM as follows:

$$\langle O_{11}\rangle_{AB} = \frac{m_Z}{e} f_{AB} \, \text{Im}\,(\tilde{d}_\tau) \tag{84}$$

and

$$\langle O_{12}\rangle_{AB} = \frac{m_Z}{e} c_{AB} \, \text{Re}\,(\tilde{d}_\tau) \tag{85}$$

Here c_{AB} and f_{AB} are dimensionless constants called sensitivities; they depend on the decay mode AB and can be calculated from the known geometry and other properties of the detector.

Two experiments have been carried out at the large electron-positron collider

(LEP) at CERN. One employed the OPAL detector (Akers et al., 1995) and accumulated 27490 $Z^0 \to \tau\tau$ decays in the years 1991–1993. Many final state channels AB (leptonic, semileptonic and hadronic) were observed. The results of the OPAL experiment are as follows:

$$\mathrm{Re}\,(\tilde{d}_\tau) = (-0.2 \pm 3.6 \pm 1.4) \cdot 10^{-18}\, e\,\mathrm{cm} \tag{86}$$

which yields

$$\mathrm{Re}\,(\tilde{d}_\tau) < 7.8 \cdot 10^{-18}\, e\,\mathrm{cm} \tag{87}$$

and

$$\mathrm{Im}\,(\tilde{d}_\tau) = (0.95 \pm 1.55 \pm .91) \cdot 10^{-17}\, e\,\mathrm{cm} \tag{88}$$

which gives:

$$\mathrm{Im}\,(\tilde{d}_\tau) < 4.5 \cdot 10^{-17}\, e\,\mathrm{cm} \tag{89}$$

Another LEP experiment was performed at the ALEPH detector (Buskelic et al., 1995). Here only the observable defined by Eq. (85) was employed, and $\tilde{d}_\tau$ was assumed to be real. Once again, many final state channels AB were analyzed. The result is:

$$\tilde{d}_\tau = (0.15 \pm 0.58 \pm 0.38) \cdot 10^{-17}\, e\,\mathrm{cm} \tag{90}$$

or

$$|\tilde{d}_\tau| < 1.5 \cdot 10^{-17}\, e\,\mathrm{cm} \tag{91}$$

The OPAL and ALEPH results may be combined to yield:

$$|\mathrm{Re}\,\tilde{d}_\tau| < 6.7 \cdot 10^{-18}\, e\,\mathrm{cm} \tag{92}$$

Escribano and Masso (1997) have carried out a precise calculation of $\Gamma_{\tau\tau}$ using an effective Lagrangian approach, to improve the above limits slightly. They obtain:

$$|\tilde{d}_\tau| < 5.8 \cdot 10^{-18}\, e\,\mathrm{cm} \tag{93}$$

They also obtain the following relation between d_τ and $\tilde{d}_\tau$:

$$|d_\tau| = \cot\theta_W |\tilde{d}_\tau| \tag{94}$$

where θ_W is the weak mixing angle (Weinberg angle). Equations (93) and (94) imply that:

$$|d_\tau| < 1.1 \cdot 10^{-17}\, e\,\mathrm{cm} \tag{95}$$

Impressive as the OPAL and ALEPH results are, the limits attained for d_τ and $\tilde{d}_\tau$ (95, 93) are still substantially larger than the predictions for d_τ and $\tilde{d}_\tau$ in most theoretical models. Considerable gains could be achieved with an e^+e^- colliding beam accelerator of very high luminosity and with $\sqrt{s}$ in the range 3.5 to 6 GeV, the so-called "Tau-Charm Factory," which has been proposed as an international collaboration and may be constructed in the next decade in Beijing. Here it might be possible to employ longitudinally polarized electrons, and thus use more convenient T,P odd correlations to place far more precise limits on d_τ (Huang et al., 1997).

VI. Magnetic and Electric Dipole Moments of Neutrinos

Neutrinos are electrically neutral and have spin 1/2, but it is not yet known whether the neutrino and antineutrino belonging to a given lepton generation are distinct particles ("Dirac" neutrino and antineutrino) or whether the neutrino is self-charge-conjugate ("Majorana" neutrino). If neutrinos are of the Dirac type, then total lepton number $L = L_e + L_\mu + L_\tau$ is conserved in all weak processes, because the charged and neutral weak current operators create as many particles as they destroy. However, if neutrinos are of the Majorana type, a charged weak current constructed from Majorana fields contains terms that change the lepton number by two units: $|\Delta L| = 2$, (Commins and Bucksbaum, 1983).

It is not known whether any or all of the neutrinos from the three lepton generations have zero rest mass. If the neutrino rest masses of at least two generations are different, the possibility exists of neutrino flavor mixing in which the neutrino mass eigenstates $\nu_{1,2,3}$ are nontrivial linear combinations of the neutrino weak eigenstates $\nu_{e,\mu,\tau}$. This could give rise to neutrino oscillations, in which for example a neutrino created in eigenstate ν_e is found at some later time in a mixture of states ν_e, ν_μ, and/or ν_τ. In such oscillations, separate lepton numbers L_e, L_μ, and L_τ are not conserved. At the present time there is no incontrovertible evidence from terrestrial laboratory experiments for neutrino oscillations, but the most widely favored explanation for the solar neutrino deficit is "matter assisted" neutrino oscillations in the solar interior, which operate according to the "MSW mechanism" (Nakamura, 1996 and references therein).

Neutrino magnetic or electric dipole moments are described by 3×3 matrices μ_{ij}, d_{ij} respectively, where the diagonal elements μ_{ii}, d_{ii} refer to the "static dipole moments" of the ith mass eigenstate, and if neutrino flavor mixing occurs, the matrices have nonzero off-diagonal elements as well. If neutrinos are of the Majorana type, the diagonal elements μ_{ii}, d_{ii} must be zero, which can be seen intuitively because Majorana neutrinos are self-charge-conjugate, whereas under charge conjugation the magnetic dipole and electric dipole operators change sign. (However, in the Majorana case, the nondiagonal matrix elements μ_{ij}, d_{ij} with $i \neq j$ are not necessarily zero.) If neutrinos are of the Dirac type and are described

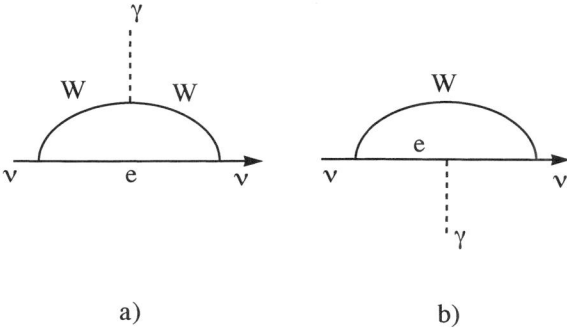

FIG. 14. The two lowest-order diagrams contributing to a neutrino magnetic moment, according to the standard model.

by the standard model, the (diagonal) neutrino magnetic dipole moment arises in lowest order from the diagrams shown in Fig. 14, and is given by the following formula:

$$\mu_{\nu j} = \frac{3 e G_F m_{\nu j}}{8 \pi^2 \sqrt{2}} = 3.2 \cdot 10^{-19} \left(\frac{m_{\nu j}}{1 \text{ eV}} \right) \mu_B \tag{96}$$

where μ_B is the electron Bohr magneton (Lee and Shrock, 1977; Fujikawa and Shrock, 1980; Shrock, 1982). Thus for a Dirac neutrino, according to the standard model, $\mu_{\nu j}$ is proportional to $m_{\nu j}$ and vanishes for zero rest mass; furthermore for any reasonable value of the mass it is extremely small. This occurs because in the standard model there are no right-handed currents to cause a neutrino spin flip. In some alternatives to the standard model (for example left-right symmetric models) the Dirac-neutrino magnetic moment could be orders of magnitude larger (Shrock, 1982). Neutrino electric dipole moments are expected to be extremely small in most theoretical models where the EDM is proportional to the neutrino mass. As in the case of the electron, the standard model yields no neutrino EDM even at the three-loop level, except for the possibility of gluonic corrections to internal quark lines.

The best direct experimental limits on neutrino magnetic moments are obtained by comparing the results of $\nu-e$ scattering experiments with expectations based on the standard model. For example, in $\bar{\nu}_e-e$ scattering (with electron-antineutrinos from a reactor) the standard model electroweak cross section arises in lowest order from W and from Z exchange (see Figs. 15 a,b). If $\bar{\nu}_e$ possesses a nonzero magnetic moment μ, there is an additional photon exchange diagram (Fig. 15c). It can be shown that this contribution adds incoherently to the neutral

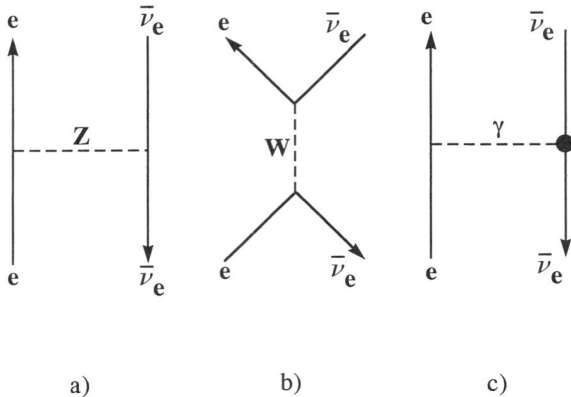

FIG. 15 (a), (b). Lowest-order weak contributions to the scattering amplitude for $\bar{\nu}_e + e \to \bar{\nu}_e + e$. (c). Additional electromagnetic contribution arising from neutrino magnetic moment. The black dot at the photon-antineutrino vertex corresponds to Figs. 14 (a),(b).

weak amplitude (there is no interference term) and that the cross section is given by the formula:

$$\left[\frac{d\sigma}{dE}\right]_{total} = \left[\frac{d\sigma}{dE}\right]_{W,Z} + \left[\frac{d\sigma}{dE}\right]_{\gamma} \tag{97}$$

where

$$\left[\frac{d\sigma}{dE}\right]_{W,Z} = \frac{2G_F^2 m_e}{\pi}\left[g_R^2 + g_L^2\left(1 - \frac{E_e}{E_\nu}\right)^2 - g_L g_R \frac{m_e E_e}{2E_\nu^2}\right] \tag{98}$$

$g_R = \sin^2\theta_W$, $g_L = \frac{1}{2} + \sin^2\theta_W$, E_ν is the energy of the incoming neutrino, and E_e is the energy of the scattered electron, whereas:

$$\left[\frac{d\sigma}{dE}\right]_\gamma = \mu^2 \pi r_0^2 \left(\frac{1}{E_e} - \frac{1}{E_\nu}\right) \tag{99}$$

where $r_0 = 2.8 \cdot 10^{-13}$ e cm is the classical electron radius. (Here for practical purposes we make no distinction between $\bar{\nu}_1$ and $\bar{\nu}_e$.) The experimental limit thus obtained (Derbin, 1994) is:

$$\mu_{\nu 1}(\text{expt}) < 1.8 \cdot 10^{-10} \mu_B \tag{100}$$

This is many orders of magnitude larger than the result obtained from Eq. (96) by inserting the limit $m_{\nu 1} < 7.3$ eV derived from tritium beta decay:

$$\mu_{\nu 1}(\text{std. mod}) < 2.3 \cdot 10^{-18} \mu_B \tag{101}$$

Similarly, from the results of $\nu_\mu - e$ and $\bar{\nu}_\mu - e$ scattering experiments using accelerator muon-neutrinos and antineutrinos with average energies of 1 GeV, one obtains the limit (Ahrens *et al.*, 1990):

$$\mu_{\nu 2}(\text{expt}) < 8.5 \cdot 10^{-10} \mu_B \tag{102}$$

whereas from (96) and the mass limit $m_{\nu 2} < 0.17$ MeV one obtains:

$$\mu_{\nu 2}(\text{std. mod}) < 5.1 \cdot 10^{-14} \mu_B \tag{103}$$

Also, from a beam dump experiment at CERN using 400-GeV muon neutrinos, in which many tau neutrinos are produced by secondary reactions, one obtains (Cooper-Sarkar *et al.*, 1992):

$$\mu_{\nu 3}(\text{expt}) < 5.4 \cdot 10^{-7} \mu_B \tag{104}$$

whereas from Eq. (96) and $m_{\nu 3} < 35$ GeV, one has:

$$\mu_{\nu 3}(\text{std. mod}) < 1.1 \cdot 10^{-11} \mu_B \tag{105}$$

A more sensitive (but also much more indirect, and therefore questionable) limit on neutrino magnetic and electric dipole moments is obtained by comparing the observed characteristics of red giant stars in globular clusters with expectations based on the theory of stellar evolution, including provision for energy loss by $\nu - \bar{\nu}$ pair emission (Raffelt, 1990). It is known that a red giant contains a dense inert core of helium outside of which exists an annular shell of hydrogen. As gravitational contraction of the core proceeds, its temperature rises, and the star's luminosity increases; (the star evolves upward in the Hertzsprung-Russell (HR) diagram.) When a certain critical core density and temperature are reached ("red giant tip" on the HR diagram), the exothermic, strongly temperature-dependent reaction:

$$3\alpha \to {}^{12}C^* \to {}^{12}C + \gamma \tag{106}$$

is suddenly ignited ("helium flash") and the star undergoes a rapid readjustment, moving quickly to a separate place on the HR diagram at lower luminosity ("horizontal branch"). It can be shown that the ratio of luminosity at the red giant tip to that on the horizontal branch is a convenient practical observable that depends on the rate at which neutrino pair emission robs the star of energy. The dominant mechanism here is the plasma process, in which a photon in the dense hot plasma of the stellar core acquires an effective mass and can thus decay via a virtual $e^+ e^-$ pair to a neutrino-antineutrino pair, (Bernstein *et al.*, 1963; Sutherland *et al.*,

1975). If the neutrinos possess magnetic and/or electric dipole moments the plasma process is enhanced by an amount proportional to:

$$\mu^2 \equiv \sum_{i,j=1}^{3} [\mu_{ij}^2 + d_{ij}^2] \qquad (107)$$

where one assumes that all mass eigenstates have masses small enough (<5 keV) to be emitted. From this analysis Raffelt obtained the following limit:

$$\mu < 3 \cdot 10^{-12} \mu_B \qquad (108)$$

Finally we mention an apparent problem that, while it lasted, stimulated considerable discussion about possible anomalous neutrino magnetic and electric dipole moments. For some years it seemed that there was an anticorrelation between the counting rate of the ^{37}Cl solar neutrino experiment and the 11.5-year solar sunspot cycle. This suggested to some authors (Voloshin and Vysotskii, 1986; Okun, 1986; Voloshin *et al.*, 1986) that the spins of neutrinos emitted from the solar core might be rotated in the large (and variable) magnetic field of the solar envelope, thereby rendering them more or less sterile for the ^{37}Cl detector as time elapsed. This could only happen if the neutrino magnetic dipole moment and/or EDM were much larger than expected on the basis of the standard model. It is possible to formulate models with charged scalar intermediate bosons in which large dipole moments can be accommodated (Fukugita and Yanazita, 1987). Inclusion of such scalars leads to subtle and interesting complications (Stockinger and Grimus, 1994). However, in the meanwhile, seven years of solar neutrino data from Kamiokande II reveal that there is no statistically significant evidence for correlation or anticorrelation between the solar neutrino flux and the solar sunspot cycle (Nakamura, 1996).

VII. Acknowledgments

The author wishes to thank W. Bernreuther, D. DeMille, V. Ezhov, E. Hinds, and B. C. Regan for many useful discussions.

VIII. References

Abdullah, K., Carlberg, C., Commins, E. D., Gould, H., and Ross, S. B. (1990). New experimental limit on the electron electric dipole moment. *Phys. Rev. Lett.* 65, 2347–2350.

Ahrens, L. A. *et al.* (1990). Determination of electroweak parameters from the elastic scattering of muon neutrinos and antineutrinos on electrons. *Phys. Rev. D* 41, 3297–3316.

Akers, R. *et al.* [OPAL Collaboration]. (1995). A test of *CP* invariance in $Z^0 \to \tau^+\tau^-$ using optimal observables. *Z. Phys. C* 66, 31–44.

Altarev, I. S., Borisov, Yu. V., Borikova, N. V., Egorov, A. I., Ivanov, S. N., Kolomensky, E. A., Lasakov, M. S., Lobashev, V. M., Nazarenko, V. A., Piroshkov, A. N., Serebrov, A. P., Sobolev, Yu. V., and Shulgina, E. V. (1996). Search for the neutron electric dipole moment. *Phys. Atomic Nucl.* 54, 1152–70.

Babu, K. S. and Barr, S. M. (1994a). Spontaneous *CP* violation in the supersymmetric Higgs sector. *Phys. Rev. D* 49, R2156–2160.

Babu, K. S. and Barr, S. M. (1994b). A solution to the small phase problem of supersymmetry. *Phys. Rev. Lett.* 72, 2831–2834.

Bailey, J. *et al.* (CERN Muon Storage Ring Collaboration). (1978). New limits on the electric dipole moment of positive and negative muons. *J.Phys. G* 4, 345–352.

Bailey, J. *et al.* [CERN-Mainz-Daresbury Collaboration]. (1979). Final report on the CERN muon storage ring including the anomalous magnetic moment and the electric dipole moment of the muon, and a direct test of relativistic time dilation. *Nucl. Phys. B* 150, 1–75.

Barbieri, R., Hall, L. J., and Romanino, A. (1997). Consequences of a $U(2)$ flavor symmetry. *Phys. Lett. B* 401, 47–53.

Barbieri, R., Hall, L. J., and Strumia, A. (1995). Violations of lepton flavor and *CP* in supersymmetric unified theories. *Nucl. Phys. B* 445, 219–251.

Barbieri, R., Romanino, A., and Strumia, A. (1996). Electric dipole moments as signals of grand unification. *Phys. Lett. B* 369, 283–288.

Barger, V., Das, A., and Kuo, C. (1997). Electric dipole moment of the muon in a 2-Higgs doublet model. *Phys. Rev. D* 55, 7099–7103.

Bargmann, V., Michel, L., and Telegdi, V. L. (1959). Precession of the polarization of particles moving in a homogeneous magnetic field. *Phys. Rev. Lett.* 2, 435–436.

Barkov, L. M., Zolotorev, M. S., and Melik-Pashaev, D. A. (1988). Amplification of a $(P+T)$-odd optical activity. *JETP Lett.* 48, 144–147.

Barr, S. M. (1992a). Measurable T and P-odd e-N Interactions from Higgs boson exchange. *Phys. Rev. Lett.* 68, 1822–1825.

Barr, S. M. (1992b). T, P-odd e-N interactions and the electric dipole moments of large atoms. *Phys. Rev. D* 45, 4148–4155.

Barr, S. M. (1993a). Magnitude of Higgs boson exchange *CP* violation in 2-doublet models with large $\tan\beta$. *Phys. Rev. D* 47, 2025–2031.

Barr, S. M. (1993b). A review of *CP* violation in atoms. *Int. J. Mod. Phys. A* 8, 209–236.

Barr, S. M. and Segre, G. (1993). Spontaneous *CP* violation and supersymmetry. *Phys. Rev. D* 48, 302–306.

Bernreuther, W. and Nachtmann, O. (1989). *CP*-violating correlations in electron-positron annihilation into τ leptons. *Phys. Rev. Lett.* 63, 2787–2790.

Bernreuther, W., Low, U., Ma, J. P., and Nachtmann, O. (1989). *CP* violation and Z boson decays. *Z. Phys. C* 43, 117–132.

Bernreuther, W., Botz, G. W., Nachtmann, O., and Overmann, P. (1991). *CP* violating effects in Z decays to τ leptons. *Z. Phys. C* 52, 567–573.

Bernreuther, W. and Suzuki, M. (1991). The electric dipole moment of the electron. *Revs. Mod. Phys.* 63, 313–340.

Bernreuther, W. (1992a). The electric dipole moment of the muon. *J. Phys. C* 56, 597.

Bernreuther, W., Schroder, T., and Pham, T. N. (1992b). *CP* violating form factors in $e^+e^- \to t\bar{t}$. *Phys. Lett. B* 279, 389–396.

Bernreuther, W., Brandenburg, A., and Overmann, P. (1997). *CP* violation beyond the standard model and tau pair production in e^+e^- collisions. *Phys. Lett. B* 391, 413–419.

Bernstein, J., Ruderman, M., and Feinberg, G. (1963). Electromagnetic properties of the neutrino. *Phys. Rev.* 132, 1227–1233.

Bijlsma, M., Verhaar, J., and Heinzen, D. (1994). Role of collisions in the search for an electron electric dipole moment. *Phys. Rev. A* 49, R4285–4287.

Bjorken, J. D. and Drell, S. D. (1964). *Relativistic Quantum Mechanics*. McGraw-Hill (New York).

Bouchiat, C. (1975). A limit on scalar-pseudoscalar weak neutral currents from a new interpretation of atomic electric dipole experiments. *Phys. Lett. B* 57, 284–288.

Bouchiat, M. A., and Bouchiat, C. (1975). Parity violation induced by weak neutral currents in atomic physics. Part II. *Jour. Phys. (Paris)* 36, 493–509.

Buskelic, D. et al. [ALEPH Collaboration]. (1995). Search for CP violation in the decay $Z \to \tau^+\tau^-$. *Phys. Lett. B* 346, 371–378.

Chu, S. (1997). Private communication.

Cho, D., Sangster, K., and Hinds, E. (1991). Search for time reversal symmetry violation in thallium fluoride using a jet source. *Phys. Rev. A* 44, 2783–2799.

Commins, E. D., and Bucksbaum, P. H. (1983). *Weak interactions of leptons and quarks*. Chap. 10. Cambridge Univ. Press (Cambridgeshire, U.K.; New York).

Commins, E. D., Ross, S. B., DeMille, D., and Regan, B. C. (1994). Improved experimental limit on the electric dipole moment of the electron. *Phys. Rev. A* 50, 2960–2977.

Conti, R. S. and Khriplovich, I. B. (1992). New limits on T-odd, P-even interactions. *Phys. Rev. Lett.* 68, 3262–3265.

Cooper-Sarkar, A. M., Sarkar, S., Guy, J., Venns, W., Hulth, P. O., and Hultqvist, K. (1992). Bound on the tau neutrino magnetic moment from the BEBC beam dump experiment. *Phys. Lett. B* 280, 153–158.

Czarnecki, A. and Krause, B.(1997). Neutron electric dipole moment in the standard model: complete three-loop calculation of the valence quark contributions. *Phys. Rev. Lett.* 78, 4339–4342.

DeMille, D. (1997). Private communication.

Derbin, A. V. (1994). Restriction on the magnetic dipole moment of reactor neutrinos. *Phys. Atomic Nucl.* 57, 222–226.

Deshpande, N. G. and He, K. G. (1994). CP violation in a multi-Higgs doublet model with flavor changing neutral currents. *Phys. Rev. D* 49, 4812–4819.

Deshpande, N. G. Dutta, B., and Keith, E. (1996). Electric dipole moments and b-τ unification in the presence of an intermediate scale in supersymmetric grand unification. *Phys. Lett. B* 388, 605–610.

Dimopoulos, S. and Hall, L. J. (1995). Electric dipole moments as a test of supersymmetric unification. *Phys. Lett. B* 344, 185–192.

Dmitriev, V. F., Khriplovich, I. B., and Telitzin, V. B. (1994). Nuclear magnetic quadrupole moments in the single particle approximation. *Phys. Rev. C* 50, 2358–2361.

Donoghue, T. (1978). T violation in $SU(2) \times U(1)$ gauge theories of leptons. *Phys. Rev. D* 18, 1632–1678.

Dzuba, V. A., Flambaum, V. V., and Sylvestrov, V. V. (1985). Bounds on electric dipole moments and T-violating weak interactions of the nucleons. *Phys. Lett. B* 154, 93–95.

Dzuba, V. A., Flambaum, V. V., Sylvestrov, P. G., and Sushkov, O. P. (1986). Shielding of an external electric field in atoms. *Phys. Lett. A* 118, 177–180.

Escribano, R. and Masso, E. (1997). Improved bounds on dipole moments of the tau lepton. *Phys. Lett. B* 395, 369–372.

Ezhov, V. (1997). Private communication.

Falk, T., Olive, K. A., and Srednicki, M. (1995). Phases in the MSSM, electric dipole moments, and cosmological dark matter. *Phys. Lett. B* 354, 99–106.

Falk, T. and Olive, K. (1996). Electric dipole moment constraints on phases in the constrained minimal supersymmetric model. *Phys. Lett. B* 375, 196–202.
Fischler, W., Paban, S., and Thomas, S. (1992). Bounds on microscopic physics from P and T violation in atoms and molecules. *Phys. Lett. B* 289, 373–380.
Flambaum, V. V. (1976). On the enhancement of the electron electric dipole moment in atoms. *Sov. J. Nucl. Phys.* 24, 199–202.
Flambaum, V. V., and Khriplovich, I. B. (1985). On the enhancement of parity nonconserving effects in diatomic molecules. *Phys. Lett. A* 110, 121–125.
Flambaum, V. V., Khriplovich, I. B., and Sushkov, O. P. (1985). Limits on the constant of T-nonconserving nucleon–nucleon interaction. *Phys. Lett. B* 162, 213–215.
Flambaum, V. V., Khriplovich, I. B., and Sushkov, O. P. (1986). On the P and T nonconserving nuclear moments. *Nucl Phys. A* 449, 750–760.
Flambaum, V. V. (1987). Thesis. Institute of Nuclear Physics, Novosibirsk, Russia. (Unpublished).
Flambaum, V. V. and Murray, D. W. (1997). Limits on the monopole polarization magnetic field from measurements of the electric dipole moments of atoms, molecules, and the neutron. *Phys. Rev. A* 55, 1736–1742.
Fortson, E. N. (1983). Sensitivity to an electric dipole moment of an electron through hyperfine coupling to the nuclear spin in an atom. *Bull. Am Phys. Soc.* 28, 1321.
Fujikawa, K. and Shrock, R. E. (1980). Magnetic moment of a massive neutrino and neutrino spin rotation. *Phys. Rev. Lett.* 45, 963–966.
Fukujita, M. and Yanagita, T. (1987). Particle physics model for the Voloshin-Vysotsky-Okun solution to the solar neutrino problem. *Phys. Rev. Lett.* 58, 1807–1809.
Garisto, R. (1994). Moderate SUSY CP violation. *Phys. Rev. D* 49, 4820–4825.
Geng, C. Q. and Ng, J. N. (1989). Fermion electric dipole moments, muon polarization in $\eta \to \mu\bar{\mu}$, $K_L^0 \to \mu\bar{\mu}$ decays, and the scalar-pseudoscalar mixing mechanism. *Phys. Rev. Lett.* 62, 2645–2647.
Gilman, F. J., Kleinknecht, K., and Renk, B. (1996). The Cabibbo-Kobayashi-Maskawa mixing matrix. In R. M. Barnett *et al.* (Particle Data Group), *Review of particle physics* (pp. 94–97). *Phys. Rev. D* 54, 1–720.
Gorshkov, V. G., Labzowsky, L. N., and Moskalev, A. N. (1979). Space and time parity nonconservation effects in the spectra of diatomic molecules. *Sov. Phys. JETP* 49, 414–421.
Gould, H. (1970). Search for an electric dipole moment in thallium. *Phys. Rev. Lett.* 24, 1091–1093.
Gould, H. (1995). Private communication.
Haber, H. E. and Kane, G. L. (1985). The search for supersymmetry: probing physics beyond the standard model. *Phys. Rep.* 117C, 75–263.
Hagedorn, R. (1963). "Relativistic kinematics: a guide to the kinematic problems of high energy physics." Benjamin (New York).
Hartley, A., Lindroth, E., and Martensson-Pendrill, A.-M. (1990). Parity nonconservation and electric dipole moments in cesium and thallium. *J. Phys. B* 23, 3417–3436.
Haxton, W. C. and Henley, E. M. (1983). Enhanced T-nonconserving nuclear moments. *Phys. Rev. Lett.* 51, 1937–1940.
Hayashi, T., Koide, Y., Matsuda, M., Tanimoto, M., and Wakaizumi, S. (1995). Electric dipole moments of neutron and electron in 2-Higgs doublet model with maximal CP violation. *Phys. Lett. B* 348, 489–495.
He, X., McKellar, B. H. J., and Pakvasa, S. (1989). The neutron electric dipole moment. *Int. J. Mod. Phys.* A4, 5011–5046. (1991). Errata and addendum. *Int. J. Mod. Phys.* A6, 1063.
He, X. and McKellar, B. (1997). Constraints on CP-violating 4-fermion interactions. *Phys. Lett. B* 390, 318–322.
Hinds, E. (1997). Private communication.

Hinds, E., Loving, C., and Sandars, P. G. H. (1976). Limits on P and T violating neutral current weak interactions. *Phys. Lett. B* 62, 97–99.

Hinds, E. and Sauer, B. (1997). Electron dipole moments. *Phys. World* 10, no. 4, 37–40.

Hoogeveen, F. (1990). The standard model prediction for the electric dipole moment of the electron. *Nucl. Phys. B* 341, 322–340.

Huang, T., Lu, W., and Tao, Z. (1997). Search for nonstandard model CP or T violation at the τ-Charm factory. *Phys. Rev. D* 55, 1643–1652.

Jacobs, J. P., Klipstein, W. M., Lamoreaux, S. K., Heckel, B. R., and Fortson, E. N. (1995). Limit on the electric dipole moment of ^{199}Hg using synchronous optical pumping. *Phys. Rev. A* 52, 3521–3540.

Jarlskog, C. (1985a). Commutator of the quark mass matrices in the standard electroweak model and a measure of maximal CP nonconservation. *Phys. Rev. Lett.* 55, 1039–1042.

Jarlskog, C. (1985b). A basis-independent formulation of the connection between quark mass matrices, CP violation, and experiment. *Z. Phys. C* 29, 491–497.

Johnson, W. R., Guo, D. S., Idrees, M., and Sapirstein, J. (1985). Weak interaction effects in heavy atomic systems. *Phys. Rev. A* 32, 2093–2099.

Johnson, W. R., Guo, D. S., Idrees, M., and Sapirstein, J. (1986). Weak interaction effects in heavy atomic systems II. *Phys. Rev. A* 34, 1043–1057.

Kanorsky, S. I., Lang, S., Lucke, S., Ross, S. B., Hansch, T. W., and Weis, A. (1996). Millihertz magnetic resonance spectroscopy of cesium atoms in body-centered-cubic ^4He. *Phys. Rev. A* 54, R1010–1013.

Khatsymovsky, V. M., Khriplovich, I. B., and Yelkhovsky, A. S. (1988). Neutron electric dipole moment, T-odd nuclear forces, and nature of CP violation. *Annals of Phys.* 186, 1–14.

Khriplovich, I. B. (1976). A bound on the proton electric dipole moment derived from atomic experiments. *Sov. Phys. JETP* 44, 25–30.

Khriplovich, I. B. (1991a). *Parity nonconservation in atomic phenomena.* Gordon and Breach (Philadelphia).

Khriplovich, I. B. (1991b). What do we know in fact about T-odd but P-even interactions? *Nucl. Phys. B* 352, 385–401.

Kizukuri, Y. and Oshimo, N. (1992). Neutron and electron electric dipole moments in supersymmetric theories. *Phys. Rev. D* 46, 3025–3033.

Kleinknecht, K. (1990). CP violation and the flavor structure of weak interactions. *Prog. Nucl. Part. Phys.* 25, 81–137.

Kobayashi, M., and Maskawa, T. (1973). CP violation in the renormalizable theory of weak interactions. *Prog. Theor. Phys. (Japan)* 49, 652–657.

Kozlov, M. G. (1985). Semi-empirical calculations of P- and P,T-odd effects in diatomic molecules-radicals. *Sov. Phys. JETP* 62, 1114–1118.

Kozlov, M. (1988). Calculations of the degree of parity nonconservation in the spin-rotational spectrum of the PbF molecule. *Sov. J. Quantum Electronics* 18, 713–715.

Kozlov, M. G., Fomichev, V. I., Dmitriev, Yu. Yu., Labzowsky, L. N., and Titov, A. V. (1988). Calculation of the P- and T-odd spin rotational Hamiltonian of the PbF molecule. *J. Phys. B* 20, 4939–4948.

Kozlov, M. G. and Ezhov, V. F. (1994). Enhancement of the electric dipole moment of the electron in the YbF molecule. *Phys. Rev. A* 49, 4502–4507.

Kozlov, M. G. and Labzowsky, L. N. (1995). Parity violation in diatomics. *J. Phys. B* 28, 1933–1961.

Kozlov, M. G. and Yashchuk, V. V. (1996). Estimate of P- and P,T-odd effects in diatomic van der Waals molecules. *JETP Lett.* 64, 709–713.

Kraftmakher, A. Ya. (1988). On the Hartree-Fock calculation of the electron electric dipole enhancement factor for the thallium atom. *J. Phys. B* 21, 2803–2813.

Kuo, C. and Xu, R. M. (1992). Charged Higgs loop contribution to the electron electric dipole moment. *Phys. Lett. B* 296, 435–439.

Landau, L. (1957). Conservation laws in weak interactions. *Sov. Phys. JETP* 5, 336–337.

Lee, B. W. and Shrock, R. E. (1977). Natural suppression of symmetry violation in gauge theories: muon and electron-number-nonconservation. *Phys. Rev. D* 16, 1444–1473.

Liu, J. (1986). Electric dipole moment of the electron in left-right models. *Nucl. Phys.* 271, 531–539.

Liu, Z. W. and Kelly, H. P. (1992). Analysis of atomic electric dipole moment of thallium by all-order calculations in many-body perturbation theory. *Phys. Rev. A* 45, R4210–4213.

Mahanta, U. (1992). Constraint on a composite 2-Higgs doublet model of *CP* violation from the electric dipole moment of the electron. *Phys. Lett. B* 281, 320–324.

Martensson-Pendrill, A. M. (1985). Calculation of a *P*- and *T*-nonconserving weak interaction in *Xe* and *Hg* with many body perturbation theory. *Phys. Rev. Lett.* 54, 1153–1155.

Martensson-Pendrill, A. M. and Oster, P. (1987). Calculations of atomic electric dipole moments. *Phys. Scripta* 36, 444–452.

Martensson-Pendrill, A. M. and Lindroth, E. (1991). Limits on a *P* and *T* violating electron–nucleon interaction. *Europhys. Lett.* 15, 155–160.

Matsuda, M. and Tanimoto, M. (1995). Explicit *CP* violation of the Higgs sector in the next-to-minimal supersymmetric model. *Phys. Rev. D* 52, 3100–3107.

Mohapatra, R. N. and Pati, J. C. (1975a). Left-right gauge symmetry and "isoconjugate" model of *CP* violation. *Phys. Rev. D* 11, 566–571.

Mohapatra, R. N. and Pati, J. C. (1975b). "Natural" left-right symmetry. *Phys. Rev. D* 11, 2558–2561.

Mohapatra, R. N. and Sidhu, D. (1977). Implications for gauge theories if search for parity nonconservation in atomic physics fails. *Phys. Rev. Lett.* 38, 667–670.

Mohapatra, R. N. and Senjanovic, G. (1980). Neutrino mass and spontaneous parity nonconservation. *Phys. Rev. Lett.* 44, 912–915.

Mohapatra, R. N. and Senjanovic, G. (1981). Neutrino masses and mixings in gauge models with spontaneous symmetry violation. *Phys. Rev. D* 23, 165–180.

Murthy, S. A., Krause, D., Li, Z. L., and Hunter, L. (1989). New limit on the electric dipole moment from cesium. *Phys. Rev. Lett.* 63, 965–968.

Nakamura, K. (1996). Solar Neutrinos. In R. M. Barnett *et al.*, (Particle Data Group), *Review of Particle Physics* (pp. 294–296). *Phys. Rev. D* 54, 1–720.

Nelson, D. F., Schupp, A. A., Pidd, R. W., and Crane, H. R. (1959). Search for an electric dipole moment of the electron. *Phys. Rev. Lett.* 2, 492–495.

Neuffer, D. V. and Commins, E. D. (1977). Calculation of parity nonconserving effects in the $6\,^2P_{1/2}$-$7\,^2P_{1/2}$ $M1$ transition in thallium. *Phys. Rev. A* 16, 844–862.

Nieves, J. F., Chang, D., and Pal, P. B. (1986). Electric dipole moment of the electron in left-right-symmetric theories. *Phys. Rev. D* 33, 3324–3328.

Nilles, H. P. (1984). Supersymmetry, supergravity, and particle physics. *Phys. Reports* 110, 1–162.

Okun, L. B. (1986). On the electric dipole moment of the neutrino. *Sov. J. Nucl. Phys.* 44, 546.

Pati, J. C. and Salam, A. (1974). Lepton number as the fourth "color." *Phys. Rev. D* 10, 275–289.

Penney, W. G. (1931). The Stark effect in band spectra. *Phil. Mag.* 11, 602–609.

Player, M. A. and Sandars, P. G. H. (1970). An experiment to search for an electric dipole moment in the $^3P_2$ metastable state of xenon. *J. Phys. B* 3, 1620–1635.

Pospelov, M. and Khriplovich, I. B. (1991). Electric dipole moment of the *W* boson and the electron in the Kobayashi-Maskawa model. *Sov. J. Nucl. Phys.* 53, 638–640.

Purcell, E. M. and Ramsey, N. F. (1950). On the possibility of electric dipole moments for elementary particles and nuclei. *Phys. Rev.* 78, 807.

Raffelt, G. (1990). New bound on neutrino dipole moments from globular cluster stars. *Phys. Rev. Lett.* 64, 2856–2858.

Ramsey, N. F. (1990). Electric dipole moment of the neutron. *Annu. Rev. Nucl. Part. Sci.* 40, 1–14.

Salpeter, E. E. (1958). Some atomic effects of an electronic electric dipole moment. *Phys. Rev.* 112, 1642–1648.

Sandars, P. G. H. (1965). The electric dipole moment of an atom. *Phys. Lett.* 14, 194–196.

Sandars, P. G. H. (1966). Enhancement factor for the electric dipole moment of the valence electron in an alkali atom. *Phys. Lett.* 22, 290–291.

Sandars, P. G. H. (1968). The electric dipole moment of an atom I. *J. Phys. B* 1, 499–510. The electric dipole moment of an atom II. *J. Phys. B* 1, 511–520.

Sandars, P. G. H. and Sternheimer, R. M. (1975). Electric dipole moment enhancement factor for the thallium atom and a new upper limit on the electric dipole moment of the electron.

Sauer, B. E., Wang, J., and Hinds, E. (1995). Anomalous spin-rotation coupling in the $X^2\Sigma^+$ state of YbF. *Phys. Rev. Lett.* 74, 1554–1557.

Sauer, B. E., Wang, J., and Hinds, E. (1996). Laser-rf double resonance spectroscopy of ^{174}YbF in the $X^2\Sigma^+$ state: spin-rotation, hyperfine interactions, and the electric dipole moment. *J. Chem Phys.* 105, 7412–7420.

Schiff, L. I. (1963). Measurability of nuclear electric dipole moments. *Phys. Rev.* 132, 2194–2200.

Semertzidis, Y. (1997). Private communication.

Shabalin, E. P. (1978). Electric dipole moment of the quark in a gauge theory with left-handed currents. *Sov. J. Nucl. Phys.* 28, 75–78.

Shabalin, E. P. (1983). Electric dipole moment of the neutron in gauge theory. *Sov. Phys. Usp.* 26, 297–310.

Shrock, R. E. (1982). Electromagnetic properties and decays of Dirac and Majorana neutrinos in a general class of gauge theories. *Nucl. Phys. B* 206, 359–379.

Smith, J. H., Purcell, E. M., and Ramsey, N. F. (1957) Experimental limit to the electric dipole moment of the neutron. *Phys. Rev.* 108, 120–122.

Smith, K. F., Crampin, N., Pendlebury, J. M., Richardson, D. J., Shiers, D., Green, K., Kilvington, A. I., Moir, J., Prosper, H. B., Thompson, D., Ramsey, N. F., Heckel, B. R., Lamoreaux, S., Ageron, P., Mampe, W., and Steyerl, A. (1990). A search for the electric dipole moment of the neutron. *Phys. Lett. B* 234, 191–195.

Steinberger, J. (1990). Experimental status of *CP* violation. In J. Tran Thanh Van (Ed.), *CP violation in particle physics and astrophysics* (pp. 55–92). Editions Frontieres (Gif-sur-Yvette, France).

Stockinger, P. and Grimus, W. (1994). Effects of scalar exchange and neutrino magnetic and electric dipole moments in ν_e-e^- scattering. *Phys. Lett. B* 327, 327–334.

Sushkov, O. P. and Flambaum, V. V. (1978). Parity violating effects in diatomic molecules. *Sov. Phys. JETP* 48, 608–613.

Sushkov, O. P., Flambaum, V. V., and Khriplovich, I. B. (1984). Possibility of investigating *P*- and *T*-odd nuclear forces in atomic and molecular experiments. *Sov. Phys. JETP* 60, 873–883.

Sutherland, P., Ng, J. N., Flowers, E., Ruderman, M., and Inman, C. (1976). Astrophysical limitations on possible tensor contributions to weak neutral current interactions. *Phys. Rev. D* 13, 2700–2704.

Van Vleck, J. H. (1932). *The theory of electric and magnetic susceptibilities*. (p. 154). Oxford University Press (Oxford, U.K.).

Voloshin, M. B. and Vysotsky, M. I. (1986). The neutrino magnetic moment and time variation of the solar neutrino flux. *Sov. J. Nucl. Phys.* 44, 544–545.

Voloshin, M. B., Vysotsky, M. I., and Okun, L. B. (1986). Neutrino electrodynamics and possible consequences for solar neutrinos. *Sov. Phys. JETP* 64, 446–452.

Weisskopf, M. C., Carrico, J. P., Gould, H., Lipworth, E., and Stein, T. S. (1968). Electric dipole moment of the cesium atom. A new upper limit to the electric dipole moment of the electron. *Phys. Rev. Lett.* 21, 1645–1648.

Wolfenstein, L. (1986). Present status of *CP* violation. *Ann. Rev. Nucl. Part. Sci.* 36, 137–170.

Wolfenstein, L. (1991). Status of *CP* violation. *Nucl. Phys. B. Proc. Suppl.* 24A, 32–36.

Wolfenstein, L. (1996). *CP* violation. In R. M. Barnett *et al.* (Particle Data Group), *Review of particle physics* (p. 103). *Phys. Rev. D* 54, 1–720.

HIGH-PRECISION CALCULATIONS FOR THE GROUND AND EXCITED STATES OF THE LITHIUM ATOM

FREDERICK W. KING
Department of Chemistry, University of Wisconsin-Eau Claire, Eau Claire, Wisconsin

I. Introduction	57
II. Computational Approaches	58
A. Convergence Considerations	62
III. Some Mathematical Issues	63
A. Integration Problems	64
IV. Nonrelativistic Energies	67
A. The Ground State	68
1. Lower Bound Estimates for E_{NR}	69
2. Distribution Functions	72
B. Excited States	72
1. The Low-Lying Excited Doublet States	72
2. The Low-Lying Quartet States	74
3. The Core-Excited Doublet States	76
4. Highly Excited Quartet States	79
V. Relativistic Corrections to the Energies	80
VI. Specific Mass Shift Correction to the Energy Levels	83
A. Transition Isotope Shifts	84
VII. Quantum Electrodynamic Corrections	88
VIII. The First Ionization Potential	90
A. Transition Energies	91
IX. Hyperfine Coupling Constants	93
X. Other Properties	100
A. Polarizabilities	100
B. Oscillator Strengths	101
C. Lifetimes	103
XI. Outlook	103
XII. Acknowledgments	105
XIII. References	105

I. Introduction

Over the past decade there has been considerable progress on high-precision calculations for the lithium atom. In this chapter, a summary is presented of some of the progress, for selected properties of the ground and excited states of *Li*. The lithium atom has long served as a test case for newly developed computational methods. This system has also been intensively investigated for its own intrinsic interest. It is a few-body system, so we might expect that calculations of very high precision are possible for the lithium atom. The helium atom and members of its isoelectronic series have been extremely popular targets for computationalists. Very high levels of precision are available for a number of calculated properties of two-electron systems. For example, the nonrelativistic ground state energy of the helium atom is known to around eighteen digits of precision (Goldman, 1998). The lithium atom, with one additional electron, turns out to be a much more complicated system to study. A number of new mathematical difficulties arise, and some of these have hampered progress on the calculation of several properties.

The lithium atom in its ground state represents the simplest atomic system for which it is possible to study valence, core, and valence-core electronic effects. It is an ideal target system for investigating the nature of the Coulomb and Fermi hole structure of highly correlated wave functions.

There has been recent renewed interest in the experimental determination of high-precision values of a number of properties of atomic lithium. These advances have provided additional stimulus for theoretical progress.

Two conventions will be employed in this chapter. Error estimates will be shown in parentheses, so that 23.4 ± 3.2 will be written as 23.4(32). Expectation values will employ an implied summation convention, so that $<r_{ij}>$ is equivalent to $\langle \sum_{i=1}^{3} \sum_{j>i}^{3} r_{ij} \rangle$. Absolute atomic units will be used for energies, with the conversion to cm^{-1} being 1 a.u. (absolute) = 219474.6313688(62) cm^{-1}.

II. Computational Approaches

A variety of computational techniques have been applied to calculate various properties of the lithium atom. These include the Hylleraas approach (HY), the configuration interaction method (CI), the hybrid CI-Hylleraas technique (CI-HY), many-body perturbation theory (MBPT), multiconfiguration Hartree-Fock methods (MCHF), and others. The CI, MBPT, and MCHF approaches have the advantage that they can be applied to systems beyond the few-electron level. The HY and CI-HY techniques are presently constrained to few-electron systems, but these techniques lead to the highest levels of precision when applied to the lithium atom. The CI-HY method has shown considerable recent promise in yielding results of very high precision. It may be feasible to extend this approach (with some restrictions on the basis functions) to atomic systems with more than four electrons.

The highest precision results for several properties have been obtained using the Hylleraas technique, and a large part of the focus of this chapter will be on this technique. The Hylleraas expansion for the $^2S_{1/2}$ ground state of Li takes the form

$$\Psi = \mathcal{A} \sum_{\mu=1}^{N} C_\mu \phi_\mu \chi_\mu \qquad (1)$$

where $\mathcal{A}$ is the three-electron antisymmetrizer, C_μ are the variationally determined expansion coefficients, and N designates the number of basis functions employed. The basis functions ϕ_μ are expanded in terms of the electron-nuclear (r_i) and electron-electron (r_{ij}) coordinates, and are defined in the following way:

$$\phi_\mu(r_1, r_2, r_3, r_{23}, r_{31}, r_{12}) = r_1^{i_\mu} r_2^{j_\mu} r_3^{k_\mu} r_{23}^{l_\mu} r_{31}^{m_\mu} r_{12}^{n_\mu} e^{-\alpha_\mu r_1 - \beta_\mu r_2 - \gamma_\mu r_3} \qquad (2)$$

The indices $\{i_\mu, j_\mu, k_\mu, l_\mu, m_\mu, n_\mu\}$ take integer values and are ≥ 0, and the orbital exponents α_μ, β_μ and γ_μ are > 0. In Eq. (1), χ_μ represents the possible spin eigenfunctions, which take one of the following two forms for the doublet states:

$$\chi = \alpha(1)\beta(2)\alpha(3) - \beta(1)\alpha(2)\alpha(3) \qquad (3)$$

or

$$\chi = 2\alpha(1)\alpha(2)\beta(3) - \beta(1)\alpha(2)\alpha(3) - \alpha(1)\beta(2)\alpha(3) \qquad (4)$$

The wave function in Eq. (1) is an eigenfunction of the spin operators S^2 and S_Z.

The importance of including both spin eigenfunctions in the wave function has been discussed in a number of papers, particularly in regard to determining precise values of properties such as the energy, and especially, precise hyperfine coupling constants (Larsson, 1968; King and Shoup, 1986; King, 1988, 1989). The majority of Hylleraas calculations on three-electron systems are carried out using just the spin eigenfunction given in Eq. (3). Omission of the second spin eigenfunction (Eq. (4)) can be compensated for by using larger basis sets with only the first spin function (Eq. (3)) included. Spin-dependent expectation values (for example, the Fermi contact term) appear to be more sensitive to the omission of the second spin eigenfunction.

The general topic of the spatial-spin form of the ground state wave function for the Li atom has been discussed by a number of authors. The interested reader might start with the articles by Slater (1970), White and Stillinger (1970) and Smolenskii and Zefirov (1993). Studies on this topic have been fruitful. The well-known determinantal wave function structure so widely employed in atomic and molecular calculations emerged from a consideration of the form of the wave function for the lithium atom (Slater, 1970).

For excited bound states, Eq. (2) is supplemented by the inclusion of appropriate spherical harmonics to account for any angular dependence of the basis terms. If excited quartet states are of interest, the spin function is taken to be

$$\chi = \alpha(1)\alpha(2)\alpha(3) \qquad (5)$$

There are several strategies that can be followed for the selection of the indices $\{i_\mu, j_\mu, k_\mu, l_\mu, m_\mu, n_\mu\}$. The earliest approach employed a selection of terms that in effect minimized the mathematical problems (James and Coolidge, 1936). Given the computer technology available at the time, this was undoubtedly the method of choice. A second procedure was to pick indices based on the expected impact in producing a good energy (Larsson 1968; King and Shoup, 1986). This procedure has been particularly successful at producing some rather precise results for the ground state energy of Li. There are two drawbacks to this approach. The first is that it depends on the experience of the theorist in choosing important terms. It is possible, because of a predisposed bias, to omit basis functions that might singly, or collectively with other terms, lead to important contributions to the energy. A second issue is that the convergence behavior of the calculation can be very difficult (if not impossible) to determine. This eliminates the possibility of trying to determine an extrapolated estimate for the energy (or other expectation values) for a basis set of infinite size. A third approach is to define an index, ω:

$$\omega = i_\mu + j_\mu + k_\mu + l_\mu + m_\mu + n_\mu \qquad (6)$$

and then add basis functions in order of increasing values of ω; $\omega = 0, 1, 2, \ldots$ (King and Shoup, 1986; McKenzie and Drake, 1991; Lüchow and Kleindienst, 1993; Yan and Drake, 1995a; Barrois et al., 1997a, 1997b, 1997c; King, 1998b). This approach avoids any bias in the selection of the basis terms. It does not provide the fastest possible convergence for the energy, but does in general lead to a reasonable monotonic convergence pattern for the energy. This has the advantage that attempts at determining extrapolated estimates of the energy are more likely to be reliable. The total number of basis functions grows significantly as ω increases. A table of the maximum number of basis functions for values of ω up to 11 is given by King and Shoup (1986).

All the Hylleraas-type calculations on three-electron systems that have been published have restricted the basis functions to integer values for the indices $\{i_\mu, j_\mu, k_\mu, l_\mu, m_\mu, n_\mu\}$. There are reasons to expect that improved wave functions can be constructed using noninteger values for the set $\{i_\mu, j_\mu, k_\mu, l_\mu, m_\mu, n_\mu\}$. These indices can be selected as arbitrary floating point values to optimize the energy. This idea has been tried by the author and Feldmann for some preliminary calculations on two-electron systems, with fairly encouraging results. For two-electron systems, there is essentially no increase in the level of difficulty for the integrals required to evaluate the energy and a variety of expectation values. The main change that occurs is that gamma functions of noninteger argument arise, but these can be readily evaluated. For three-electron systems, the switch from integer to noninteger exponent indices leads to a significantly more involved integration problem, when the required matrix elements are evaluated.

The orbital exponent parameters α_μ, β_μ and γ_μ are usually selected in one of the following ways. The simplest choice has been to use fixed values of the orbital exponents, that is,

$$\left.\begin{array}{l}\alpha_\mu = \alpha \\ \beta_\mu = \alpha \\ \gamma_\mu = \gamma\end{array}\right\} \text{all } \mu \qquad (7)$$

The values of α and γ can be determined in one of several ways, such as using the appropriate screened nuclear charge to fix these values, or by optimizing these values by using small to modest size wave functions, and then employing the optimized exponents for larger basis sets. A significant advantage of using Eq. (7) is that it is feasible to store all the calculated integrals. Using modern desktop workstations, with relatively inexpensive memory, it is possible to store almost all the required integrals (about one to two hundred megabytes) needed for the construction of a wave function of considerable size (one to two thousand terms). The same integral file can be employed in the evaluation of a large range of expectation values.

An alternative approach is to optimize the orbital exponents to obtain the minimum energy. This can be done either on a term-by-term basis or by optimizing blocks of terms. The lowest energies have been obtained using approaches involving orbital exponent optimization. In the term-by-term optimization approach, the fraction of matrix elements being reevaluated at each step scales as $2(N + 1)^{-1}$ of the total number of matrix elements. With present computer technology, it is not feasible to optimize simultaneously all the nonlinear parameters of a large number of basis functions.

There are two drawbacks to schemes involving optimization procedures. There is effectively very little gained by attempting to carry out any type of integral store and retrieve strategy. The cpu requirements also increase very significantly in this approach. An additional factor that the reader needs to be aware of is that the energy surface in the multidimensional parameter space $\{\alpha_\mu, \beta_\mu, \gamma_\mu\}$ typically has a number of local minima at any stage of the construction of the wave function. Getting the best energy can then be dependent on making skillful choices for the starting values for $\{\alpha_\mu, \beta_\mu, \gamma_\mu\}$, with the hope that these lead to an energy close to the global minimum. An alternative approach that the author has employed is to use a stochastic global optimization algorithm. The advantage of this approach is that there is minimal bias in locating the best starting set of exponents to determine the energy minimum. The drawback is the significant cpu resources required by this method.

It is possible to supplement the basis functions shown in Eq. (2) with a factor of the form $\exp(-\alpha_{12} r_{12} - \alpha_{23} r_{23} - \alpha_{13} r_{13})$, where α_{12}, α_{23}, and α_{13} are constants that may be separately optimized for each term. This leads to extra flexibility in the basis functions, but at the cost of introducing additional complexity into the integrals that must be evaluated. Methods have been developed to evaluate the integrals that emerge (Fromm and Hill, 1987), but no results appear to have been published for any three-electron species based on these more elaborate basis functions. For two-electron atomic systems, incorporation of logarithmic functions of the electronic coordinates allows relatively compact wave functions of rather high quality (in the energetic sense) to be constructed. Basis functions of this type have not been employed for calculations on three-electron systems. The integration problems that emerge when logarithmic functions are employed to treat three-electron systems are rather severe, and there are no published procedures to solve the integrals that are required.

A. CONVERGENCE CONSIDERATIONS

All the theoretical methods that have been applied to carry our high-precision calculations on the lithium atom, converge rather slowly as the size of the basis set is increased. A central issue is how well the basis sets employed describe the electron cusps in the wave function. Because the exact analytic structure of the N-electron ($N \geq 2$) nonrelativistic wave function is unknown, the theorist must resort to educated guesswork to select basis functions that will mimic both the shape characteristics of the exact wave function, as well as the form of the wave function at the coalescence points, which occur where the electron-nuclear separation or the interelectronic distances are zero.

There have been a number of studies on the general convergence characteristics of variational calculations (Schwartz, 1962, 1963; Lakin, 1965; Klahn and Bingel, 1977; Klahn and Morgan, 1984; Hill, 1985, 1995; Kutzelnigg, 1985; Klahn, 1985; Kutzelnigg and Klopper, 1991; Kutzelnigg and Morgan, 1992). From these works it is possible to gain insight into the expected rates of convergence of some of the computational techniques currently employed, and to anticipate why the r_{ij}-dependent basis sets lead to particularly good convergence when the nonrelativistic energy is calculated. There has also been some effort expended on investigating the behavior of the electronic wave function in the neighborhood of the coalescence points (Kato, 1957; Hoffmann-Ostenhof and Seiler, 1981; Johnson, 1981; Hoffmann-Ostenhof et al., 1992). With the notable exception of the work of White and Stillinger (1970, 1971), very little attention has been directed toward finding the analytic structure of three-electron wave functions near the singular points of the potential. White and Stillinger (1971) find a logarithmic dependence on the three electronic coordinates, reminiscent of what had been discovered earlier by Fock (1954) for two-electron atoms.

The conventional CI approach does rather poorly at describing the electron-electron cusps in the wave function. The convergence of the CI calculations for the nonrelativistic ground state energy (E_{NR}) of Li is very slow (Chung, 1991), and it is necessary to resort to a careful evaluation of the errors arising from basis set truncation. Armed with these estimates, an extrapolated value for E_{NR} can be obtained. A serious drawback in this extrapolation approach is that the final result is no longer guaranteed to be a strict upper bound estimate for E_{NR}. There are several recently published results for the ground and several excited states of *Li*, where the reported energies are *below* the expected "exact" results for E_{NR}. The conventional CI approach is clearly not the method of choice to obtain the highest precision results for the *Li* atom. When fairly high precision is not required for the calculation of a property, the CI approach is a very viable computational technique. The principal advantage of the CI approach, in contrast to some of the other methods discussed in this section, is that it can be readily extended to treat larger electronic systems. For a recent review of MCHF and CI calculations of atomic properties, see Godefroid *et al.* (1996).

The hybrid CI-Hylleraas technique (Sims and Hagstrom, 1971a) explicitly incorporates r^n_{ij}-dependent terms in the basis set. Early users of this technique restricted the basis functions to a maximum of one r^n_{ij} factor per term. Recent applications of the technique to the *Li* atom (Lüchow and Kleindienst, 1992b, 1994) have removed this restriction, at a greater cost in terms of the integral complexity of the calculations. A very significant improvement in the rate of convergence for the calculation of E_{NR} is found using the CI-HY technique, relative to what is known from standard CI calculations.

The Hylleraas technique (Hylleraas, 1929), which incorporates factors of r_{ij} into the basis terms, shows the fastest convergence rate for the calculation of E_{NR}. This improved convergence is tied in part to the better description of the electron-electron cusps of the exact wave function. The faster convergence is also due to a superior description of the wave function over a more extended region of configuration space (Gilbert, 1963). These gains are counterbalanced by two other features of the general approach. The first is that the Hylleraas technique has not been extended to systems with more than four electrons. Even for four-electron atomic systems, there are significant unresolved integration problems when Hylleraas basis sets are employed (see, for example, King, 1993). For three-electron systems, there are several expectation values—for example, the relativistic kinetic energy correction and the electron-electron orbital correction—that require integrals that are very difficult to deal with, when a general Hylleraas expansion is employed. Despite these rather severe drawbacks of the Hylleraas approach, the technique has proved to be of considerable value, and has yielded the highest precision and fastest convergence for a number of properties of the lithium atom. Several of these properties are discussed in later sections.

III. Some Mathematical Issues

The essential feature of any high-precision calculation of atomic properties is the interplay between the choice of the mathematical form of the basis functions that are expected to give a good description of the electronic distribution, and the tractable nature of the integrals arising in the evaluation of the required expectation values. Because we have only very limited knowledge about the functional form of the atomic wave function for an N-electron ($N > 2$) system in the nonrelativistic approximation, the construction of quality wave functions is still, to a large degree, a matter of trial and error (trial and success!). Brute force procedures are prevalent.

The significance of explicitly incorporating factors of the interelectronic coordinates has long been recognized. The importance of the r_{ij} factors for improving the cusp characteristics of the wave function has been discussed in numerous papers (see, for example, Morgan, 1989). The inclusion of r_{ij}-dependent terms in basis sets is becoming increasingly common for atomic and molecular calculations, which in turn leads to a number of interesting integration problems.

A. INTEGRATION PROBLEMS

For basis functions of the form given in Eq. (2) (which are appropriate for an S-state of a three-electron atomic system), it is straightforward to show that all the integrals required for the determination of the energy and a number of other expectation values reduce to the form

$$I(i, j, k, \ell, m, n, a, b, c) = \int r_1^i r_2^j r_3^k r_{23}^\ell r_{31}^m r_{12}^n e^{-ar_1-br_2-cr_3} \, d\mathbf{r}_1 d\mathbf{r}_2 d\mathbf{r}_3 \quad (8)$$

For an energy determination, only integral cases with $\ell, m, n \geq -1$ arise. Various cases of Eq. (8) have been discussed extensively in the literature (Huang, 1946; Szász, 1961; Öhrn and Nordling, 1963; Hinze and Pitzer, 1964; Bonham, 1965; Burke, 1965; Byron and Joachain, 1966; Roberts, 1966a, 1966b; Larsson, 1968; Perkins, 1968, 1969; Ho and Page, 1975; Berk et al., 1986; King, 1991c; Remiddi, 1991; King et al., 1992; Lüchow and Kleindienst, 1993; Porras and King, 1994; Drake and Yan, 1995; Yan and Drake, 1997a; King, 1998a; Pelzl and King, 1998). The integrals appearing in Eq. (8) can be simplified by using the Sack expansion (Sack, 1964) for the interelectronic coordinate to obtain the auxiliary function

$$W(I, J, K, \alpha, \beta, \gamma) = \int_0^\infty x^I e^{-\alpha x} \, dx \int_x^\infty y^J e^{-\beta y} \, dy \int_y^\infty z^K e^{-\gamma z} \, dz \quad (9)$$

These auxiliary functions have been well studied in the literature (James and Coolidge, 1936; Öhrn and Nordling, 1963; McKoy, 1965; Burke, 1965; Larsson, 1968; Berk et al., 1986; Drake and Yan, 1995; Frolov and Smith, 1997).

The most difficult I-integral cases that arise occur when $\ell, m,$ and n are all odd. This case is typically evaluated by truncation of an infinite summation, which behaves roughly like $\Sigma_{k=1}^\infty k^{-6}$ (the convergence of the sum depends explicitly on the values of $\ell, m,$ and $n,$ but the worst case convergence of any I-integral in an energy calculation is as indicated). As remarked in Section II, when fixed exponents are employed (Eq. (7)), an integral store and retrieve strategy becomes very effective for dealing with the I-integrals, but this is not practical (or useful) when the exponent parameters are optimized. Ideally, it would be highly desirable if the I-integrals could be expressed in the form

$$I(i, j, k, \ell, m, n, a, b, c) \Rightarrow \sum_u f_u(a, b, c) g_u(i, j, k, \ell, m, n) \quad (10)$$

The advantage of this simplification is that the function g_u can be calculated and stored as a large array. This is feasible given the relatively inexpensive cost of memory on currently available workstations. The amount of computational activity is thereby significantly reduced when orbital exponent optimization is carried out. The closest form to Eq. (10) that the author has found is

$$I(i, j, k, \ell, m, n, a, b, c) = \sum_u f_u(a, b, c) \sum_v g_{uv}(i, j, k, \ell, m, n) \quad (11)$$

This expression is obviously less suitable because the g-function has an additional index dependence.

An extended form of the I-integrals given in Eq. (8) has been investigated where the additional factor $\exp(-\alpha_{12} r_{12} - \alpha_{23} r_{23} - \alpha_{13} r_{13})$ is included, and α_{12}, α_{13}, and α_{23} are constants (Fromm and Hill, 1987; Remiddi, 1991; Harris, 1997). Solution of these integrals allows for increased flexibility in the choice of basis functions, by the inclusion of additional exponential factors like $\exp(-\alpha_{12} r_{12})$. Including this type of exponential term in the basis set may offer the advantage of ensuring that the wave function approximately satisfies the electron-electron cusp condition.

We note parenthetically that for four-electron atomic systems, the generalization of Eq. (8) now includes up to six interelectronic coordinates. Although there has been some effort directed at solving these four-electron integrals (Sims and Hagstrom, 1971b; King, 1993; Kleindienst et al., 1995), there are a number of important cases still unresolved. There are clearly formidable mathematical problems to be overcome if Hylleraas-type expansions are to be routinely employed beyond three-electron atomic systems.

For three-electron systems, there are two important problems where more complex I-integrals arise. The first is the evaluation of $\langle \Psi | H_{NR}^2 | \Psi \rangle$, where H_{NR} is the nonrelativistic Hamiltonian. This expectation value is required for the determination of lower bounds to the nonrelativistic energies. The second problem is the evaluation of certain relativistic corrections to the energy. For these expectation values, I-integrals arise with at least one of the indices $\ell, m, n = -2$. For both of these problems, it is possible to select the basis functions in such a way that these very difficult integral cases are avoided. However, the resulting wave functions turn out to be of rather poor quality, particular in the near-nuclear region of configuration space. The I-integrals with one (or two) factors of r_{ij}^{-2} are much harder to evaluate, and these integrals have received considerably less attention in the literature (King, 1991c; King et al., 1992; Lüchow and Kleindienst, 1993; Porras and King, 1994; Yan and Drake, 1997a; King, 1998a). The increased complexity can be appreciated by observing the form for r_{ij}^{-2} (Pauli and Kleindienst, 1984; King, 1991c; Lüchow and Kleindienst, 1992a; Porras and King, 1994), which can be expressed as (King, 1991c)

$$\frac{1}{r_{12}^2} = \sum_{l=0}^{\infty} \left[\frac{2l+1}{2}\right] 4^{-l} \left[\ell n \left|\frac{r_1+r_2}{r_1-r_2}\right| \sum_{\kappa=0}^{l} r_1^{2\kappa-l-1} r_2^{l-2\kappa-1} \right.$$
$$\sum_{\nu=0}^{\min[\kappa,l-\kappa]} (-4)^{\nu} \binom{l}{\nu}\binom{2l-2\nu}{l}\binom{l-2\nu}{\kappa-\nu} -2 \sum_{\kappa=0}^{l-1} r_1^{-l+2\kappa} r_2^{l-2\kappa-2} \quad (12)$$
$$\left. \sum_{j=0}^{\min[\kappa,l-\kappa-1]} 4^j \binom{l-2j-1}{\kappa-j} \sum_{\nu=0}^{j} \frac{(-1)^{\nu}\binom{l}{\nu}\binom{2l-2\nu}{l}}{2j-2\nu+1} \right] P_l(\cos\theta_{12})$$

where P_l denotes a Legendre polynomial and $\binom{a}{b}$ is a binomial coefficient. This expansion can be contrasted with the familiar expansion of r_{12}^{-1} in terms of Legendre polynomials. The expansion terms for r_{12}^{-2} involve a logarithmic function of the coordinates r_1 and r_2, which adds significantly to the complexities of the integral evaluations. The author (King, 1991c) has succeeded in reducing a large number of integrals of the form $I(i, j, k, -2, m, n, a, b, c)$ to two-electron integrals, which can be computed relatively quickly.

A particularly difficult case to evaluate is $\ell = -2$, and m and n both odd. The methods that are currently available for this case lead to limited precision (about 12–16 digits) for the integral evaluations, which in turn becomes a factor controlling the precision of the required expectation values, when very large basis set expansions are employed. If a general Hylleraas expansion is employed, then the considerably more complicated case $\ell = -2, m = -2$ arises. Although methods are available to deal with these integrals (Lüchow and Kleindienst, 1993; Porras and King, 1994), the precision available is limited. For practical calculations, it appears possible to delete basis functions that lead to these $\ell = -2, m = -2$ cases, without any appreciable loss of quality for the overall wave function.

When excited states of *Li* are under investigation, the additional angular factors in the basis functions lead to more tedious integration problems. A number of the required integrals involving additional angular factors have been evaluated. For a detailed account see, for example, Barrois et al. (1997b, 1997c) and Yan and Drake (1997a).

There are three additional mathematical issues that emerge, particularly when large basis sets are employed. As the size of the basis set grows, significant problems associated with linear dependence in the basis set arise. These are numerical problems that are difficult to circumvent, because increasing the precision level of the calculation can be a very expensive proposition. Using double-precision arithmetic with a 64-bit word or quadruple precision with a 32-bit word will generally allow well over a thousand terms to be incorporated in a Hylleraas-type expansion, though the actual maximum size will obviously depend on the particular basis functions employed. To proceed to several thousand terms would most likely require that extended precision arithmetic be employed. Although Fortran source codes have been developed to deal with this, the high cpu costs would make the routine use of such codes prohibitive.

For large basis sets, particularly when ω (Eq.(6)) reaches double-digit values, the precision level of calculated expectation values can significantly decline. This can be easily monitored by separately evaluating the positive and negative contributions to the expectation value of interest. For large values of ω, individual matrix elements can be very large, and consequently a sum of several hundred thousand contributions can lead to a value of considerable size. When the positive and negative components of an expectation value are very large, and the value of the particular expectation value is not, then very significant loss of precision occurs. If the positive and negative components are not tracked separately, then this precision loss can be easily overlooked, particularly if the convergence of the expectation value is nonmonotonic.

Large Hylleraas expansions often lead to many local energy minima in the $\{\alpha_\mu, \beta_\mu, \gamma_\mu\}$ parameter space. There can be significant differences between these minima, so an effective optimization scheme needs to be able to handle a global strategy in a cost-effective manner. Global strategies that perform some type of stochastic sampling require the evaluation of a significant number of matrix elements, so the bottleneck in the calculation of the wave function becomes the evaluation of the three-electron I-integrals (Eq.(8)). Because the all-odd $\{\ell, m, n\}$ I-integral case is significantly slower than the other cases, developing new mathematical approaches for the evaluation of these integrals would have an important impact on progress in the area of high-precision calculations on three-electron atomic systems.

IV. Nonrelativistic Energies

The nonrelativistic Hamiltonian conventionally employed is (in atomic units)

$$H_{\text{NR}} = \sum_{i=1}^{3} \left(-\frac{1}{2} \nabla_i^2 - \frac{Z}{r_i} \right) + \sum_{i=1}^{3} \sum_{j>1}^{3} \frac{1}{r_{ij}} \quad (13)$$

where Z is the nuclear charge ($= 3$ for Li). The specific mass shift (mass polarization correction) is not included in Eq. (13). Several high-precision calculations have been carried out with this factor incorporated in H_{NR} (King, 1986; Lüchow and Kleindienst, 1994; Yan and Drake, 1995a), but it is more common to evaluate the specific mass shift using first-order perturbation theory.

For the S states of a three-electron atom, it is possible to work with the Hamiltonian in the form

$$H_{NR} = -\sum_{i=1}^{3} \left(\frac{1}{2} \frac{\partial^2}{\partial r_i^2} + \frac{1}{r_i} \frac{\partial}{\partial r_i} + \frac{3}{r_i} + \frac{\partial^2}{\partial u_i^2} + \frac{2}{u_i} \frac{\partial}{\partial u_i} - \frac{1}{u_i} \right)$$
$$- \frac{1}{2} \sum_{\mathcal{P}_{ijk}} \left(\frac{r_i^2 + u_k^2 - r_j^2}{u_k r_i} \frac{\partial^2}{\partial r_i \partial u_k} + \frac{u_i^2 + u_k^2 - u_j^2}{2 u_i u_k} \frac{\partial^2}{\partial u_i \partial u_k} \right) \quad (14)$$

In Eq. (14), $\mathcal{P}_{ijk}$ signifies that the summation is over the six permutations ($\begin{smallmatrix} 1 & 2 & 3 \\ i & j & k \end{smallmatrix}$), and the notational simplifications $u_1 = r_{23}$, $u_2 = r_{31}$, and $u_3 = r_{12}$ are employed. The form of H_{NR} in Eq. (14) is very convenient to use, because of the structure of the basis functions employed in the Hylleraas expansion (see Eq. (2)). It is also straightforward to show, using this simplified form of H_{NR}, that the matrix elements required for the determination of the energy reduce to integrals of the form given in Eq. (8).

A. The Ground State

A considerable amount of attention has been directed at the problem of determining high precision estimates for the nonrelativistic energies of the lithium atom, and much of this activity has focused on the ground state (King and Shoup, 1986; King, 1989, 1995; Kleindienst and Beutner, 1989; King and Bergsbaken, 1990; Jitrik and Bunge, 1991; Chung, 1991; McKenzie and Drake, 1991; Pipin and Bishop, 1992; Lüchow and Kleindienst, 1992b, 1994; Tong et al., 1993; Yan and Drake, 1995a; Jitrik and Bunge, 1997). Recent work with the Monte Carlo technique is discussed by Alexander and Coldwell (1997). A summary of some of the highest precision values is presented in Table 1. An extensive tabulation of earlier calculations of the ground state E_{NR} for Li is given by King (1997). Not surprisingly, progress in obtaining higher precision values of E_{NR} has largely mirrored advances in computer technology.

In Table 1, the last digits for each energy that have been determined by either extrapolation procedures or by estimation of basis set truncation errors are shown in italics. The reader needs to note that neither of these procedures necessarily leads to a strict upper bound for E_{NR}. Except for the CI calculations of Chung (1991), all the E_{NR} values for the ground state presented in Table 1 are based on wave functions involving explicit dependence on the interelectronic coordinates. Chung's result includes a sizeable basis set truncation correction of 134.6 μhartree. Calculations based on the CI-HY technique (Pipin and Bishop, 1992; Lüchow and Kleindienst, 1992b, 1994) lead to rather precise values for

TABLE 1
High-Precision Estimates of the Nonrelativistic Energies of the Low-Lying Doublet States of Li.

State	$E_{\rm NR}$ (a.u.)	Reference
2 ^2S	−7.478058	King and Shoup (1986)
($1s^22s$)	−7.478059	King (1989)
	−7.4780595	King and Bergsbaken (1990)
	−7.4780597	Chung (1991)
	−7.478060326	McKenzie and Drake (1991)
	−7.4780601	Pipin and Bishop (1992)
	−7.478060252	Lüchow and Kleindienst (1992b)
	−7.4780603208	Lüchow and Kleindienst (1994)
	−7.478060	King (1995)
	−7.47806032310	Yan and Drake (1995a)
	[−7.47806035(12)]	This chapter
3 ^2S	−7.354030	Pipin and Woznicki (1983)
($1s^23s$)	−7.354076	King (1991)
	−7.3540978	Lüchow and Kleindienst (1992b)
	−7.3540980	Wang, Zhu, and Chung (1992a)
	−7.354098369	Lüchow and Kleindienst (1994)
	−7.354098	King (1998c)
	[−7.3540981(2)]	This chapter
4 ^2S	−7.318491	King (1992)
($1s^24s$)	−7.318525	Lüchow and Kleindienst (1992b)
	−7.3185303	Wang, Zhu, and Chung (1992a)
	−7.318530665	Lüchow and Kleindienst (1994)
	−7.318529	King (1998c)
	[−7.3185306(2)]	This chapter
5 ^2S	−7.303439	King (1991)
($1s^25s$)	−7.303547	Lüchow and Kleindienst (1992b)
	−7.3035508	Wang, Zhu, and Chung (1992a)
	−7.303546	King (1998c)
	[−7.3035515(2)]	This chapter
6 ^2S	−7.29583	Lüchow and Kleindienst (1992b)
($1s^26s$)	−7.295846	King (1998c)
	[−7.295859(1)]	This chapter
2 ^2P	−7.410106	Pipin and Woznicki (1983)
($1s^22p$)	−7.4101554	Pipin and Bishop (1992)
	−7.4101541	Wang, Zhu, and Chung (1993)
	−7.4101565218	Yan and Drake (1995a)
	−7.41015652	Barrois, Kleindienst, and Lüchow (1997b)
	−7.410156531763	Yan and Drake (1997b)
	[−7.41015645(11)]	This chapter
3 ^2P	−7.337059	Pipin and Woźnicki (1983)
($1s^23p$)	−7.3371503	Wang, Zhu, and Chung (1993)
	−7.33715170	Barrois, Kleindienst, and Lüchow (1997b)
	[−7.3371516(2)]	This chapter

TABLE 1 (*Continued*)

State	E_{NR} (a.u.)	Reference
4 ^2P	−7.311736	Sims and Hagstrom (1975)
($1s^24p$)	−7.311*8881*	Wang, Zhu, and Chung (1993)
	−7.3118888	Barrois, Kleindienst, and Lüchow (1997b)
	[−7.3118887(2)]	This chapter
5 ^2P	−7.300142	Sims and Hagstrom (1975)
($1s^25p$)	−7.3002875	Wang, Zhu, and Chung (1993)
	[−7.3002883(2)]	This chapter
3 ^2D	−7.3355231	Pipin and Bishop (1992)
($1s^23d$)	−7.3355*239*	Wang, Zhu, and Chung (1992b)
	−7.33552354*110*	Yan and Drake (1995a)
	[−7.3355234(2)]	This chapter
4 ^2D	−7.311187	Pipin and Bishop (1992)
($1s^24d$)	−7.311*190*	Wang, Zhu, and Chung (1992b)
	[−7.3111896(2)]	This chapter
5 ^2D	−7.299928	Wang, Zhu, and Chung (1992b)
($1s^25d$)	[−7.2999277(2)]	This chapter
4 ^2F	−7.311*1687*	Wang, Zhu, and Chung (1992b)
($1s^24f$)	[−7.3111668(2)]	This chapter
5 ^2F	−7.2999*171*	Wang, Zhu, and Chung (1992b)
($1s^25f$)	[−7.2999159(2)]	This chapter

Digits based on extrapolation or estimates of the basis set truncation errors are indicated in italics. Estimates of the nonrelativistic energies are shown in [ ].

E_{NR}. All the other ground state values of E_{NR} shown in Table 1 were determined using Hylleraas-type expansions.

1. Lower Bound Estimates for E_{NR}

When the convergence of the variational calculation of E_{NR} is slow, it is particularly advantageous to have access to precise lower bounds for the nonrelativistic energies. The lower bounds supplement the upper bound estimates obtained via the standard variational approach. Relatively little work has been published on finding lower bounds for the energies of the lithium atom. The principal reason for this is the extremely difficult nature of the integration problems that arise. In fact, almost all of the progress that has been made on finding lower bounds for E_{NR} has been restricted to one- and two-electron atomic and molecular systems. The problem of determining a lower bound for the ground state E_{NR} has been approached using two rather different methods. The first, based on intermediate Hamiltonian techniques (Bazley, 1959, 1960; Bazley and Fox, 1961) has been applied to the lithium atom (Reid, 1972, 1974; Fox and Sigillito, 1972a, b, c; Fox, 1972; Russell and Greenlee, 1985). This approach has not led to precise estimates for any lower bounds to the nonrelativistic energies of the lithium atom. The second method is based on the more familiar classical lower bound formulas derived by Temple (1928),

$$E_0 \geq E_T = \langle \Psi|H|\Psi \rangle - \frac{\sigma}{E_1 - \langle \Psi|H|\Psi \rangle} \quad (15)$$

by Weinstein (1934),

$$E_0 \geq E_W = \langle \Psi|H|\Psi \rangle - \sigma^{1/2} \quad (16)$$

and by Stevenson (1938) and Stevenson and Crawford (1938),

$$E_0 \geq E_S = \alpha - (\alpha^2 - 2\alpha\langle \Psi|H|\Psi \rangle + \langle \Psi|H^2|\Psi \rangle)^{1/2}$$
$$= \alpha - [\sigma + (\alpha - \langle \Psi|H|\Psi \rangle)^2]^{1/2} \quad (17)$$

In Eqs. (15–17), E_0 denotes the exact nonrelativistic ground state energy, and E_T, E_W, and E_S designate, respectively, the Temple, Weinstein, and Stevenson lower bound estimates to E_0. The variance, σ, is defined by

$$\sigma = \langle \Psi|H^2|\Psi \rangle - \langle \Psi|H|\Psi \rangle^2 \quad (18)$$

As Ψ approaches the exact solution of the Schrödinger equation, $\sigma \to 0$. These lower bound formulas have been discussed extensively in the literature (Kato, 1949; Caldow and Coulson, 1961; Fröman and Hall, 1961; Wilson, 1965; Switkes, 1967; Schmid and Schwagner, 1968; Delves, 1972; Coulson and Haskins, 1973; Cohen and Feldmann, 1979; Scrinzi, 1992). Further references on application of these formulas to one- and two-electron systems can be found in King (1995).

There are two high-precision estimates of a lower bound for E_{NR} (Lüchow and Kleindienst, 1994; King, 1995) and an earlier more approximate estimate (Conroy, 1964). The high-precision estimates are

$$-7.478176 \text{ a.u.} < E_0 \quad (19)$$

obtained using a 920-term CI-HY wave function (Lüchow and Kleindienst, 1994) and

$$-7.47830 \text{ a.u.} < E_0 \quad (20)$$

using a 600-term Hylleraas wave function (King, 1995). These estimates are approximately .12 and .24 millihartrees too low, based on the extrapolated value given for E_0 in Table 1.

The lower bound calculations converge at a significantly slower rate than the corresponding variational calculations of the upper bound estimates of E_0. The key expectation value required in the lower bound evaluation, $\langle H^2 \rangle$, samples the region of configuration space close to the nucleus. This region is less well described in the standard variational approach. To significantly improve upon the results indicated in Eqs. (19, 20), the strategy most likely to be successful is to build a wave function that has been optimized to minimize σ (Eq. (18)) directly, or as an auxiliary constraint in the standard variational approach. This idea has been in the literature for a considerable time (see, for example, Preuss, 1961).

2. Distribution Functions

The radial electronic density function for the ground state of Li has been evaluated in closed form, starting from a Hylleraas-type wave function (King and Dressel, 1989). To obtain the density in a compact analytic form, it is necessary to make some restrictions on the basis functions employed in the Hylleraas expansion. For example, terms with three odd powers of the interelectronic coordinates must be excluded, otherwise a finite expansion for the density does not appear possible. The moments $\langle r_i^n \rangle$, for $n = -2$ to 6, generated from the calculated radial density, are found to be in good agreement with other calculations using much larger basis sets (King, 1995). For the excited states of Li, there has been relatively little work devoted to determining precise densities. King (1991b) has evaluated radial densities for some excited S states using Hylleraas basis sets of modest size.

Whereas the spin density at the nucleus has received a considerable amount of attention, the radial dependence of the spin density has received almost no attention. The most precise results available for the radial dependence of the spin density are due to Esquivel *et al.* (1991), who used large-scale CI calculations to study this function.

Starting from a Hylleraas-type basis set, Dressel and King (1994) have managed to determine a compact analytic expression for the electron-electron distribution function. Once again, it was necessary to make some simplifications on the possible terms in the basis set, otherwise a compact and finite series expansion could not be obtained. Extension of this work to allow for more general basis set expansions would be desirable, as the electron-electron distribution function plays an important role in discussions of the Coulomb hole.

B. EXCITED STATES

The calculation of nonrelativistic energies for the excited states of lithium has received considerable attention. Progress in this area can be summarized most easily in terms of the following four groups: single valence-electron excited doublet states, low-lying quartet states, core-excited doublet states, and doubly core-excited quartet states. The next four subsections discuss each of these states.

1. The Low-Lying Excited Doublet States

The calculation of E_{NR} for the low-lying ^2S states has been carried out with a variety of computational techniques (Larsson, 1972; Perkins, 1972; Sims and

Hagstrom, 1975; Sims et al., 1976a; Pipin and Woźnicki, 1983; Hijikata et al., 1987; King, 1991a, 1998c; Lüchow and Kleindienst, 1992b; Wang et al., 1992a; Yan and Drake, 1995a). The determination of the nonrelativistic energies for the low-lying ^2P states has also attracted considerable attention (Ahlenius and Larsson, 1973, 1978; Sims and Hagstrom, 1975; Sims et al., 1976a; Muszyńska et al., 1980; Pipin and Woźnicki, 1983; Hijikata et al., 1987; Pipin and Bishop, 1992; Chung and Zhu, 1993; Wang et al., 1993; Yan and Drake, 1995a; Yan and Drake, 1997b). The low-lying ^2D and ^2F states have received far less attention, but there has been some very recent progress on the calculation of E_{NR} for some of these states (Wang et al., 1992b; Pipin and Bishop, 1992; Yan and Drake, 1995a). A selection of the highest precision results for the nonrelativistic energies of these states is presented in Table 1. Also included in Table I is an estimate of the nonrelativistic energy, which was derived using

$$E_{NR}(^2X) = E_{NR}(Li^+) + \Delta E_{REL} + \Delta E_{MASS} + \Delta E_{QED} - I_1 + \Delta E(^2X) \quad (21)$$

where

$$\Delta E_{REL} = E_{REL}(Li^+) - E_{REL}(^2X) \quad (22)$$

$$\Delta E_{MASS} = E_{MASS}(Li^+) - E_{MASS}(^2X) \quad (23)$$

$$\Delta E_{QED} = E_{QED}(Li^+) - E_{QED}(^2X) \quad (24)$$

and I_1 designates the first ionization potential of the ground state of Li, and $\Delta E(^2X)$ is the transition energy from the ground state to the ^2X state of interest. REL, MASS, and QED refer to the relativistic correction, nuclear mass dependent correction, and quantum electrodynamic shift, respectively. The excitation energies have been taken from Radziemski et al. (1995) or Sansonetti et al. (1995). All the relativistic corrections are taken from the work of Wang et al. (1992a, 1992b, 1993), except the result for the 2 ^2S state, which is taken from King (1997). The latter result is a combination of individual relativistic corrections taken from Chung (1991), King (1995) and Yan and Drake (1995a). The mass correction includes both the Bohr mass shift and the specific mass shift. The specific mass shifts used to evaluate ΔE_{MASS} were taken from Yan and Drake (1995a) and Wang et al. (1992a, 19992b, 1993). For the 2 ^2S state, the value of ΔE_{QED} based on the work of Feldman and Fulton (1995) was employed, and for the 3 ^2S state, ΔE_{QED} was taken from Wang et al. (1992a). For the other doublet states, ΔE_{QED} has been set to zero, which should be a satisfactory approximation, at least for the higher lying states, based on the values calculated from the standard one-electron formula for ΔE_{QED} (which is given later in Section VII). The results for Li^+ that enter Eqs. (21-24) are taken from Yan and Drake (1995a), Pekeris (1958, 1962), Johnson and Soff (1985) and Drake (1988). The semiempirical estimates for E_{NR} are in particularly good agreement with the results obtained from variational calculations. The error estimates are only rough, as the uncertainties for the $E_{REL}(^2X)$ values are not available, and have been estimated at two in the last quoted digit of the published calculations. Also, the error associated with ΔE_{QED} is rather difficult to gauge.

2. The Low-Lying Quartet States

There has been renewed interest in high precision calculations on the low-lying quartet states of the lithium atom (Fischer, 1990; Hsu et al., 1991, 1994; Lüchow et al., 1993; Barrois et al., 1996, 1997a; King, 1998d). Efforts to improve model-potential calculations for these states have also attracted recent attention (Chen, 1996). A summary of some of the higher precision calculations is presented in Table 2. All the states presented in Table 2 lie below the $1s2s\ ^3S$ state of Li^+. There has been extensive experimental work on these states (see Feldman and Novick, 1967; and for reviews, Berry, 1975; Mannervik, 1989).

A semiempirical estimate of the nonrelativistic energy of a number of the ^4X states can be determined using

$$E_{NR}(^4X) = E_{NR}(Li^+, 1s2s\ ^3S) + \Delta E_{MASS}(^4X) \quad (25)$$
$$+ \Delta E_{REL}(^4X) + \Delta E_{QED}(^4X) - I_4 + \Delta E(^4X)$$

where $\Delta E(^4X)$ is the transition energy from the $1s2s2p\ ^4P$ state, and

$$\Delta E_{MASS}(^4X) = E_{MASS}(1s2s,\ ^3S\ Li^+) - E_{MASS}(^4X) \quad (26)$$

$$\Delta E_{REL}(^4X) = E_{REL}(1s2s,\ ^3S\ Li^+) - E_{REL}(^4X) \quad (27)$$

$$\Delta E_{QED}(^4X) = E_{QED}(1s2s,\ ^3S\ Li^+) - E_{QED}(^4X) \quad (28)$$

Experimental values of $\Delta E(^4X)$ have been taken from the work of Mannervik and Cederquist (1983). In Eq. (25), I_4 designates the ionization potential of the $1s2s2p$ ^4P state (to yield Li^+ $(1s2s)\ ^3S$). This ^4P state is the lowest in energy of the quartet states, and is metastable against both radiative decay and autoionization. The values of E_{NR}, E_{SMS} (a part of E_{MASS}) and E_{REL} for Li^+ $(1s2s\ ^3S)$ are taken from Pekeris (1962). There are several estimates available for the ionization potential of $1s2s2p\ ^4P$. These include an experimental determination of 56473(5) cm^{-1} (Mannervik and Cederquist, 1983) and five theoretically derived estimates of 56460.6 cm^{-1} (Bunge, 1981a), 56459.6(5)cm^{-1}, and 56460.1(2)cm^{-1} (Hsu et al., 1991), 56461.7 cm^{-1} (Barrois et al., 1997a) and 56462.2(2) cm^{-1} (King, 1998d). The experimental result appears too high; it leads to estimates of $E_{NR}(^4X)$ that are significantly different from the results of some recent well-converged calculations. The estimates of Hsu et al. are based on nonrelativistic energies of the states $1s2s2p\ ^4P$ and $1s2s3s\ ^4S$, which are too high by approximately 2.0 cm^{-1}. The value of E_{NR} for the $1s2s3s\ ^4S$ state used by Bunge is too high by about 1.6 cm^{-1}.

TABLE 2
Nonrelativistic Energies of the Low-Lying Quartet States of Li.

State	E_{NR} (a.u.)	Reference
$(1s2s3s)\ ^4S$	−5.212396	Larsson (1972)
	−5.21259	Larsson, Crossley, and Ahlenius (1979)
	−5.212737	Bunge and Bunge (1978a)
	−5.212741	Bunge (1981a)
	−7.212739	Hsu, Chung, and Huang (1991)
	−5.212748246	Barrois, Lüchow, and Kleindienst (1996)
	−5.212748	King (1998d)
	[−5.212748(1)]	This chapter
$(1s2s4s)\ ^4S$	−5.15823	Larsson and Crossley (1982)
	−5.15839345	Barrois, Lüchow, and Kleindienst (1996)
	−5.158391	King (1998d)
	[−5.15844(2)]	This chapter
$(1s2s5s)\ ^4S$	−5.13816	Larsson and Crossley (1982)
	−5.1384624	Lüchow, Barrois, and Kleindienst (1993)
	−5.138460	King (1998d)
	[−5.13845(2)]	This chapter
$(1s2s6s)\ ^4S$	−5.12829	Larsson and Crossley (1982)
	−5.128880	Lüchow, Barrois, and Kleindienst (1993)
	−5.128872	King (1998d)
	[−5.12893(4)]	This chapter
$(1s2s2p)\ ^4P^o$	−5.367948	Bunge and Bunge (1978b)
	−5.36783	Larsson and Crossley (1982)
	−5.367917	Fischer (1990)
	−5.368001	Hsu, Chung, and Huang (1991)
	−5.3680059	Hsu, Chung, and Huang (1994)
	−5.36801014	Barrois, Bekavac, and Kleindienst (1997a)
	[−5.368013(2)]	This chapter
$(1s2s3p)\ ^4P^o$	−5.186742	Glass (1978)
	−5.18687	Larsson and Crossley (1982)
	−5.187278	Bunge (1981b)
	−5.1872793	Hsu, Chung, and Huang (1994)
	−5.18728815	Barrois, Bekavac, and Kleindienst (1997a)
	[−5.18731(2)]	This chapter
$(1s2s4p)\ ^4P^o$	−5.14338	Larsson and Crossley (1982)
	−5.149722	Bunge (1981b)
	−5.1497361	Hsu, Chung, and Huang (1994)
$(1s2s5p)\ ^4P^o$	−5.134454	Bunge (1981b)
	−5.1344767	Hsu, Chung, and Huang (1994)[a]
$(1s2p3s)\ ^4P^o$	−5.1195222	Hsu, Chung, and Huang (1994)
	[−5.119533(6)]	This chapter
$(1s2s3d)\ ^4D$	−5.1730806	Hsu, Chung, and Huang (1994)
	[−5.173086(2)]	This chapter
$(1s2s4f)\ ^4F$	−5.142818	Galán and Bunge (1981)

[a] Additional higher states in the quartet P series were studied by these authors.

An accurate way to determine the ionization potential of the $1s2s2p\ ^4P$ state is to use the relationship

$$I_4 = E(1s2s\ ^3S\ Li^+) - E(1s2s3s\ ^4S) + \Delta E(1s2s3s \rightarrow 1s2s2p) \quad (29)$$

where $\Delta E(1s2s3s \rightarrow 1s2s2p)$ is the energy for the transition $1s2s3s\ ^4S \rightarrow 1s2s2p\ ^4P$, and $E(1s2s,\ ^3S\ Li^+)$ and $E(1s2s3s\ ^4S)$ are the energies for the lowest $^3S\ Li^+$ state and the lowest ^4S state, respectively. The transition energy $\Delta E(1s2s3s \rightarrow 1s2s2p)$ was reported (as a then unassigned line) by Herzberg and Moore (1959) in their study of the spectrum of Li^+. It was later suggested that several of the unassigned lines in the observed spectrum of Li^+ were actually transitions between quartet levels of Li I (Feldman and Novick, 1963; Garcia and Mack, 1965). The series of lines at approximately 2934 Å observed by Herzberg and Moore were assigned to be the $1s2s3s\ ^4S \rightarrow 1s2s2p\ ^4P$ transition by Holøien and Geltman (1967) on the basis of calculations, and confirmed by Feldman et al. (1968) and Levitt and Feldman (1969). The latter authors gave for this transition energy the value $\Delta E(1s2s3s \rightarrow 1s2s2p) = 34071.91(5)$ cm^{-1}. If this value for ΔE is employed in Eq. (29) along with $E(1s2s,\ ^3S\ Li^+) = 1121722.13(1)$ cm^{-1} (Accad et al., 1971) and the calculated values for E_{NR} (King, 1998d) and E_{REL} (Hsu et al., 1991) for the $1s2s3s\ ^4S$ state, then $I_4 = 56462.25\ (22)$ cm^{-1}. The error estimate is determined primarily by the uncertainty in the value of E_{REL}. This is the value of I_4 that has been employed in Eq. (25). Other values of $E_{REL}\ (^4X)$ were taken from Hsu et al., (1991, 1994).

The semiempirical estimates of E_{NR} reported in Table 2 are in satisfactory agreement with the results from recent high-precision calculations. The error estimates are rather large, particularly in comparison with those given in Table 1. The experimental transition energies between quartet levels are not known with the same high precision as the transition energies between the low-lying doublet states. There are significant uncertainties in the relativistic corrections for some quartet states, and there are difficulties associated with pinning down precise estimates of $\Delta E_{QED}(^4X)$.

There has been considerable interest in the modes of decay of some of the low-lying quartet states. The reader can pursue this avenue of research starting with the following works: Manson (1971), Nicolaides and Aspromallis (1986, 1988), Mannervik and Cederquist (1986), Davis and Chung (1987, 1988), and Sonnek and Mannervik (1990).

3. The Core-Excited Doublet States

Core-excited doublet states of the Li atom have received considerable attention. These states play an important role in electron scattering and various collision experiments, as well as certain photoabsorption processes. The core-excited doublet states have energies above the first ionization energy of the neutral atom. They are not discrete states in the same sense as the bound excited doublet states. These core-excited states are in most cases coupled to the $1s^2\epsilon\ell$ continua via interelectron Coulomb interactions. The standard variational method cannot be *directly* applied to treat these autoionizing states, the exceptions being states like $1s2p^2$ $^2P^e$ (or more generally, $1s2p\,\epsilon\ell\,^2L\,(L=\ell)$), which are bound metastable core-excited states.

A variety of theoretical techniques have been employed to treat a number of the core-excited resonances (see for example, Ho, 1983; Chung and Davis, 1985). Most of the standard techniques available have been applied to various core-excited states of the lithium atom (Bhatia, 1978; Bunge, 1979; Wakid *et al.*, 1980; Chung, 1981a, 1981b, 1982; Jáuregui and Bunge, 1981; Woźnicki *et al.*, 1983; Davis and Chung, 1984, 1985, 1990b; Jaskólska and Woźnicki, 1989a, 1989b; Chung and Gou, 1995; Barrois *et al.*, 1997c; Chung, 1997a, 1997b). The majority of these calculations deal only with the nonrelativistic energy contribution. The only workers that have attempted any evaluation of the relativistic contributions to the energy are Chung and coworkers. Much of this work has employed the CI technique. There has been experimental interest in these states for many years (see, for example, Ederer *et al.*, 1970; Berry *et al.*, 1972; Pegg *et al.*, 1975; Ziem *et al.*, 1975; Cantù *et al.*, 1977; Rassi *et al.*, 1977; McIlrath and Lucatorto, 1977; Rødbro *et al.*, 1979; Cederquist and Mannervik, 1985; Mannervik and Cederquist, 1985; Mannervik *et al.*, 1986; Meyer *et al.*, 1987; Kiernan *et al.*, 1996). Most recently, interest has focused on "hollow" atomic states of *Li*. These states of *Li* have an empty *K* shell. The reader interested in this avenue of work might start with the work of Journel *et al.* (1996).

The $^2P^e$ states have attracted attention from several authors (Bunge, 1979; Chung, 1982; Woźnicki *et al.*, 1983 and Barrois *et al.*, 1997c). For the lowest three $^2P^e$ terms, Barrois *et al.* (1997c) have used large scale CI-HY wave functions to obtain high-precision estimates of E_{NR}. The E_{NR} values reported by these authors have converged to better than 1 μhartree. Barrois *et al.* have also evaluated the expectation value $\langle \nabla_i \cdot \nabla_j \rangle$, which allows the specific mass shift correction to the energy levels to be calculated.

Some of the theoretical energies for the core-excited states are collected in Table 3. For most of the other core excited doublets that have been investigated theoretically, the nonrelativistic energies have not been determined to the same level of precision as the best results for the $^2P^e$ terms. The precision available for the experimental energies for the core-excited states is around 10 cm^{-1} (or better) (Mannervik, 1989), and clearly does not rival the results available, particularly the most recent ones, for the bound low-lying excited doublet states. Generally, the agreement between the theoretically determined energies and experimental results is satisfactory.

TABLE 3
ENERGIES FOR THE LOW-LYING CORE EXCITED DOUBLET STATES OF Li.

State	E_{NR} (a.u.)	E_{TOTAL} (a.u.)	Reference
$(1s2s^2)\ ^2S$	−5.405219	−5.405833	Davis and Chung (1984)
$[(1s2s)\ ^3S\ 3s]\ ^2S$	−5.199641	−5.200237	Davis and Chung (1985)
$[1s(2s2p)\ ^3P]\ ^2P^o$	−5.312761	−5.313056	Davis and Chung (1985)
	−5.312936		Jaskólska and Woźnicki (1989a)
	−5.313212	−5.313312	Chen and Chung (1994)
$[1s(2s2p)\ ^1P]\ ^2P^o$	−5.256864	−5.257464	Davis and Chung (1985)
	−5.257499		Jaskólska and Woźnicki (1989a)
	−5.258351	−5.258471	Chen and Chung (1994)
$1s2p^2\ ^2P^e$	−5.21365		Bunge (1979)
	−5.213702	−5.213734	Chen and Chung (1994)
	−5.21373920		Barrois, Lüchow, and Kleindienst (1997c)
$[(1s2s)\ ^3S\ 3p]\ ^2P^o$	−5.183387	−5.183993	Davis and Chung (1985)
	−5.183842		Jaskólska and Woźnicki (1989a)
	−5.184006	−5.184057	Chen and Chung (1994)
$[(1s2s)\ ^3S\ 4p]\ ^2P^o$	−5.149599		Jaskólska and Woźnicki (1989a)
	−5.149695	−5.149725	Chen and Chung (1994)
$[(1s2s)\ ^3S\ 5p]\ ^2P^o$	−5.134334		Chung (1981a)
	−5.134940		Jaskólska and Woźnicki (1989a)
$[(1s2p)\ ^3S\ 6p]\ ^2P^o$	−5.126683		Chung (1981a)
	−5.127315		Jaskólska and Woźnicki (1989a)
$[(1s2p)\ ^3P\ 3p]\ ^2P^e$	−5.10429		Bunge
	−5.104364	−5.104374	Chen and Chung (1994)
	−5.10438176		Barrois, Lüchow, and Kleindienst (1997c)
$[(1s2p)\ ^3P\ 4p]\ ^2P^e$	−5.07012		Bunge (1979)
	−5.070284	−5.070305	Chen and Chung (1994)
	−5.0703159		Barrois, Lüchow, and Kleindienst (1997c)
$[(1s2pS)\ ^1P\ 3p]\ ^2P^e$	−5.06214		Bunge (1979)
	−5.061841		Jaskólska and Woźnicki (1989a)
$(1s2p^2)\ ^2D^e$	−5.233703	−5.234200	Davis and Chung (1985)
	−5.233789		Jaskólska and Woźnicki (1989b)
	−5.234138	−5.234236	Chen and Chung (1994)
$[(1s2s)\ ^3S\ 3d]\ ^2D^e$	−5.166023	−5.166619	Davis and Chung (1985)
	−5.166187		Jaskólska and Woźnicki (1989b)
	−5.166434	−5.166475	Chen and Chung (1994)
$[(1s2s)\ ^3S\ 4d]\ ^2D^e$	−5.141919		Jaskólska and Woźnicki (1989b)
	−5.142194	−5.142220	Chen and Chung (1994)
$[1s2p)\ ^3P\ 3d]\ ^2D^o$	−5.08929		Jáuregui and Bunge (1981)
	−5.089285	−5.089293	Chen and Chung (1994)
$[(1s2p)\ ^3P\ 4d]\ ^2D^o$	−5.06163		Jáuregui and Bunge (1981)
	−5.061584	−5.061594	Chen and Chung (1994)
$[(1s2p)\ ^1P\ 3d]\ ^2D^o$	−5.05367		Jáuregui and Bunge (1981)
	−5.053853	−5.053884	Chen and Chung (1994)

For the CI results shown in Table 3 (Chung and coworkers, Bunge, and others), the estimates of basis set truncation errors are typically ten or more μhartrees. The work of Barrois et al. (1997c), using the CI-HY technique, provides a valuable check on the reliability of these estimates of the CI basis set truncation errors.

Not all the term designations in Table 3 are straightforward. For example, $^2P^e(4)$, denoted in Table 3 as $[(1s2p)\ ^1P\ 3p]\ ^2P^e$ has a significant admixture of $[(1s2p)\ ^3P\ 4p]\ ^2P^e$, with CI coefficients of magnitude 0.71 for the latter configuration, and 0.66 for the former configuration (Bunge, 1979). Bunge (1979) reports that the $^2P^e(3)$ state has similar principal CI configurations and coefficients.

4. Highly Excited Quartet States

The higher lying quartet states of Li have attracted some theoretical attention (see, for example, Davis and Chung, 1990a, 1990b; Chung and Gou, 1995). For a number of the triply excited states, interelectron correction effects would be expected to be of importance. Theoretical results for the energies for some of the high-lying states are presented in Table 4. These results are all based on large-scale CI calculations. No Hylleraas-type calculations appear to have been carried out on these states.

The $2p^3\ ^4S^o$ state is particularly interesting. This is a bound metastable state; it has an energy lying below the $2p^2\ ^3P$ threshold of Li^+. There has been theoretical interest in this state of Li going back many years (Wu and Shen, 1944). The $2p^3\ ^4S$ state does not couple to the $1s2s\epsilon s\ ^4S$ continuum, but can decay via radiative autoionization to the $1s2p\ ^3P^o$ continuum. For the decay process $2p^3\ ^4S^o \rightarrow 1s2p^2\ ^4P$, the theoretically determined wave length was found to be 145.009 Å, which is in close agreement with the experimental result of 145.02(5) Å (Agentoft et al., 1984). A more recent experimental result yields 145.016(6) Å (Mannervik et al., 1989), and a subsequent theoretical reevaluation gives 145.019 Å (Davis and Chung, 1990b), which is in very close agreement with the aforementioned experimental measurement.

TABLE 4
ENERGIES FOR SOME DOUBLY AND TRIPLY CORE-EXCITED QUARTET STATES OF Li.

State	E_{NR}	E_{TOTAL}	Reference
$(1s3s3p)\ ^4P^o$	−4.878651	−4.879214	Davis and Chung (1990a)
$(1s3p3p)\ ^4P^e$	−4.846711	−4.847261	Davis and Chung (1990a)
$2s2p2p\ ^4P$	−2.239379	−2.239559	Chung and Gou (1995)
$2s2p3p\ ^4P$	−1.961782	−1.961972	Chung and Gou (1995)
$2p^3\ ^4S^o$	−2.103588	−2.103684	Davis and Chung (1990b)
$2p^23p\ ^4S^o$	−1.873415	−1.873516	Davis and Chung (1990b)
$2p^24p\ ^4S^o$	−1.835481	−1.835583	Davis and Chung (1990b)

The $2p^3$ $^4S^o$ state is of special interest for the isoelectronic species He^-. It is one of the few bound (metastable) states of this anion, and as a result has been subject to a fair amount of attention (Beck and Nicolaides, 1978; Chung, 1979; Nicolaides et al., 1981; Nicolaides and Komninos, 1981).

V. Relativistic Corrections to the Energies

The standard approach by which the relativistic corrections to the energy are incorporated is to use a first-order perturbation theoretic procedure using the relativistic Breit-Pauli Hamiltonian

$$H_{rel} = H_{mass} + H_{enD} + H_{eeD} + H_{ssc} + H_{oo}, \tag{30}$$

where the various terms in Eq. (30) are (in atomic units)

$$H_{mass} = -\frac{\alpha^2}{8} \sum_{i=1}^{3} \nabla_i^4 \tag{31}$$

$$H_{enD} = \frac{1}{2}\alpha^2 Z\pi \sum_{i=1}^{3} \delta(\mathbf{r}_i) \tag{32}$$

$$H_{eeD} = -\pi\alpha^2 \sum_{i=1}^{3} \sum_{j>i}^{3} \delta(\mathbf{r}_{ij}) \tag{33}$$

$$H_{ssc} = -\frac{8\pi\alpha^2}{3} \sum_{i=1}^{3} \sum_{j>i}^{3} \mathbf{s}_i \cdot \mathbf{s}_j \delta(\mathbf{r}_{ij}) \tag{34}$$

$$H_{oo} = \frac{1}{2}\alpha^2 \sum_{i=1}^{3} \sum_{j>i}^{3} \left(\frac{\nabla_i \cdot \nabla_j}{r_{ij}} + \frac{\mathbf{r}_{ij} \cdot (\mathbf{r}_{ij} \cdot \nabla_i)\nabla_j}{r_{ij}^3} \right) \tag{35}$$

The fine structure constant is denoted by α, $\delta(\mathbf{r})$ is a Dirac delta function, and $\mathbf{s}_i$ is an electron spin operator. H_{mass} represents the kinetic energy mass correction, H_{enD} is the electron-nuclear Darwin term, H_{eeD} denotes the electron-electron Darwin term, H_{ssc} is the spin-spin contact interaction, and H_{oo} designates the electron-electron orbit interaction. Only the nonfine-structure contributions have been shown in Eq. (30). In addition to these terms, there are fine-structure contributions that include spin-orbit, spin-other-orbit, and spin-spin interactions.

Almost all the estimates of the relativistic corrections for the energies of the ground and excited states of the lithium atom have been calculated by Chung and coworkers using the CI approach (see Tables 2, 3, 4, and 5 for specific references). Some Hylleraas calculations have been carried out for parts of H_{rel} for the ground state of Li and a few excited states. A summary of some of the relativistic results is presented in Table 5. For the ground state of Li, Hylleraas-type calculations of the Breit-Pauli terms given in Eqs. (31–35) have just been completed by King

TABLE 5
Relativistic Corrections for Some Low-Lying Doublet States of Li in Atomic Units.

State	$\langle H_{\text{mass}} \rangle$	$\langle H_{\text{enD}} \rangle$	$\langle H_{\text{mass}} + H_{\text{enD}} \rangle$	$\langle H_{\text{eeD}} \rangle$	$\langle H_{\text{eeD}} + H_{\text{ssc}} \rangle$	$\langle H_{\text{oo}} \rangle$	Reference
$2\,^2S$	-4.18317×10^{-3}						King (1995)
	-4.18769×10^{-3}						Esquivel et al. (1992)
		3.4734×10^{-3}					King and Shoup (1986)
		3.47348×10^{-3}					King (1989)
		3.47370×10^{-3}					King and Bergsbaken (1990)
		3.473663×10^{-3}		-9.10630×10^{-5}			Yan and Drake (1995a)
			-7.0748×10^{-4}		9.5340×10^{-5}	-2.3331×10^{-5}	Chung (1991)[a]
			-7.0942×10^{-4}		9.1154×10^{-5}	-2.3201×10^{-5}	Chung (1991)[a]
$3\,^2S$		3.4457×10^{-3}					King (1991a)
			-6.968×10^{-4}		9.43×10^{-5}	-2.30×10^{-5}	Wang et al. (1992a)[a]
$4\,^2S$		3.4397×10^{-3}					King (1991a)
			-6.947×10^{-4}		9.40×10^{-5}	-2.30×10^{-5}	Wang et al. (1992a)
$5\,^2S$		3.4378×10^{-3}					King (1991a)
			-6.940×10^{-4}		9.40×10^{-5}	-2.30×10^{-5}	Wang et al. (1992a)
$2\,^2P$		3.431887×10^{-3}		-8.90484×10^{-5}			Yan and Drake (1995a)
$3\,^2P$			-6.933×10^{-4}		9.36×10^{-5}	-2.13×10^{-5}	Wang et al. (1993)
			-6.935×10^{-4}		9.38×10^{-5}	-2.24×10^{-3}	Wang et al. (1993)
$4\,^2P$			-6.939×10^{-4}		9.39×10^{-5}	-2.27×10^{-5}	Wang et al. (1993)
$5\,^2P$			-6.934×10^{-4}		9.39×10^{-5}	-2.28×10^{-5}	Wang et al. (1993)
$3\,^2D$		3.438817×10^{-3}		-8.92896×10^{-5}			Yan and Drake (1995a)
$4\,^2D$			-6.956×10^{-4}		8.93×10^{-5}	-2.28×10^{-5}	Wang et al. (1992b)
			-6.954×10^{-4}		8.93×10^{-5}	-2.28×10^{-5}	Wang et al. (1992b)
$5\,^2D$			-6.954×10^{-4}		8.93×10^{-5}	-2.28×10^{-5}	Wang et al. (1992b)

[a] Chung and Wang et al. report only $\langle H_{\text{mass}} + H_{\text{enD}} \rangle$ and $\langle H_{\text{eeD}} + H_{\text{ssc}} \rangle$.

et al. (1998e). The limited number of Hylleraas results available provides a useful check on some of the CI calculations.

There has been limited application of the Hylleraas technique to calculating relativistic corrections for the three-electron systems, primarily because of the difficult nature of the integrals that arise. Along with cases $\ell = -2$ (or the more difficult case, $\ell = -2, m = -2$) in Eq. (8), integrals such as

$$I_1(i, j, k, \ell, m, n, a, b, c) = \int r_1^i r_2^j r_3^k (r_1^2 - r_2^2) r_{23}^\ell r_{31}^m r_{12}^{-3} e^{-ar_1 - br_2 - cr_3} \, d\mathbf{r}_1 d\mathbf{r}_2 d\mathbf{r}_3 \tag{36}$$

and

$$I_2(i, j, k, \ell, m, n, a, b, c) = \int r_1^i r_2^j r_3^k (r_{23}^2 - r_{31}^2) r_{23}^\ell r_{31}^m r_{12}^{-3} e^{-ar_1 - br_2 - cr_3} \, d\mathbf{r}_1 d\mathbf{r}_2 d\mathbf{r}_3 \tag{37}$$

also arise. The I_1 and I_2 integrals cannot be separated into the obvious two parts, because the separate contributions are divergent. These integrals have recently been studied by Feldmann *et al.* (1998).

For the ground state of *Li*, a comparison of theoretical methods is possible for the principal part of the relativistic correction to the energy, that is, the contribution $\langle H_{\text{mass}} + H_{\text{enD}} \rangle$. Chung (1991) calculates $\langle H_{\text{mass}} + H_{\text{enD}} \rangle = -7.0748 \times 10^{-4}$ a.u., and he also evaluates this quantity for the $1s^2$ ^1S state of Li^+. He finds a difference with the results of Pekeris (1958, 1962) for Li^+, and accordingly adopts a core correction procedure. When this core correction is included for the *Li* ground state, Chung finds $\langle H_{\text{mass}} + H_{\text{enD}} \rangle = -7.0942 \times 10^{-4}$ a.u. If the Hylleraas results for $\langle H_{\text{mass}} \rangle$ (King, 1995) and $\langle H_{\text{enD}} \rangle$ (Yan and Drake, 1995a) are combined, the value found for $\langle H_{\text{mass}} + H_{\text{enD}} \rangle$ is -7.0951×10^{-4} a.u., which is in fairly close agreement with Chung's result. This close comparison does validate Chung's core correction approach, at least for $\langle H_{\text{mass}} + H_{\text{enD}} \rangle$. The most complicated contribution to evaluate is H_{oo}, and there are no published Hylleraas results available to check the CI results of Chung and coworkers.

The level of precision of the relativistic corrections calculated using the CI technique is generally adequate to aid in the assignment of spectral lines arising from a wide variety of states. However, when a high-precision theoretical calculation of the ionization potential is the target, for example for the *Li* ground state, six-digit accuracy is required for the relativistic corrections in order to match up with the currently available experimental result, for which eight digits of precision are available. As higher precision spectroscopic work continues for *Li*, there will be increased interest in knowledge of higher precision values for the relativistic corrections.

VI. Specific Mass Shift Correction to the Energy Levels

The extension of Eq. (13) to incorporate the effect of finite nuclear mass is

$$H_M = \sum_{i=1}^{3}\left[-\frac{1}{2\mu}\nabla_i^2 - \frac{3}{r_i}\right] - \frac{1}{M}\sum_{i=1}^{3}\sum_{j>i}^{3}\nabla_i\cdot\nabla_j + \sum_{i=1}^{3}\sum_{j>i}^{3}\frac{1}{r_{ij}} \quad (38)$$

where μ is the reduced mass,

$$\mu = \frac{m_e M}{M + m_e} \quad (39)$$

and M and m_e denote the nuclear mass and the electron mass, respectively. The effect of finite nuclear mass involves two principal contributions. The normal mass shift (also referred to as the Bohr mass shift) can be determined using

$$\Delta E_{\text{Bohr}} = \frac{-\mu\ ^\infty E}{M} \quad (40)$$

where $^\infty E$ is the state energy computed in the infinite nuclear mass approximation. The second contribution is specific mass shift, ΔE_{sms} (also referred to as the mass polarization correction). Two methods have been used to evaluate ΔE_{sms}. The most commonly employed approach is to evaluate ΔE_{sms} using the first-order perturbation theory formula

$$\Delta E_{\text{sms}} = -\frac{\mu}{M}\left\langle\Psi\left|\sum_{i=1}^{3}\sum_{j>i}^{3}\nabla_i\cdot\nabla_j\right|\Psi\right\rangle \quad (41)$$

where Ψ is the approximate solution of the infinite nuclear mass Schrödinger equation. An alternative operator form is available (Vinti, 1932, 1940), which gives

$$\Delta E_{\text{sms}} = \left(\frac{\mu}{M}\right)\left\{\frac{1}{2}\left\langle\Psi\left|\sum_{i=1}^{3}\sum_{j>i}^{3}\frac{1}{r_{ij}}\right|\Psi\right\rangle\right.\\
\left.+ Z\left\langle\Psi\left|\sum_{i=1}^{3}\sum_{j>i}^{3}\mathbf{r}_i\cdot\mathbf{r}_j\left(\frac{1}{r_i^3}+\frac{1}{r_j^3}\right)\right|\Psi\right\rangle\right\} \quad (42)$$

The result given in Eq. (42) has been less frequently employed, but for an application to Li, see Tong et al. (1993).

A different approach that has been investigated for Li is to evaluate ΔE_{sms} using the result

$$\Delta E_{\text{sms}} = -\frac{1}{M}\left\langle\Psi_M\left|\sum_{i=1}^{3}\sum_{j>i}^{3}\nabla_i\cdot\nabla_j\right|\Psi_M\right\rangle \quad (43)$$

where Ψ_M is the nuclear mass-dependent approximate solution of the Schrödinger equation using H_M (Eq. (38)). High-precision calculations using this procedure have been carried out by King (1986), Lüchow and Kleindienst (1994), and Yan and Drake (1995a).

A selection of high-precision results for ΔE_{sms} for various low-lying states of Li is given in Table 6. These results are all based on the use of Eq. (41), and the values $\mu/M = 7.8202022(6) \times 10^{-5}$ for ^7Li, and $\mu/M = 9.1216762(8) \times 10^{-5}$ for ^6Li have been employed. These values of μ/M are calculated from the nuclear masses of 7.0143584(5) amu for ^7Li, and 6.0134766(5) amu for ^6Li (Audi and Wapstra, 1993, 1995). For the most precise results for the $2\,^2S$ ground state, using the full H_M approach, a significant fraction of the uncertainty in ΔE_{sms} comes from the error in determining the nuclear masses of ^6Li and ^7Li.

There are two smaller mass-dependent contributions to the energy. The first are the nuclear mass-dependent relativistic corrections. No high-precision calculations of these contributions have been carried out for the lithium atom. A second correction is the field shift contribution (also called the volume shift) (King, 1984). This correction arises from the electric field generated by the nuclear charge distribution. For light atoms, this contribution is usually regarded as negligible. However, the accuracy of recent isotope shift measurements (Sansonetti et al., 1995) suggests that a high-precision calculation of this field shift correction would be of value. There is relatively little published work on this correction for the lithium atom. Veseth (1985) has determined values of 0.02168 cm^{-1} for the ground state of ^7Li, 0.02143 cm^{-1} for the $2\,^2P$ state of ^7Li and 0.02147 cm^{-1} for $^7Li^+(1s^2)$. There is, not unexpectedly, a significant cancellation of these contributions when transitions such as $^7Li(2s) \rightarrow {}^7Li(2p)$ and $^7Li(1s^22s) \rightarrow {}^7Li^+(1s^2) + e^-$ are considered. Improvements in the experimental precision of isotope shift measurements will provide some significant challenges for theorists in this area.

A. Transition Isotope Shifts

The transition isotope shift (TIS) for a transition from state X to state Y for isotopes with mass numbers A_1 and A_2 ($A_1 > A_2$) is

$$\Delta E_{TIS} = [E(^{A_1}Y) - E(^{A_1}X)] - [E(^{A_2}Y) - E(^{A_2}X)] \quad (44)$$
$$= E(^{A_1}Y) - E(^{A_2}Y) - [E(^{A_1}X) - E(^{A_2}X)]$$

where $E(^{A_1}Y)$ is the energy of state Y for the isotope of mass A_1. If this energy is factored into a value that is computed in the infinite nuclear mass approximation and a mass correction shift to the energy,

$$E(^{A_1}Y) = {}^\infty E(Y) + \Delta E_{MASS}(^{A_1}Y) \quad (45)$$

then the mass-dependent form of ΔE_{TIS} is

$$\Delta E_{TIS,mass} = \Delta E_{mass}(^{A_1}Y) - \Delta E_{mass}(^{A_2}Y) - [\Delta E_{mass}(^{A_1}X) - \Delta E_{mass}(^{A_2}X)] \quad (46)$$

TABLE 6
Specific Mass Shifts for the Ground and Selected Excited States of Li.

State	$\left\langle \sum_{i=1}^{3} \sum_{j>i}^{3} \nabla_i \cdot \nabla_j \right\rangle$ (absolute a.u.)	ΔE_{sms} (μhartree) ^6Li	ΔE_{sms} (μhartree) ^7Li	Reference
$2\,^2S$	−0.30185	27.531	23.603	King (1986)
	−0.3018467	27.53097	23.60318	King (1989)
	−0.3018436	27.53068	23.60293	King and Bergsbaken (1990)
	−0.30180	27.527	23.600	Chung (1991)
	−0.301842799[a]	27.530611	23.602871	Lüchow and Kleindienst (1994)
	−0.301842809[a]	27.530612	23.602872	Yan and Drake (1995a)
$3\,^2S$	−0.29212	26.644	22.843	King (1991a)
	−0.292039995[a]	26.636513	22.836332	Lüchow and Kleindienst (1994)
$4\,^2S$	−0.29033	26.481	22.703	King (1991a)
	−0.2901575[a]	26.46481	22.68913	Lüchow and Kleindienst (1992b)
$5\,^2S$	−0.28969	26.422	22.653	King (1991a)
	−0.289540	26.4085	22.6408	Lüchow and Kleindienst (1992b)
$6\,^2S$	−0.28942	26.398	22.631	Lüchow and Kleindienst (1992b)
$2\,^2P$	−0.24673781[a]	22.50457	19.29389	Yan and Drake (1995a)
	−0.24674181	22.50494	19.29420	Barrois et al. (1997b)
$3\,^2P$	−0.27589098	25.16357	21.57355	Barrois et al. (1997b)
$3\,^2D$	−0.288928837[a]	26.352749	22.59305	Yan and Drake (1995a)
$1s2s3s\,^4S$	−0.019098739	1.741966	1.493443	Barrois et al. (1996)
$1s2s4s\,^4S$	−0.018619609	1.698266	1.455977	Barrois et al. (1996)
$1s2s5s\,^4S$	−0.01791668	1.634151	1.401011	Lüchow et al. (1993)
$1s2s6s\,^4S$	−0.0175922	1.60456	1.37564	Lüchow et al. (1993)
$1s2s2p\,^4P$	0.1975568	−18.01885	−15.44813	Barrois et al. (1997a)
$1s2s3p\,^4P$	0.02001852	−1.825858	−1.565366	Barrois et al. (1997a)
$1s2p2p\,^2P^e$	−0.15493607	14.13148	12.11537	Barrois et al. (1997c)
$1s2p3p\,^2P^e$	0.23991361	−21.88215	−18.76026	Barrois et al. (1997c)

[a] A more precise value calculated using the finite mass Hamiltonian is available for this state (see Lüchow and Kleindienst, 1994; and Yan and Drake, 1995a).

There are two principal components to ΔE_{TIS}. The Bohr mass shift (put MASS = Bohr in Eq. (46)) is straightforward to calculate (see Eq (40)). Of greater interest to theorists, is the specific mass shift contribution to the transition isotope shift,

$$\Delta E_{TIS,sms} = \Delta E_{sms}(^{A_1}Y) - \Delta E_{sms}(^{A_2}Y) - [\Delta E_{sms}(^{A_1}X) - 1\Delta E_{sms}(^{A_2}X)] \quad (47)$$

This quantity is a sensitive measure of the adequate description of correlation effects (it is zero in the Hartree-Fock approximation), and can be compared directly with experimental results.

For transitions of the type $Li(^2X) \rightarrow Li^+(1s^2) + e^-$, there has been considerable theoretical interest (Prasad and Stewart, 1966; Mårtensson and Salomonson, 1982; Chambaud et al. 1984; King, 1986, 1989; King and Bergsbaken, 1990; Lüchow and Kleindienst, 1994; Yan and Drake, 1995a; Barrois et al. 1997b). These authors either explicitly calculate ΔE_{TIS} or provide the necessary expectation values to determine it. There has also been a good deal of experimental interest in $\Delta E_{TIS,sms}$ for the same process (Hughes, 1955; Mariella, 1979; Lorenzen and Niemax, 1982; Goy et al., 1986; Vadla et al., 1987; Sansonetti et al., 1995; Radziemski et al., 1995). A summary of some of the higher precision theoretical results is presented in Table 7. In most cases, the theoretical results fall within the error limits of the experimental results. Unfortunately, the error limits are rather large in a number of cases where the theoretical precision is high. The precision of the best theoretical results for $\Delta E_{TIS,sms}$ is limited by the present uncertainties in the nuclear masses of ^6Li and ^7Li.

A small but notable discrepancy between theory and experiment occurs for the 2 ^2P state of Li. Radziemski et al. (1995) report values of $-3.6100(6)$ GHz for 2 $^2P_{1/2}$ and $-3.6103(5)$ GHz for 2 $^2P_{3/2}$. Two high-precision theoretical estimates are -3.61635 GHz and -3.61601 GH$_z$ (Yan and Drake, 1995a; Barrois et al., 1997b); which are both smaller than the aforementioned experimental results. The specific mass shift contribution to the 2 ^2P_J level is given by

$$\Delta E_{TIS,sms}(2p_j) = \Delta E_{TIS,sms}(2s) - \Delta E_{TIS,sms}(2s \rightarrow 2p_j) \quad (48)$$

Using the value of the $2s \rightarrow 2p$ TIS from Sansonetti et al. (1995), the Bohr shift contribution as 5.813 GHz, and the shift for the ground state of 1.109(8) GHz (Vadla et al., 1987), leads to a shift for the 2 ^2P state of $-3.612(8)$ GHz. If the alternative value of 1.111(6) GHz for the specific mass shift of the 2 ^2S state (Lorenzen and Niemax, 1982) is used, then the shift for the 2 ^2P state is $-3.609(6)$ GHz. The error bars are too large to distinguish any difference between the 2 $^2P_{1/2}$ and 2 $^2P_{3/2}$ states. The first estimate, $-3.612(8)$ GHz, is close to the results of high-precision calculations, and the second estimate, $-3.609(6)$ GHz, almost overlaps the theoretical results. The error bars on these two values are too large for these results to provide a tight check on the theoretical calculations.

The specific mass shift contribution for a transition between any pair of levels

TABLE 7
Specific Mass Shift Contribution to the Transition Isotope Shifts (TIS) for the Ground and Selected Excited States of Li.

State	Shift for ^6Li–^7Li (GHz)	Specific mass shift contribution to the TIS (GHz)[a]	Experimental specific mass shift (GHz)	Reference for theoretical calculation
2 ^2S	25.844	1.102	1.108(8)[b]	King (1986)
	25.8436	1.1020	1.111(6)[c]	King (1989)
	25.84336	1.10172		King and Bergsbaken (1990)
	25.84329(3)	1.10165(4)		Lüchow and Kleindienst (1994)
	25.84329(3)	1.10165(4)		Yan and Drake (1995a)
3 ^2S	25.011	0.269	0.276(26)[b]	King (1991a)
	25.00399	0.26235	0.260(30)[d]	Lüchow and Kleindienst (1994)
4 ^2S		0.088	0.111(12)[c]	Mårtensson and Salomonson (1982)
	24.858	0.116	0.094(30)[d]	King (1991a)
	24.84281	0.10117		Lüchow and Kleindienst (1994)
5 ^2S		0.042	0.053[b,e]	Mårtensson and Salomonson (1982)
	24.803	0.061	0.027(30)[d]	King (1991a)
	24.7899	0.0483		Lüchow and Kleindienst (1992b)
6 ^2S		0.024	0.029[b,e]	Mårtensson and Salomonson (1982)
	24.780	0.038	−0.046(120)[d]	Lüchow and Kleindienst (1992b)
2 ^2P	21.12529	−3.61635	−3.596(26)[b]	Yan and Drake (1995a)
	21.12563	−3.61601	−3.608(8)[f]	Barrois et al. (1997b)
			−3.611(6)[g]	
			−3.6100(6)[h,i]	
			−3.6103(5)[h,j]	
			−3.603(15)[d,i]	
			−3.603(15)[d,j]	
3 ^2P		−1.034	−1.105(8)[b]	Mårtensson and Salomonson (1982)
	23.62134	−1.12030	−1.116(30)[d]	Barrois et al. (1997b)
4 ^2P		−0.442	−0.504(45)[d]	Mårtensson and Salomonson (1982)
5 ^2P		−0.227	−0.308(60)[d]	Mårtensson and Salomonson (1982)
3 ^2D		−0.00205	−0.011(45)[d]	Mårtensson and Salomonson (1982)
	24.73762	−0.00402		Yan and Drake (1995a)
4 ^2D		−0.00058	−0.024(45)[d]	Mårtensson and Salomonson (1982)

[a] The shift for $1s^2\ ^6Li^+ - 1s^2\ ^7Li^+$ has been taken as 24.74164(3) GHz, which has been computed from the available value of $\langle \nabla_i \cdot \nabla_j \rangle$ (Lüchow and Kleindienst, 1994; Yan and Drake, 1995a) and the conversion factor 1 a.u. = 85.61837(7) GHz.
[b] Results from Vadla et al. (1987).
[c] Result from Lorenzen and Niemax (1982).
[d] Results from Radziemski et al. (1995).
[e] These experimental results are derived by extrapolation using scaling formulas.
[f] Result from Mariella (1979).
[g] Result from Fuchs and Rubahn (1986).
[h] Results from Radziemski et al. (1995) based on measurements of Sansonetti et al. (1995).
[i] Result for 2 $^2P^o_{1/2}$.
[j] Result for 2 $^2P^o_{3/2}$.

can be evaluated using the level shift information for the various states given in Table 7, and the result

$$\Delta E_{\text{TIS,sms}}(^2X_J \to {}^2Y_J') = \Delta E_{\text{TIS,sms}}(^2X_J) - \Delta E_{\text{TIS,sms}}(^2Y_J') \tag{49}$$

Precise experimental results for a number of transitions for Li can be found in Mariella (1979), Fuchs and Rubahn (1986), Vadla *et al.* (1987), Windholz *et al.* (1990), and Sansonetti *et al.* (1995).

VII. Quantum Electrodynamic Corrections

A high-precision calculation of transition energies or ionization potentials requires a determination of the Lamb shift correction, ΔE_{QED}. For the lithium atom there has been limited work in this area. One approach to calculating ΔE_{QED} for a transition from the $1s^2 2s$ ground state of Li, is to ignore the $1s^2$ core, effectively reducing the problem to a one-electron correction. ΔE_{QED} can then be calculated using (Bethe and Salpeter, 1977)

$$\Delta E_{\text{QED}}(n, \ell) = \frac{4Z_{\text{eff}}^4 \alpha^3}{3\pi n^3} \left\{ \delta_{\ell,0} \left[\frac{31}{120} - 2\ell n(\alpha Z_{\text{eff}}) \right] \right. \tag{50}$$
$$\left. + \frac{3}{8} \frac{c_{\ell j}}{(2\ell + 1)} - \ell n \left[\frac{k_0(n\ell)}{Z_{\text{eff}}^2 R_\infty} \right] \right\}$$

and the dependence of $c_{\ell j}$ on the quantum numbers ℓ and j is given by

$$c_{\ell j} = (\ell + 1)^{-1} \delta_{j,\ell+1/2} - \ell^{-1} \delta_{j,\ell-1/2} \tag{51}$$

In Eq. (51), R_∞ denotes the infinite nuclear mass Rydberg constant, Z_{eff} is the effective nuclear charge, and $\delta_{m,n}$ is a Kronecker delta. Values of the Bethe logarithm $\ell n[k_0(n\ell)/R_\infty]$ have been tabulated as a function of the quantum numbers n and ℓ (Drake and Swainson, 1990).

The principal problem with the use of Eq. (50) for transitions from the Li ground state is that Z_{eff} is not known with any precision. Using a value of Z_{eff} that would be characteristic of a $Z = 3$ nucleus screened by a pair of $1s$ electrons leads to the value $\Delta E_{\text{QED}}(2, 0) = -0.08$ cm^{-1}. This value is about one-third the size of estimates based on more refined calculations (McKenzie and Drake, 1991; Feldman and Fulton, 1995). The other drawback of the application of Eq. (50) is that the many-electron nature of the correction is lost.

For transitions from higher excited doublet states, $\Delta E_{\text{QED}}(n, \ell)$ makes a negligible contribution to the transition energy, based on the current levels of precision available experimentally. For such transitions, the QED effects for the $1s^2$ core effectively cancel for the two states in question.

HIGH-PRECISION CALCULATIONS OF THE LITHIUM ATOM

For a precise theoretical determination of the first ionization energy of Li, an improved estimate of ΔE_{QED} based on the three-electron nature of the problem is required. This can be determined using

$$\Delta E_{QED} = E_L(1s^2) - E_L(1s^2 2s) \tag{52}$$

where $E_L(1s^2)$ can be calculated from

$$E_{L,1}(1s^2) = \frac{4Z\alpha^3}{3} \langle \delta(\mathbf{r}_i) \rangle \left\{ -2\ell n\alpha - \ell n(k_0/R_\infty) + \frac{19}{30} + 2.2962\pi\alpha Z \right\} \tag{53}$$

and

$$E_{L,2}(1s^2) = \alpha^3 \left[\langle \delta(\mathbf{r}_{12}) \rangle \left\{ \frac{14}{3} \ell n\alpha + \frac{164}{15} \right\} - \frac{7}{6\pi} \lim_{a \to 0} \{ \langle r_{12}^{-3}(a) \rangle \right.$$
$$\left. + 4\pi(\gamma + \ell na)\delta(\mathbf{r}_{12}) \} \right] \tag{54}$$

where

$$r_{12}^{-3}(a) = \begin{cases} 0 & r_{12} \leq a \\ r_{12}^{-3} & r_{12} > a \end{cases} \tag{55}$$

and γ is Euler's constant. Similarly, $E_L(1s^2 2s)$ can be determined from

$$E_{L,1}(1s^2 2s) = Z\alpha^3 \left\{ F(1s^2 2s) \right.$$
$$\left. \cdot \left\langle \sum_{i=1}^{3} \delta(\mathbf{r}_i) \right\rangle - \frac{4}{3} \ell n[(Z-\sigma)/Z]^2 \left\langle \sum_{i=1}^{3} \delta(\mathbf{r}_i) \right\rangle \right\} \tag{56}$$

and

$$E_{L,2}(1s^2 2s) = \alpha^3 \left[\left(\frac{14}{3} \ell n\alpha + \frac{164}{15} \right) \left\langle \sum_{i=1}^{3} \sum_{j>i}^{3} \delta(\mathbf{r}_{ij}) \right\rangle \right.$$
$$\left. - \frac{7}{6\pi} \lim_{a \to 0} \left\{ \left\langle \sum_{i=1}^{3} \sum_{j>i}^{3} [r_{ij}^{-3}(a) + 4\pi(\gamma + \ell na)\delta(\mathbf{r}_{ij})] \right\rangle \right\} \right] \tag{57}$$

In Eq. (56), $F(1s^2 2s)$ denotes a combination of one-electron functions $F(1s)$ and $F(2s)$, which can be written as a sum of one-electron quantum electrodynamic corrections (Drake, 1993; Johnson and Soff, 1985), and σ is a screening constant. Feldman and Fulton (1995) find a different result in place of Eq. (57); the factor 164/15 is found as (129/15) − (3π/2).

The correction to the ionization energy of Li using Eqs. (52–57) is −0.22(2) cm^{-1} (McKenzie and Drake, 1991) or -0.24 cm^{-1} (Feldman and Fulton, 1995). The uncertainty in this correction is a major component in the error associated with the theoretical determination of the first ionization potential of *Li*. Further progress in this area will be needed as higher precision experimental data becomes available for the *Li* atom.

The previous discussion has focused on transitions involving the low-lying doublet states. Complications arise when more excited states are considered. No detailed calculations appear to have been published. For a state like [$(1s2p)$ $^3P,3d$] $^2D^o$, an estimate of the QED contribution to the term energy can be made by combining the ΔE_{QED} contributions to the ionization potentials of the $1s^2$ Li^+, $1s^2 2s$ Li and $1s2p$ ^3P Li^+ states (Chen and Chung, 1994). Implicit in this type of calculation is the assumption that the contribution from the $3d$ electron is negligible. This can be verified to be a reasonable assumption using the one-electron formula, Eq. (50). A more problematic situation arises for states like $1s2s2p$ ^4P. It would probably be an inadequate approximation to estimate the QED contribution to the term energy of this state, using a combination of the ΔE_{QED} contributions to the ionization potentials of the $1s^2$ Li^+, $1s^2 2s$ Li and $1s2s$ ^3S Li^+ states. The estimate could be improved by trying to determine the QED contribution of the $2p$ electron using the one-electron formula, but this would be a rather rough approximation in this case. For states such as $1s2s2p$ ^4P, where there would be expected to be significant correlation effects in the valence shell, the QED contribution should therefore be evaluated using the many-electron expression for ΔE_{QED}.

VIII. The First Ionization Potential

The calculation of the first ionization potential of *Li* has attracted considerable attention over many years. The ionization potential has been a benchmark property to test different computational techniques, some of which include many-body perturbation theory (Lindgren, 1985; Johnson *et al.*, 1987, 1988; Blundell *et al.*, 1989), CI (Chung, 1991; Weiss, 1992; Morrison *et al.*, 1996), CI-HY (Pipin and Bishop, 1992), MCHF (Tong *et al.*, 1993), and HY (Yan and Drake, 1995a; King *et al.*, 1998e).

The first ionization potential, I_1, can be determined from the formula

$$I_1 = E_{NR}(Li^+) - E_{NR}(Li) + \Delta E_{REL} + \Delta E_{MASS} + \Delta E_{QED} \tag{58}$$

where the various terms in Eq. (58) have been defined previously in Eq. (22–24). A breakdown of the component contributions has been given recently by King (1997) and King *et al.* (1998e). The theoretical value of I_1 (in absolute a.u.) is 0.1981420(1) (43487.14(2) cm^{-1}), which compares closely with the experimental value of 0.19814203(2) a.u. (43487.150(5) cm^{-1}) (Johansson, 1959). The major sources of error in the theoretical determination of I_1 lie with ΔE_{REL} and ΔE_{QED}. A combined experimental–theoretical approach has been suggested by Yan and

Drake (1995a) to evaluate I_1. Because the experimental $2\,^2S$–$2\,^2P$ and $2\,^2P$–$3\,^2D$ transition energies are known to high precision, combining these values with the theoretical ionization energy of the $3\,^2D$ state leads to the precise value $I_1 = 43487.167(4)$ cm^{-1} (Yan and Drake, 1995a). A recalculation using more recent experimental results (Sansonetti et al., 1995; Radziemski et al., 1995) yields the value $I_1 = 43487.163(5)$ cm^{-1} (King, 1997). An essential advantage of this approach is that ΔE_{QED} for the $3\,^2D$ state is negligibly small.

A. Transition Energies

In Table 8, precise theoretical estimates are presented for the term energies of the lower-lying doublet states relative to the Li ground state. The term energies (relative to the ground state energy) are obtained theoretically using

$$T(^2X) = E_{\text{NR}}(^2X) - E_{\text{NR}}(1s^22s) + \Delta E_{\text{REL}} + \Delta E_{\text{MASS}} + \Delta E_{\text{QED}} \quad (59)$$

with

$$\Delta E_{\text{REL}} = E_{\text{REL}}(^2X) - E_{\text{REL}}(1s^22s) \quad (60)$$

$$\Delta E_{\text{MASS}} = E_{\text{MASS}}(^2X) - E_{\text{MASS}}(1s^22s) \quad (61)$$

$$\Delta E_{\text{QED}} = E_{\text{QED}}(^2X) - E_{\text{QED}}(1s^22s) \quad (62)$$

An alternative approach is to use a rearranged form of Eq. (21),

$$T(^2X) = I_1 + E_{\text{NR}}(^2X) - E_{\text{NR}}(Li^+) - \Delta E_{\text{MASS}} - \Delta E_{\text{REL}} - \Delta E_{\text{QED}} \quad (63)$$

with ΔE_{MASS}, ΔE_{REL} and ΔE_{QED} defined in Eqs. (22–24). Equation (63) has the advantage that the relativistic and quantum electrodynamic corrections need to be explicitly evaluated for only one three-electron state, rather than the two required for Eq. (59). This is partially offset by the need for a high-precision value of I_1. If the experimental result for I_1 is employed, we have a combined experimental–theoretical determination of $T(^2X)$, with the error resulting principally from the uncertainties in ΔE_{REL} and ΔE_{QED}. For a higher lying doublet state, ΔE_{REL} and ΔE_{QED} can be evaluated from one-electron formulas, with a corresponding reduction in the estimated uncertainty for $T(^2X)$ when Eq. (63) is used.

The E_{NR} values used to construct the theoretical entries in Table 8 are taken from Table 1 (the least-upper-bound result for each term energy was used). The relativistic corrections were taken from the references cited in Section V and the mass corrections (the E_{sms} component) were taken from Table 6. The QED corrections were estimated from the one-electron formula, Eq. (50), with Z_{eff} determined from

$$Z_{\text{eff}} = n[2\{E_{\text{NR}}(1s^2, Li^+) - E_{\text{NR}}(1s^2nx, ^2X)\}]^{1/2} \quad (64)$$

TABLE 8
Level Energies above the Ground State for the
Low-Lying Doublet States of ^7Li.

State	Team energy (cm^{-1})	
	Experimental[a,b]	Theoretical[c]
$2s\ ^2S_{1/2}$	0.0000	
$3s\ ^2S_{1/2}$	27206.0952(10)	27206.09(5)
$4s\ ^2S_{1/2}$	35012.0326(10)	35012.05(5)
	35012.0337(7)[d]	
$5s\ ^2S_{1/2}$	38299.4627(10)	38299.6(2)
$6s\ ^2S_{1/2}$	39987.586(3)	39994(7)
$2p\ ^2P_{1/2}$	14903.648130(14)[e]	
$2p\ ^2P_{3/2}$	14903.983468(14)[e]	
	[14903.871689(20)]	14903.86(11)
$3p\ ^2P_{1/2}$	[30925.5530(10)]	
$3p\ ^2P_{3/2}$	30925.6494(10)	
	[30925.6173(14)]	30925.60(11)
$4p\ ^2P_{1/2}$	36469.7542(15)	
$4p\ ^2P_{3/2}$	36469.7943(15)	
	[36469.7809(21)]	36469.76(12)
$5p\ ^2P_{1/2}$	39015.6988(20)	
$5p\ ^2P_{3/2}$	39015.7199(20)	
	[39015.7129(28)]	39015.89(25)
$6p\ ^2P_{1/2}$	40391.283(10)	
$6p\ ^2P_{3/2}$	40391.295(10)	
	[40391.291(14)]	
$3d\ ^2D_{3/2}$	31283.0505(10)	
	31283.0496(7)[d]	
$3d\ ^2D_{5/2}$	31283.0866(10)	
	31283.0856(7)[d]	
	[31283.0722(14)]	31283.05(11)
$4d\ ^2D_{3/2}$	36623.3360(10)	
	36623.3444(7)[d]	
$4d\ ^2D_{5/2}$	36623.3511(10)	
	36623.3596(7)[d]	
	[36623.3451(14)]	36623.2(2)
$5d\ ^2D_{3/2}$	39094.861(10)	
$5d\ ^2D_{5/2}$	39094.869(10)	
	[39094.866(14)]	39094.8(2)
$6d\ ^2D_{3/2,5/2}$	40437.220(20)	
$4f\ ^2F_{5/2}$	36628.329(3)	
$4f\ ^2F_{7/2}$	36628.336(3)	
	[36628.333(4)]	36627.9(7)
$5f\ ^2F_{5/2}$	39097.499(15)	
$5f\ ^2F_{7/2}$	39097.503(15)	
	[39097.501(21)]	39097.2(7)
$6f\ ^2F_{5/2,7/2}$	40438.90(5)	

[a] Experimental data from Radziemski et al. (1995), except where noted.

[b] Level values are determined from the center of gravity of the hyperfine structure of the ground state. Center of gravity estimates are given in [ ].

[c] Theoretical values are calculated from Eq. (59) or Eq. (63).

[d] From Lorenzen and Niemax (1983).

[e] From Sansonetti et al. (1995).

For the P, D, and F states, the quantity to compare with the theoretical result is the center of gravity of the two level energies, that is

$$E_{CG} = \frac{\sum_{j=1}^{2}(2J_j + 1)E_j}{\sum_{j=1}^{2}(2J_j + 1)} \qquad (65)$$

The agreement between theory and the recent experimental results (Windholz and Umfer, 1994; Sansonetti *et al.*, 1995; Radziemski *et al.*, 1995) is, in general, rather good. Because there are often no published uncertainties for $E_{REL}(^2X)$, we have made rough estimates of the errors, based on the numbers of digits the authors have quoted. The uncertainties for ΔE_{QED} for the lowest lying ^2X states are also very difficult to gauge, and an error estimate based on the difference between the result calculated using a one-electron approximation and more precise results for the ^2S ground state has been employed. The uncertainties for the other quantities that determine $T(^2X)$ are much less important, and do not have an impact on the final error estimates. The final error estimates reported in Table 8 are believed to be generous, but do involve some rough estimation.

Using Table 8, the theoretically determined transition energies between different levels—actually, between center of gravity estimates—are found to be in very close agreement with experimental results. To progress beyond the current precision levels will require a high-precision determination of E_{REL} for the states involved. Progress on this front has been limited by mathematical difficulties, a topic addressed earlier in Section V. Further progress on the calculation of ΔE_{QED} for systems beyond the two-electron level will also be required.

IX. Hyperfine Coupling Constants

The lithium atom has long served as a benchmark for testing various theoretical methodologies for calculating precise hyperfine coupling constants. High-precision experimental data are available for the coupling constants of the ground states of both ^6Li and ^7Li, and this serves as a valuable comparison point for the theoretical work.

The principal part of the magnetic Hamiltonian describing hyperfine interactions is

$$H_{hfs} = H_c + H_d + H_o + H_Q = H_{mhfs} + H_Q \qquad (66)$$

where

$$H_c = \frac{8\pi}{3} g_J g_I \mu_B \mu_N \mathbf{I} \cdot \sum_{i=1}^{3} \mathbf{s}_i \delta(\mathbf{r}_i) \qquad (67)$$

$$H_d = g_J g_I \mu_B \mu_N \mathbf{I} \cdot \sum_{i=1}^{3} [3(\mathbf{s}_i \cdot \mathbf{r}_i)\mathbf{r}_i - r_i^2 \mathbf{s}_i] r_i^{-5} \tag{68}$$

$$H_o = 2 g_I \mu_B \mu_N \mathbf{I} \cdot \sum_{i=1}^{3} \boldsymbol{\ell}_i r_i^{-3} \tag{69}$$

$$H_Q = -\sum_{p=1}^{3} \sum_{i=1}^{3} (r_p^2 r_i^{-3}) P_2(\cos \theta_{ip}) \tag{70}$$

In Eqs. (67–70), g_J is the electronic g-factor, g_I is the nuclear g-factor, μ_B is the Bohr magneton, μ_N is the nuclear magneton, $\mathbf{I}$ is the nuclear spin operator, $\mathbf{s}_i$ is the electron spin operator for electron i, $\delta(\mathbf{r}_i)$ is a Dirac delta function, $\boldsymbol{\ell}_i$, is the orbital angular momentum operator for electron i, P_n is a Legendre polynomial, and the p-summation is over protons. Equations (67–70) represent the Fermi contact hyperfine interaction, the spin dipolar hyperfine term, the orbital contribution, and the electric quadrupole interaction, respectively.

For a given J, it is possible to write effective operator forms:

$$H_c = a_{cJ} \mathbf{I} \cdot \mathbf{J} \tag{71}$$

$$H_o = a_{oJ} \mathbf{I} \cdot \mathbf{J} \tag{72}$$

$$H_d = a_{dJ} \mathbf{I} \cdot \mathbf{J} \tag{73}$$

with an effective magnetic hyperfine operator defined by

$$H_m = A_J \mathbf{I} \cdot \mathbf{J} \tag{74}$$

where

$$A_J = a_{cJ} + a_{oJ} + a_{dJ} \tag{75}$$

To match up with theoretical calculations, the following connections to the various expectation values are employed:

$$f = \left\langle \Psi \left| 4\pi \sum_{i=1}^{3} \sigma_{zi} \delta(\mathbf{r}_i) \right| \Psi \right\rangle \tag{76}$$

$$d = \left\langle \Psi \left| \sum_{i=1}^{3} \sigma_{zi} r_i^{-3} P_2(\cos \theta_i) \right| \Psi \right\rangle \tag{77}$$

$$\ell = \left\langle \Psi \left| \sum_{i=1}^{3} r_i^{-3} \ell_{zi} \right| \Psi \right\rangle \tag{78}$$

$$q = 2 \left\langle \Psi \left| \sum_{i=1}^{3} r^{-3} P_2(\cos \theta_i) \right| \Psi \right\rangle \tag{79}$$

with

$$a_{cJ} = \gamma_I \frac{\langle \mathbf{S} \cdot \mathbf{J} \rangle}{SJ(J+1)} \left(\frac{g_J}{6}\right) f \tag{80}$$

$$a_{dJ} = \gamma_I \frac{[3\langle \mathbf{S} \cdot \mathbf{L} \rangle \langle \mathbf{L} \cdot \mathbf{J} \rangle - L(L+1)\langle \mathbf{S} \cdot \mathbf{J} \rangle]}{SL(2L-1)J(J+1)} \left(\frac{g_J}{2}\right) d \tag{81}$$

$$a_{oJ} = \gamma_I \frac{\langle \mathbf{L} \cdot \mathbf{J} \rangle}{LJ(J+1)} \left(\frac{g_J}{6}\right) \ell \tag{82}$$

and

$$\gamma_I = \frac{2\mu_B \mu_N \mu_I}{h a_0^3 I} \tag{83}$$

In Eq. (83), h is Planck's constant, a_0 is the Bohr radius, I is the nuclear spin, and μ_I is the nuclear magnetic moment. The expectation values in Eqs. (76–78) are also sometimes identified notationally with a_c, a_d, and a_o, respectively. In Eq. (76), σ_{zi} is the Pauli spin operator and satisfies $\sigma_{zi}\alpha(i) = \alpha(i)$ and $\sigma_{zi}\beta(i) = -\beta(i)$.

For the $^2S_{1/2}$ states of Li

$$A_{1/2} = a_{c1/2} \tag{84}$$

It is convenient to simplify the notation by dropping the J dependence of a_λ, and express $A_{1/2}$ and $A_{3/2}$ in terms of $a_{\lambda 3/2}$. For the state $^2P_{3/2}$

$$A_{3/2} = a_c + a_o + a_d \tag{85}$$

and for the $^2P_{1/2}$ state

$$A_{1/2} = -a_c + 2a_o - 10a_d \tag{86}$$

The connection between $a_{\lambda 3/2}$ and $a_{\lambda 1/2}$ can be obtained directly from Eqs. (80–82). A third hyperfine coupling parameter, $A_{3/2,1/2}$, arises as an off-diagonal component,

$$A_{J,J-1} = \langle JI, M_JI|H_{\text{mhfs}}|(J-1)I, M_JI \rangle, \tag{87}$$

leading to

$$A_{3/2,1/2} = -a_{c3/2} + \frac{1}{2}a_{o3/2} + \frac{5}{4}a_{d3/2} \tag{88}$$

Slightly different definitions of a_λ can be found in the literature (see, for example, Lindgren and Rosén, 1974), with the result that Eqs. (85, 86, 88) will also appear in a slightly different form.

The electric quadrupole constant is defined by

$$b_q = \langle LSI, LSI | H_Q | LSI, LSI \rangle \tag{89}$$

The determination of the hyperfine constant for the $^2S_{1/2}$ state of Li has attracted extensive theoretical attention (Larsson, 1968; Lindgren, 1985; King and Shoup, 1986; King, 1989; Panigrahy et al. 1989; Blundell et al., 1989; King and Bergsbaken, 1990; Mårtensson-Pendrill and Ynnerman, 1990; Sundholm and Olsen, 1990; Esquivel et al., 1991; Carlsson et al., 1992; Tong et al., 1993; Shabaev et al., 1995; Bieroń et al., 1996; Yan et al., 1996b; King, 1998b). A collection of references to earlier work on the Li ground state is given by King (1997). For the excited S states of Li, the most precise theoretical values available have been determined by Blundell et al. (1989), King (1991), Jönsson et al. (1995), and Yan et al. (1996b). There has been considerable interest in the theoretical determination of the hyperfine coupling constants of the $^2P_{1/2}$ and $^2P_{3/2}$ states (Ahlenius and Larsson, 1973, 1978; Garpman et al., 1975, 1976; Glass and Hibbert, 1976; Lindgren, 1985; Johnson et al., 1987; Blundell et al., 1989; Mårtensson-Pendrill and Ynnerman, 1990; Sundholm and Olsen, 1990; Carlsson et al., 1992; Tong et al., 1993; Bieroń et al., 1996; Yan et al., 1996b). Other excited states of Li have received far less attention; with only more approximate results being available (Goddard, 1968; Ladner and Goddard, 1969; Lunell, 1973).

The availability of a number of experimental hyperfine coupling constants has undoubtedly been a stimulus for theoretical developments. A good review of the earlier experimental work has been given by Arimondo et al. (1977). Experimental results are available for the $2\,^2S_{1/2}$, $3\,^2S_{1/2}$, and $4\,^2S_{1/2}$ states (Beckmann et al., 1974, Vadla et al., 1987; Stevens et al., 1995; Kowalski et al., 1978), and the $2\,^2P_{1/2}$, $2\,^2P_{3/2}$, $3\,^2P_{1/2}$, $3\,^2P_{3/2}$, and $4\,^2P_{3/2}$ states (Ritter, 1965; Brog et al., 1967; Budick et al., 1966; Isler et al., 1969; Orth et al., 1975 ; Nagourney et al., 1978; Shimizu et al., 1987; Carlsson and Sturesson, 1989). The high precision measurement of $A_{1/2}$ for the ground state of Li presents a significant computational challenge. For the ground state of the isoelectronic ion Be^+, the experimental precision is higher still, and the difference between experiment (Wineland et al., 1983) and theory (King, 1988; Yan et al., 1996b) is even more pronounced.

Table 9 presents a summary of some of the high-precision values of the Fermi contact term (Eq. (76)) and the hyperfine coupling constants for the ^2S states of Li. These values are mostly based on nonrelativistic calculations in the infinite nuclear mass approximation. To determine the coupling constant $A_{1/2}$ for the $^2S_{1/2}$ states, the following relationship is employed:

$$A_{1/2} = \left(\frac{\mu_0 \mu_B \mu_N}{2\pi h a_0^3}\right)\left(\frac{g_J \mu_I}{3I}\right) f \tag{90}$$

TABLE 9
EXPECTATION VALUES AND HYPERFINE COUPLING CONSTANTS FOR THE LOW-LYING ^2S STATES OF ^7Li.

State	f_{NR} (a.u.)	f (a.u.)	$A_{1/2}$ (MHz)	Reference
$2\,^2S$	2.906	2.907	401.9	Larsson (1968)
	2.9041	2.905	401.6	King and Shoup (1986)
	2.9064	2.9072	401.91	King (1989)
	2.9071	2.9079	402.01	King and Begrsbaken (1990)
	2.9039	2.9047	401.56	Sundholm and Olsen (1990)
	2.9047	2.9055	401.67	Carlsson et al. (1992)
	2.9051	2.9059	401.73	Tong et al. (1993)
		2.904	401.5	Shabaev et al. (1995)
		2.90578	401.714	Bieroń et al. (1996)
	2.905922(50)	2.90575(22)	401.71(3)	Yan et al. (1996b)
		expt.	401.7520433(5)	Beckmann et al. (1974)
$3\,^2S$			93.24(2)	Blundell et al. (1989)
	0.670	0.670	92.7	King (1991a)
	0.67335	0.67372	93.139	Jönsson et al. (1995)
	0.673405(50)	0.673368(86)	93.091(12)	Yan et al. (1996b)
		expt.	94.68(22)	Stevens et al. (1995)
$4\,^2S$	0.254	0.254	35.	King (1991a)
	0.25327	0.25336	35.026	Jönsson et al. (1995)
		expt.	36.4(40)	Kowalski et al. (1978)
$5\,^2S$	0.11	0.11	16	King (1991a)

where

$$C = \left(\frac{\mu_0 \mu_B \mu_N}{2\pi h a_0^3}\right)$$
$$= \alpha^2 c R_\infty (m_e/m_p) \qquad (91)$$
$$= 95.410673(9) \text{ MHz}$$

In Eq. (91), m_p is the proton mass, c is the speed of light, and μ_0 is the vacuum permeability.

Using the most recent estimates for the fundamental constants (Cohen and Taylor, 1987), and the revised values for the nuclear moments of ^6Li and ^7Li (King, 1997), Eqs. (80–82) can be expressed for the $^2P_{3/2}$ state of ^7Li as,

$$a_c = 69.123175(44) f/J \qquad (92)$$

$$a_d = 207.36953(13) d/J \qquad (93)$$

$$a_o = 207.13122(11) \ell/J \qquad (94)$$

and for $^2P_{3/2}$ state of ^6Li:

$$a_c = 26.174020(20) f/J \quad (95)$$

$$a_d = 78.522059(60) d/J \quad (96)$$

$$a_o = 78.431823(53) \ell/J \quad (97)$$

In Eqs. (92–97), f, d, and ℓ are in a.u. and a_c, a_d and a_o are in MHz.

To employ Eq. (90), the f value must be corrected for the finite nuclear mass, for relativistic effects and for quantum electrodynamic corrections, that is

$$f = f_{NR} + \Delta f_{MASS} + \Delta f_{REL} + \Delta f_{QED} \quad (98)$$

The correction for finite nuclear mass is handled by multiplying f_{NR} by $(1 - \mu/M)^3$, leading to the result

$$\Delta f_{MASS} = -\frac{3\mu}{M}\left[1 - \frac{\mu}{M} + \frac{1}{3}\left(\frac{\mu}{M}\right)^2\right] f_{NR} \approx -\frac{3\mu}{M} f_{NR} \quad (99)$$

For ^7Li this correction is -0.000682 a.u. Δf_{REL} has been estimated in several ways. One approach involves a comparison of MCHF and MCDF calculations using different basis sets, and then estimating Δf_{REL} using an extrapolation procedure (Tong et al., 1993). An alternative approach is based on a one-electron relativistic correction to $|\Psi(0)|^2$ (Yan et al., 1996b). For the $^2S_{1/2}$ ground state of Li, several estimates of Δf_{REL} lead to approximately 0.0017 a.u., with the error in the second significant digit being roughly estimated as ± 3, based on the spread of the calculated values. Relatively little work is available on the determination of Δf_{QED}. Three rather different estimates for the ground state of Li are available: -0.0002 a.u. (Panigrahy et al., 1989), 0.00336 a.u. (Bieroń et al., 1996), and $-0.000918(47)$ (Yan et al., 1996b). It should be clear from the preceding remarks on Δf_{REL} and Δf_{QED} that these two contributions are the principal factors that prevent a more precise theoretical determination of the hyperfine coupling constants from being made.

A list of calculated $A_{1/2}$ values for the ^2S states of Li is presented in Table 9. For the ^2S ground state, the corrections $\Delta f_{REL} = 0.0017$ a.u., $\Delta f_{QED} = -0.0002$ a.u., and the mass correction given above have been applied, unless the authors included values for these contributions. For the excited ^2S states, the Δf_{QED} correction was ignored, and $\Delta f_{REL} \approx 0.0006$ a.u. (Yan et al., 1996b) was employed for the $3\,^2S$ state and ignored for the $4\,^2S$ state. The precision of the experimental results for the excited ^2S states is not sufficiently high to provide a test of the calculated relativistic and QED corrections to f. Only modest agreement between theory and experiment is found for the hyperfine constants of the low-lying excited doublet S states.

In Table 10 a summary is presented for some of the higher precision theoretical

EXPECTATION VALUES (EQS. (76–79)) AND HYPERFINE COUPLING CONSTANTS FOR THE LOW-LYING ^2P STATES OF ^7Li.

State	f (a.u.)	d (a.u.)	ℓ (a.u.)	q (a.u.)	$A_{1/2}$ (MHz)[a]	$A_{3/2}$ (MHz)[a]	$A_{3/2,1/2}$ (MHz)[a]	Reference
$2\,^2P$	−0.214619	−0.013525	0.063218	−0.022824	45.667	−2.992	11.965	Nesbet (1970)
	−0.2162	−0.0134	0.0634	−0.0202	46.0	−3.07	12.0	Ahlenius and Larsson (1973)
	−0.2086	−0.0135	0.0628	−0.0224	45.6	−2.81	11.6	Ahlenius and Larsson (1978)
	−0.2210(3)[b]	−0.013476(2)[b]	0.06308(1)[b]	−0.022664(4)[b]	46.24	−3.337	12.21	Lindgren (1985)
					45.96(1)	−3.03		Blundell et al. (1989)
	−0.2158(15)	−0.01346(2)	0.06304(8)	−0.02253(8)	45.96	−3.100	11.97	Sundholm and Olsen (1990)
	−0.2159(15)[b]	−0.01346(2)[b]	0.06303(8)[b]	−0.02253(8)[b]	45.96	−3.106	11.98	Sundholm and Olsen (1990)
					45.789	−2.879		Mårtensson-Pendrill and Ynnerman (1990)
	−0.2155	−0.01346	0.06305	−0.02255	45.95	−3.085	11.96	Carlsson et al. (1992)
	−0.21705	−0.01341	0.06308	−0.02187	45.96	−3.145	12.040	Tong et al. (1993)
					45.989[c]	−3.1060[c]		Bieroń et al. (1996)
					45.977[b]	−3.058[b]		Godefroid et al. (1996)
	−0.214860 (extrapolated estimate)							Yan et al. (1996b)
	−0.214783(50)							
					45.914(25)	−3.055(14)	11.85(35)	Yan et al. (1996b)
$3\,^2P$	−0.0677	−0.00354	0.0177	−0.00708 expt.	12.9	−1.16	3.73	Orth et al. (1975)
		−0.003988	0.01868	−0.006832				Lunell (1973)
				expt.	13.5(2)	−0.96(13)		Garpman et al. (1975)
				expt.		−0.965(20)		Budick et al. (1966)
				expt.	13.7(12)	−1.036(16)		Isler et al. (1969)
								Nagourney et al. (1978)
$4\,^2P$	−0.0289	−0.00149	0.00746	−0.00298	5.45	−0.51	1.59	Lunell (1973)
		−0.001682	0.007858	−0.002892				Garpman et al. (1975)
				expt.		−0.41(2)		Isler et al. (1969)

[a] Hyperfine coupling constants have been recomputed using Eqs. (85, 86, 88, 92–94).
[b] Includes relativistic and mass corrections.
[c] Includes QED, relativistic, and mass corrections.

results for the terms f, d, ℓ, and q, along with the hyperfine coupling constants $A_{1/2}$, $A_{3/2}$, $A_{3/2,1/2}$, computed from Eqs. (85, 86, 88). The precision of the experimental results for the P-states is currently not very high, and as a consequence, the relativistic and QED corrections to the hyperfine constants of these excited states have received less attention, relative to the efforts expended on the ^2S ground state.

The quadrupole moment, Q, is related to the electric field gradient (q) at the nucleus. There is a scatter of the computed results derived from different theoretical approaches: see Diercksen et al., (1988) for a tabulation of results. Precise theoretical estimates for Q in barns (1 $b = 1 \times 10^{-28}$ m^2) are: $Q(^7Li) = -0.04055(80)$ b (Diercksen et al., 1988) and $Q(^6Li) = -0.00083$ b (Sundholm et al., 1984). The result for ^7Li is in good agreement with the experimental result $Q(^7Li) = -0.041(6)$ b (Orth et al., 1975). Unfortunately, the uncertainty in the experimental result is too large to provide a severe test for different computational approaches.

X. Other Properties

Space limitations do not allow us to discuss all the recent progress on calculating the different properties of the lithium atom. There are, however, three areas where very significant progress has been made, and these are discussed briefly next.

A. Polarizabilities

The static dipole polarizability of the ground state of Li has been of theoretical interest for a long time. A summary of the more significant calculations performed over the past thirty years is given by King (1997). Several relatively recent calculations (Pipin and Bishop, 1992; Wang and Chung, 1994; Yan et al., 1996a) settle on the value of 164.1 a.u. (Yan et al. 1996a quote a value 164.116(2) a.u.). This result is in agreement with the best experimental value available, 164.0(34) a.u. (Molof et al., 1974); unfortunately, the error associated with this experimental work is too large to provide a severe test for the theoretical calculations. A significant reduction in the error of the experimental measurement will be needed to induce theoretical developments beyond the nonrelativistic level.

There has been continuing interest in the calculation of the quadrupole polarizability and the hyperpolarizability for the ground state of Li (for a recent summary of progress see King, 1997). Unfortunately, there are no experimental measurements for Li to compare with the results of calculations. The hyperpolarizability is a particularly difficult property to determine. It is only recently (Pipin and Bishop, 1992; Kassimi and Thakkar, 1994; Jaszunski and Rizzo, 1996) that both the correct sign and the magnitude have been determined.

There has been some theoretical attention directed at evaluating the polarizabilities of some of the excited states of *Li*, including high-lying Rydberg states (Schmieder *et al.*, 1971; Shestakov *et al.*, 1972; Adelman and Szabo, 1973; Manakov *et al.*, 1975; Sims *et al.*, 1976b; Beck and Nicolaides, 1977; Redmon and Browne, 1977; Davydkin and Zon, 1982; Davydkin and Ovsiannikov, 1986; Chung, 1992; Pipin and Bishop, 1993; Themelis and Nicolaides, 1992, 1995; Ponomarenko and Shestakov, 1993; Mérawa *et al.*, 1994). The principal experimental result available is a measurement of the Stark shift of the lithium D_1 line ($1s^2 2s\ ^2S_{1/2} - 1s^2 2p\ ^2P_{1/2}$) (Hunter *et al.*, 1991; Windholz *et al.*, 1992), which is directly related to the difference in the scalar polarizabilities (α_0) of the two states involved in the transition. This experimental result offers a valuable check on the theoretically determined $\alpha_0(^2P)$. The Stark shift of the D_2 line ($1s^2 2s\ ^2S_{1/2} - 1s^2 2p\ ^2P_{3/2}$) has also been measured (Windholz *et al.*, 1992) and this allows the tensor polarizability, α_2, for the $^2P_{3/2}$ state to be determined. The best calculations of α_0 and α_2 (Pipin and Bishop, 1993) are found to be in very good agreement with the experimental results.

B. Oscillator Strengths

For an electric dipole transition, the oscillator strength is given by

$$f_{if,\ell} = \frac{2\Delta E}{3g_i} \left| \left\langle \Psi_i \left| \sum_{j=1}^{3} \mathbf{r}_j \right| \Psi_f \right\rangle \right|^2 \tag{100}$$

$$f_{if,v} = \frac{2}{3g_i \Delta E} \left| \left\langle \Psi_i \left| \sum_{j=1}^{3} \nabla_j \right| \Psi_f \right\rangle \right|^2 \tag{101}$$

$$f_{if,a} = \frac{2Z^2}{3g_i (\Delta E)^3} \left| \left\langle \Psi_i \left| \sum_{j=1}^{3} \frac{\mathbf{r}_j}{r_j^3} \right| \Psi_f \right\rangle \right|^2 \tag{102}$$

In Eqs. (100–102), ΔE denotes the transition energy between the initial and final states expressed in a.u., and g_i is the statistical weight of the initial state. An implicit summation over degeneracies is assumed for both states involved in the transition. Equations (100–102) represent the dipole length, dipole velocity, and dipole acceleration forms, respectively. These relationships are all mathematically equivalent when exact eigenfunctions are employed. Refinements to Eqs. (100–102) are necessary when the effects of finite nuclear mass are considered (Yan and Drake, 1995b). When approximate wave functions are utilized, agreement between the different forms is sometimes taken as a sign of a better calculation, but there are known risks associated with drawing this inference. The dipole acceleration form is infrequently evaluated. The transition moment involved in the evaluation of $f_{if,a}$ is sensitive to the near-nuclear region of configuration space, and this is a region that is more difficult to account for theoretically in the standard variational approach.

The other property commonly reported is the multiplet line strength, S. This is connected to the oscillator strength by the relationship

$$f_{if} = \frac{2}{3g_i} \Delta E \, S_{if} \qquad (103)$$

where ΔE and S_{if} are both expressed in a.u.

The calculation of a precise value for the $1s^2 2s \; ^2S_{1/2} - 1s^2 2p \; ^2P_{1/2}$ oscillator strength has attracted considerable attention. Gaupp et al. (1982) reported a measurement of this oscillator strength, $f = 0.7416(12)$, with an uncertainty of 0.16%, making it one of the most precise measurements of its kind. A number of theoretical results emerged shortly thereafter, using a diverse variety of techniques, including MBPT, CI, CI-HY, MCHF, HY (Fischer, 1988; Peach et al., 1988; Blundell et al., 1989; Mårtensson-Pendrill and Ynnerman, 1990; Weiss, 1992; Pipin and Bishop, 1992; Tong et al., 1993; Chung, 1993; Liaw and Chiou, 1994; Brage et al., 1994; Yan and Drake, 1995b), together with some earlier results (Ahlenius and Larsson, 1973; Sims et al., 1976a), made it apparent that despite the quoted precision, the experimental result of Gaupp et al. was slightly low. Most of the theoretical results were in the range 0.7467–0.748, with the highest precision result being 0.7469572(10) in the infinite nuclear mass approximation, and 0.7467871(10) for ^7Li (Yan and Drake, 1995b). Two later measurements by Carlsson and Sturesson (1989) and McAlexander et al. (1995) gave results 0.7439(55) and 0.7502(44), respectively. These were in closer agreement with the theoretical results, but the larger uncertainties left the issue unresolved. The most recent measurement of Volz and Schmoranzer (1996) gives the value 0.7467(16), which is in excellent agreement with a number of the theoretical calculations. The outstanding work of Yan and Drake (1995b) and of the most recent experimental measurements will probably end discussion of this discrepancy between theory and experiment.

Quantum Monte Carlo (QMC) calculations by Barnett et al. (1992, 1995) were found to support the experimental result of Gaupp et al. The most recent result of these authors was 0.7431(6). The current applications of the QMC method do not reach the precision levels obtainable by the HY technique, so it appears that the error limit is too optimistic in these QMC calculations. Because the QMC approach is a useful computation technique, particularly for larger electronic systems, there should be some interest in using the D_1 line oscillator strength of Li as a test property to refine the technique.

Less attention has been directed at the calculation of oscillator strengths for other transitions of Li. The interested reader can pursue this topic with the following works (Lunell, 1975; Caves, 1975; Sims et al., 1976a; Martin and Wiese, 1976; Pipin and Bishop, 1992; Yan and Drake, 1995b).

C. Lifetimes

Closely tied to the previous discussion is the topic of lifetimes. The radiative lifetime of an excited level k is given by

$$\tau_k = \left(\sum_n A_{kn}\right)^{-1} \tag{104}$$

where A_{kn} is a transition probability and the summation is over all levels of the atom that have an energy less than E_k. If only one decay channel is possible (say i), Eq. (104) simplifies to

$$\tau_k = A_{ki}^{-1} \tag{105}$$

The transition probability can be calculated from

$$A_{ki} = \frac{8\pi^2\mu_0 e^2 c}{m_e \lambda^2} \frac{g_i}{g_k} f_{ik} \tag{106}$$

where e is the absolute value of the electronic charge, λ is the wavelength of the transition, and the other symbols have been introduced previously. Using values of the fundamental constants from Cohen and Taylor (1987), Eq. (106) can be simplified to

$$A_{ki} = 0.66702532(44)(\Delta E)^2 \frac{g_i}{g_k} f_{ik} \tag{107}$$

where ΔE is the transition energy expressed in cm^{-1}.

A number of theoretical and experimental lifetimes for various excited levels of *Li* have been tabulated by Theodosiou and coworkers (1984, 1991). The lifetime of the $2\,^2P_{1/2}$ level of *Li* has been well studied experimentally and theoretically. The most precise theoretical result for the $2\,^2P$ term is 27.117301(36) ns (Yan and Drake, 1995b), which is in excellent agreement with the very recent experimental measurement of Volz and Schmoranzer (1996), who find a value of 27.11(6) ns, and McAlexander *et al.* (1996) who obtained 27.102(7) ns. The latter results improve the earlier experimental value of Gaupp *et al.* (1982), who obtained 27.29(4) ns. Relativistic effects were not accounted for in the calculations of Yan and Drake, but this appears not to be an important issue given the present accuracy levels of the most recent experimental work.

XI. Outlook

The lithium atom will continue to play the role of a benchmark system in testing new computational methodologies. For many properties, there are now available a number of high-precision estimates, which can serve as valuable reference points for testing different theoretical approaches. The precision level of the calculations for certain properties has progressed to the point where the theoretical results can serve as both a guide and a calibration marker for some experimental measurements.

The lithium atom will also continue to be studied for its own intrinsic interest. There are several areas where theoretical progress is desirable and likely to occur in the next few years. Improved precision determination of the relativistic corrections to the energy levels is a priority problem. Advances in this area will be directly tied to solving various recalcitrant integration problems. Resolution of these mathematical issues will improve the precision of a number of calculated ionization potentials, the accuracy of which is currently limited in part by the uncertainties associated with the relativistic corrections.

Improved calculations of the hyperpolarizability will pose a significant challenge. With the recent progress on high-precision calculations of the low-lying excited states, we might be optimistic that considerable progress can be made on the theoretical evaluation of several of the polarizabilities. There are a number of experimental opportunities available for these properties. An improved precision measurement of the static polarizability would provide an important check for some of the high-precision theoretical results of this property that have become available in the last few years.

Recently, there has been a substantial increase in the precision level for the experimental determination of some of the low-lying energy levels for ^6Li and ^7Li. These measurements provide a stimulus for a theoretical examination of some of the smaller contributions to the term energies. Currently, a significant fraction of the uncertainty for the theoretical determination of the first ionization potential arises from the correction ΔE_{QED}. There is a clear need for additional work by the QED theorists to improve the precision level of calculations of ΔE_{QED} for many-electron atoms. The nonrelativistic calculation of the low-lying energy levels has progressed significantly over the past ten years. There has been recent theoretical (Drachman and Bhatia, 1995; Bhatia and Drachman, 1997) and continuing experimental advances (Liang *et al.,* 1986; Day *et al.,* 1994; Rothery *et al.,* 1995; Hoogenraad *et al.,* 1995; Stevens *et al.,*1996; Storry *et al.,* 1997) in the study of the Rydberg states of Li. The high-lying states will provide the next computational challenge. Brute force application of the standard variational technique is not likely to be very successful for the higher lying Rydberg levels.

Improved calculations of the hyperfine coupling constants for the ground and excited states will require combined efforts on several fronts. The determination of precise Fermi contact contributions requires wave functions of high quality in the near nuclear region. Improved procedures to build wave functions that are highly accurate in this region, rather than just relying on the output from the standard variational approach, appear to be needed. Improved ways to determine the relativistic and QED contributions must be found. For the $2\,^2S_{1/2}$ ground state of ^6Li and ^7Li, the experimentalists have measured results of high precision; it is now up to the theorists to accept the challenge of calculating these constants to high precision. Two related properties, the hyperfine anomaly and the hyperfine pressure shift, have received very little theoretical attention for the lithium atom. A theoretical study of the hyperfine anomaly might provide an avenue for the determination of useful nuclear structure information.

Over the past twenty-five years there has been a close interplay between theoretical and experimental studies of the properties of the ground and excited states of atomic lithium. This trend will likely be maintained in the foreseeable future. Advances in computer technology have played a pivotal role in recent theoretical progress, and continued technological progress will be an integral component of the theoretical advances that occur in the future.

XII. Acknowledgments

Support from the National Science Foundation (grant No. PHY-9600926) is greatly appreciated. Acknowledgment is also made to the Donors of the Petroleum Research Fund, administered by the American Chemical Society, for financial support. Some of the author's results reported in this work were obtained with support from the Camille and Henry Dreyfus Foundation and from Cray Research, Inc.

XIII. References

Accad, Y., Pekeris, C. L., and Schiff, B. (1971). *Phys. Rev. A* 4, 516.
Adelman, S. A. and Szabo, A. (1973). *J. Chem. Phys.* 58, 687.
Agentoft, M., Andersen, T., and Chung, K. T. (1984). *J. Phys. B: At. Mol. Phys.* 17, L433.
Ahlenius, T. and Larsson, S. (1973). *Phys. Rev. A* 8, 1.
Ahlenius, T. and Larsson, S. (1978). *Phys. Rev. A* 18, 1329.
Alexander, S. A. and Coldwell, R. L. (1997). *Int. J. Quantum Chem. 63,* 1001.
Arimondo, E., Inguscio, M., and Violino, P. (1977). *Rev. Mod. Phys.* 49, 31.
Audi, G. and Wapstra, A. H., (1993). *Nucl. Phys.* A 565, 1.
Audi, G. and Wapstra, A. H., (1995). *Nucl. Phys.* A 595, 409.
Barnett, R. N., Reynolds, P. J., and Lester, W. A., Jr. (1992). *Int. J. Quantum Chem.* 42, 837.
Barnett, R.N., Johnson, E. M., and Lester, W. A., Jr. (1995). *Phys. Rev. A* 51, 2049.
Barrois, R., Lüchow, A., and Kleindienst, H. (1996). *Chem. Phys. Lett.* 249, 249.
Barrois, R., Bekavac, S., and Kleindienst, H. (1997a). *Chem. Phys. Lett.* 268, 531.
Barrois, R., Kleindienst, H., and Lüchow, A. (1997b). *Int. J. Quantum Chem.* 61, 107.
Barrois, R., Lüchow, A., and Kleindienst, H. (1997c). *Int. J. Quantum Chem.* 62, 77.
Bazley, N. W. (1959). *Proc. Natl. Acad. Sci.* (U.S.A.) 48, 850.
Bazley, N. W. (1960). *Phys. Rev.* 120, 144.

Bazley, N. W. and Fox, D. W. (1961). *Phys. Rev.* 124, 483.
Beck, D. R. and Nicolaides, C. A. (1977). *Chem. Phys. Lett.* 49, 357.
Beck, D. R. and Nicolaides, C. A. (1978). *Chem. Phys. Lett.* 59, 525.
Beckmann, A., Böklen, K. D., and Elke, D. (1974). *Z. Phys.* 270, 173.
Berk, A., Bhatia, A. K., Junker, B. R., and Temkin, A. (1986). *Phys. Rev. A* 34, 4591.
Berry, H. G. (1975). *Physica Scr.* 12, 5.
Berry, H.G., Pinnington, E. H., and Subtil, J. L. (1972). *J. Opt. Soc. Amer.* 62, 767.
Bethe, H. A. and Salpeter, E. E. (1977). *Quantum mechanics of one- and two-electron atoms* (p. 103). Plenum (New York).
Bhatia, A. K. (1978). *Phys. Rev. A* 18, 2523.
Bhatia, A. K. and Drachman, R. J. (1997). *Phys. Rev. A* 55, 1842.
Bieroń, J., Jönsson, P., and Fischer, C. F. (1996). *Phys. Rev. A* 53, 2181.
Blundell, S. A., Johnson, W. R., Liu, Z. W., and Sapirstein, J. (1989). *Phys. Rev. A* 40, 2233.
Bonham, R. A. (1965). *J. Mol. Spectros.* 15, 112.
Brage, T., Fischer, C. F., and Jönsson, P. (1994). *Phys. Rev. A* 49, 2181.
Brog, K. C., Eck, T. G., and Wieder, H. (1967). *Phys. Rev.* 153, 91.
Budick, B., Bucka, H., Goshen, R. J., Landman, A., and Novick, R. (1966). *Phys. Rev.* 147, 1.
Bunge, C. F. (1979). *Phys. Rev. A* 19, 936.
Bunge, C. F. (1981a). *Phys. Rev. A* 23, 2060.
Bunge, C. F. (1981b). *J. Phys. B: At. Mol. Phys.* 14, 1.
Bunge, C. F. and Bunge, A. V. (1978a). *Phys. Rev. A* 17, 816.
Bunge, C. F. and Bunge, A. V. (1978b). *Phys. Rev. A* 17, 822.
Burke, E. A. (1965). *J. Math. Phys.* 6, 1691.
Byron, Jr., F. W. and Joachain, C. J. (1966). *Phys. Rev.* 146, 1.
Caldow, G. L. and Coulson, C. A. (1961). *Proc. Camb. Philos. Soc.* 57, 341.
Cantù, A. M., Parkinson, W. H., Tondello, G., and Tozzi, G. P. (1977). *J. Opt. Soc. Amer.* 67, 1030.
Carlsson, J. and Sturesson, L. (1989). *Z. Phys. D* 14, 281.
Carlsson, J., Jönsson, P., and Fischer, C. F. (1992). *Phys. Rev. A* 46, 2420.
Caves, T. C. (1975). *J. Quant. Spectros. Radiat. Transfer.* 15, 439.
Cederquist, H. and Mannervik, S. (1985). *Phys. Rev. A* 31, 171.
Chambaud, G., Lévy, B., and Stacey, D. N. (1984). *J. Phys. B: At. Mol. Phys.* 17, 4285.
Chen, M.-K. (1996). *J. Phys. B: At. Mol. Opt. Phys.* 29, 2179.
Chen, M.-K. and Chung, K. T. (1994). *Phys. Rev. A* 49, 1675.
Chung, K. T. (1979). *Phys. Rev. A* 20, 724.
Chung, K. T. (1981a). *Phys. Rev. A* 23, 2957.
Chung, K. T. (1981b). *Phys. Rev. A* 24, 1350.
Chung, K. T. (1982). *Phys. Rev. A* 25, 1596.
Chung, K. T. (1991). *Phys. Rev. A* 44, 5421.
Chung, K. T. (1992). *J. Phys. B: At. Mol. Opt. Phys.* 25, 4711.
Chung, K. T. (1993). In P. Richard, M. Stöckli, C. L. Cocke, and C. D. Lin (Eds.). *AIP Conf. Proc. 274, VIth International Conference on the Physics of Highly Charged Ions* (p. 381). AIP (New York).
Chung, K. T. (1997a). *Phys. Rev. Lett.* 78, 1416.
Chung, K. T. (1997b). *Phys. Rev. A* 56, R3330.
Chung, K. T. and Davis, B. F. (1985). In A. Temkin (Ed.). *Autoionization: Recent developments and applications* (p. 73). Plenum Press (New York).
Chung, K. T. and Gou, B. (1995). *Phys. Rev. A* 52, 3669.
Chung, K. T. and Zhu, X.-W. (1993). *Physica Scr.* 48, 292.
Cohen, E. R. and Taylor, B. N. (1987). *Rev. Mod. Phys.* 59, 1121.
Cohen, M. and Feldmann, T. (1979). *J. Phys. B: At. Mol. Phys.* 12, 2771.

Conroy, H. (1964). *J. Chem. Phys.* 41, 1336.
Coulson, C. A. and Haskins, P. J. (1973). *J. Phys. B: At. Mol. Phys.* 6, 1741.
Davis, B. F. and Chung, K. T. (1984). *Phys. Rev. A* 29, 1878.
Davis, B. F. and Chung, K. T. (1985). *Phys. Rev. A* 31, 3017.
Davis, B. F. and Chung, K. T. (1987). *Phys. Rev. A* 36, 1948.
Davis, B. F. and Chung, K. T. (1988). *Phys. Rev. A* 37, 111.
Davis, B. F. and Chung, K. T. (1990a). *Phys. Rev. A* 41, 5844.
Davis, B. F. and Chung, K. T. (1990b). *Phys. Rev. A* 42, 5121.
Davydkin, V. A. and Ovsiannikov, V. D. (1986). *J. Phys. B: At. Mol. Phys.* 19, 2071.
Davydkin, V. A. and Zon, B. A. (1982). *Opt. Spectros. (USSR)* 52, 359.
Day, J. C., Ehrenreich, T., Hansen, S. B., Horsdal-Pedersen, E., Mogensen, K. S., and Taulbjerg, K. (1994). *Phys. Rev. Lett.* 72, 1612.
Delves, L. M. (1972). *J. Phys. A: Gen. Phys.* 5, 1123.
Diercksen, G. H. F., Sadlej, A. J., Sundholm, D., and Pyykkö, P. (1988). *Chem. Phys. Lett.* 143, 163.
Drachman, R. J. and Bhatia, A. K. (1995). *Phys. Rev. A* 51, 2926.
Drake, G. W. F. (1988). *Can. J. Phys.* 66, 586.
Drake, G. W. F. (1993). AIP Conf. Proc. (275) *At. Phys.* 13, 3.
Drake, G. W. F. and Swainson, R. A. (1990). *Phys. Rev. A* 41, 1243.
Drake, G. W. F. and Yan, Z.-C. (1995). *Phys. Rev. A* 52, 3681.
Dressel, P. R. and King, F. W. (1994). *J. Chem. Phys.* 100, 7515.
Ederer, D. L., Lucatorto, T., and Madden, R. P. (1970). *Phys. Rev. Lett.* 25, 1537.
Esquivel, R. O., Bunge, A. V., and Núñez, M. A. (1991). *Phys. Rev. A* 43, 3373.
Esquivel, R. O., Tripathi, A. N., Sagar, R. P., and Smith, V. H., Jr. (1992). *J. Phys. B: At. Mol. Opt. Phys.* 25, 2925.
Feldman, G. and Fulton, T. (1995). *Ann. Phys.* 240, 315.
Feldman, P. and Novick, R. (1963). *Phys. Rev. Lett.* 11, 278.
Feldman, P. and Novick, R. (1967). *Phys. Rev.* 160, 143.
Feldman, P. Levitt, M., and Novick, R. (1968). *Phys. Rev. Lett.* 21, 331.
Feldmann, D. M., Pelzl, P. J., and King, F. W. (1998). Submitted for publication.
Fischer, C. F. (1988). *Nucl. Instrum. Meth. Phys. Res. B* 31, 265.
Fischer, C. F. (1990). *Phys. Rev. A* 41, 3481.
Fock, V. (1954). *Izvest. Acad. Nauk USSR Ser. Fiz.* 18, 161. For an English translation see *Kgl. Norske Videnskab. Selskabs Forh.* 31, 138 (1958).
Fox, D. W. (1972). *SIAM J. Math. Anal.* 3, 617.
Fox, D. W. and Sigillito, V. G. (1972a). *Chem. Phys. Lett.* 13, 85.
Fox, D. W. and Sigillito, V. G. (1972b). *Chem. Phys. Lett.* 14, 583.
Fox, D. W. and Sigillito, V. G. (1972c). *J. Appl. Math. Phys.* 23, 392.
Frolov, A. M. and Smith, V. H., Jr. (1997). *Int. J. Quantum Chem.* 63, 269.
Fröman, A., and Hall, G. G. (1961). *J. Mol. Spectros.* 7, 410.
Fromm, D. M. and Hill, R. N. (1987). *Phys. Rev. A* 36, 1013.
Fuchs, M. and Rubahn, H.-G. (1986). *Z. Phys. D* 2, 253.
Galán, M. and Bunge, C. F. (1981). *Phys. Rev. A* 23, 1624.
Garcia, J. D. and Mack, J. E. (1965). *Phys. Rev.* 138, A987.
Garpman, S., Lindgren, I., Lindgren, J., and Morrison, J. (1975). *Phys. Rev. A* 11, 758.
Garpman, S., Lindgren, I., Lindgren, J., and Morrison, J. (1976). *Z. Phys. A* 276, 167.
Gaupp, A., Kuske, P., and Andrä, H. J. (1982). *Phys. Rev. A* 26, 3351.
Gilbert, T. L. (1963). *Rev. Mod. Phys.* 35, 491.
Glass, R. (1978). *J. Phys. B: At. Mol. Phys.* 11, 3469.
Glass, R. and Hibbert, A. (1976). *J. Phys. B: At. Mol. Phys.* 9, 875.
Goddard, W. A., III. (1968). *Phys. Rev.* 176, 106.

Godefroid, M. R., Fischer, C. F., and Jönsson, P. (1996). *Physica Scr.* T 65, 70.
Goldman, S. P. (1998). *Phys. Rev. A* 57, R 677.
Goy, P., Liang, J., Gross, M., and Haroche, S. (1986). *Phys. Rev. A* 34, 2889.
Harris, F. E. (1997). *Phys. Rev. A* 55, 1820.
Herzberg, G. and Moore, H. R. (1959). *Can. J. Phys.* 37, 1293.
Hijikata, K., Matsubara, I., and Maruyama, M. (1987). In J. Avery, J. P. Dahl, and A. E. Hansen (Eds.). *Understanding Molecular Properties* (p. 503). D. Reidel (Dordrecht, Holland).
Hill, R. N. (1985). *J. Chem. Phys.* 83, 1173.
Hill, R. N. (1995). *Phys. Rev. A* 51, 4433.
Hinze, J. and Pitzer, K. S. (1964). *J. Chem. Phys.* 41, 3484.
Ho, Y. K. (1983). *Phys. Rep.* 99, 1.
Ho, Y. K. and Page, B. A. P. (1975). *J. Comput. Phys.* 17, 122.
Hoffmann-Ostenhof, M., and Seiler, R. (1981). *Phys. Rev. A* 23, 21.
Hoffmann-Ostenhof, M., Hoffmann-Ostenhof, T., and Stremnitzer, H. (1992). *Phys. Rev. Lett.* 68, 3857.
Holøien, E. and Geltman, S. (1967). *Phys. Rev.* 153, 81.
Hoogenraad, J. H., Vrijen, R. B., van Amersfoort, P. W., van der Meer, A. F. G., and Noordam, L. D. (1995). *Phys. Rev. Lett.* 75, 4579.
Hsu, J.-J., Chung, K. T., and Huang, K.-N. (1991). *Phys. Rev. A* 44, 5485.
Hsu, J.-J., Chung, K. T., and Huang, K.-N. (1994). *Phys. Rev. A* 49, 4466.
Huang, K. (1946). *Phys. Rev.* 70, 197.
Hughes, R. H. (1955). *Phys. Rev.* 99, 1837.
Hunter, L. R., Krause, D., Jr., Berkeland, D. J., and Boshier, M. G. (1991). *Phys. Rev. A* 44, 6140.
Hylleraas, E. A. (1929). *Zeit. Phys.* 54, 347.
Isler, R. C., Marcus, S., and Novick, R. (1969). *Phys. Rev.* 187, 76.
James, H. M. and Coolidge, A. S. (1936). *Phys. Rev.* 49, 688.
Jaskólska, B. and Woźnicki, W. (1989a). *Physica Scr.* 39, 230.
Jaskólska, B. and Woźnicki, W. (1989b). *Physica Scr.* 39, 234.
Jaszuński, M. and Rizzo, A. (1996). *Int. J. Quantum Chem.* 60, 487.
Jáuregui, R. and Bunge, C. F. (1981). *Phys. Rev. A* 23, 1618.
Jitrik, O. and Bunge, C. F. (1991). *Phys Rev. A* 43, 5804.
Jitrik, O. and Bunge, C. F. (1997). *Phys. Rev. A* 56, 2614.
Johansson, I. (1959). *Arkiv Fysik* 15, 169.
Johnson, B. R. (1981). *Phys. Rev. A* 24, 2339.
Johnson, W. R. and Soff, G. (1985). *At. Data Nucl. Data Tables* 33, 405.
Johnson, W. R., Blundell, S. A., and Sapirstein, J. (1988). *Phys. Rev. A* 37, 2764.
Johnson, W. R., Idrees, M., and Sapirstein, J. (1987). *Phys. Rev. A* 35, 3218.
Jönsson, P., Fischer, C. F., and Bieroń, J. (1995). *Phys. Rev. A* 52, 4262.
Journel, L., Cubaynes, D., Bizau, J.-M., Al Moussalami, S., Rouvellou, B., Wuilleumier, F. J., VoKy, L., Faucher, P., and Hibbert, A. (1996). *Phys. Rev. Lett.* 76, 30.
Kassimi, N. E. and Thakkar, A. J. (1994). *Phys. Rev. A* 50, 2948.
Kato, T. (1949). *J. Phys. Soc. Jpn.* 4, 334.
Kato, T. (1957). *Commun. Pure Appl. Math.* 10, 151.
Kiernan, L. M., Lee, M.-K., Sonntag, B. F., Zimmermann, P., Costello, J. T., Kennedy, E. T., Gray, A., and Vo Ky, L. (1996). *J. Phys. B: At. Mol. Opt. Phys.* 29, L181.
King, F. W. (1986). *Phys. Rev. A* 34, 4543.
King, F. W. (1988). *Phys. Rev. A* 38, 6017.
King, F. W. (1989). *Phys Rev. A* 40, 1735.
King, F. W. (1991a). *Phys. Rev. A* 43, 3285.
King, F. W. (1991b). *Phys. Rev. A* 44, 3350.

King, F. W. (1991c). *Phys. Rev. A* 44, 7108.
King, F. W. (1993). *J. Chem. Phys.* 99, 3622.
King, F. W. (1995). *J. Chem. Phys.* 102, 8053.
King, F. W. (1997). *J. Mol. Struct. (Theochem)* **400**, 7.
King, F. W. (1998a). Submitted for publication.
King, F. W. (1998b). To be submitted for publication.
King, F. W. (1998c). To be submitted for publication.
King, F. W. (1998d). To be submitted for publication.
King, F. W., Ballegeer, D. G., Larson, D. J., Pelzl, P. J., Nelson, S. A., Prosa, T. J., and Hinaus, B. M. (1998e). Submitted for publication.
King, F. W. and Bergsbaken, M. P. (1990). *J. Chem. Phys.* 93, 2570.
King, F. W. and Dressel, P. R. (1989). *J. Chem. . Phys.* 90, 6449.
King, F. W. and Shoup, V. (1986). *Phys. Rev. A* 33, 2940.
King, F. W., Dykema, K. J., and Lund, A. D. (1992). *Phys. Rev. A* 46, 5406.
King, W. H. (1984). *Isotope shifts in atomic spectra*. Plenum (New York).
Klahn, B. (1985). *J. Chem. Phys.* 83, 5754.
Klahn, B. and Bingel, W. A. (1977). *Theor. Chim. Acta* 44, 9, 27.
Klahn, B. and Morgan, J. D., III. (1984). *J. Chem. Phys.* 81, 410.
Kleindienst, H. and Beutner, S. (1989). *Chem. Phys. Lett.* 164, 291.
Kleindienst, H., Büsse, G., and Lüchow, A. (1995). *Int. J. Quantum Chem.* 53, 575.
Kowalski, J., Neumann, R., Suhr, H., Winkler, K., and zu Putlitz, G. (1978). *Z. Phys. A* 287, 247.
Kutzelnigg, W. (1985). *Theor. Chim. Acta* 68, 445.
Kutzelnigg, W. and Klopper, W. (1991). *J. Chem. Phys.* 94, 1985.
Kutzelnigg, W. and Morgan, J. D., III. (1992). *J. Chem. Phys.* 96, 4484. Erratum: *J. Chem. Phys.* 97, 8821 (1992).
Ladner, R. C. and Goddard, W. A., III. (1969). *J. Chem. Phys.* 51, 1073.
Lakin, W. (1965). *J. Chem. Phys.* 43, 2954.
Larsson, S. (1968). *Phys. Rev.* 169, 49.
Larsson, S. (1972). *Phys Rev. A* 6, 1786.
Larsson, S. and Crossley, R. (1982). *Int. J. Quantum Chem.* 22, 837.
Larsson, S., Crossley, R., and Ahlenius, T. (1979). *J. Phys. Colloq.* 40, C1.
Levitt, M. and Feldman, P. D. (1969). *Phys. Rev.* 180, 48.
Liang, J., Gross, M., Goy, P., and Haroche, S. (1986). *Phys. Rev. A* 33, 4437.
Liaw, S.-S. and Chiou, F.-Y. (1994). *Phys. Rev. A* 49, 2435.
Lindgren, I. (1985). *Phys. Rev. A* 31, 1273.
Lindgren, I. and Rosén, A. (1974). *Case Studies At. Phys.* 4, 93.
Lorenzen, C.-J. and Niemax, K. (1982). *J. Phys. B: At. Mol. Phys.* 15, L139.
Lorenzen, C.-J. and Niemax, K. (1983). *Physica Scr.* 27, 300.
Lüchow, A. and Kleindienst, H. (1992a). *Int. J. Quantum Chem.* 41, 719.
Lüchow, A. and Kleindienst, H. (1992b). *Chem. Phys. Lett.* 197, 105.
Lüchow, A. and Kleindienst, H. (1993). *Int. J. Quantum Chem.* 45, 445.
Lüchow, A. and Kleindienst, H. (1994). *Int. J. Quantum Chem.* 51, 211.
Lüchow, A., Barrois, R., and Kleindienst, H. (1993). *Chem. Phys. Lett.* 216, 359.
Lunell, S. (1973). *Phys. Rev. A* 7, 1229.
Lunell, S. (1975). *Physica Scr.* 12, 63.
Lyons, J. D., Pu, R. T., and Das, T. P. (1969). *Phys. Rev.* 178, 103.
Manakov, N. L., Ovsyannikov, V. D., and Rapoport, L. P. (1975). *Opt. Spectros. (USSR)* 38, 115.
Mannervik, S. (1989). *Physica Scr.* 40, 28.
Mannervik, S., Cederquist, H., and Träbert, E. (1986). *Physica Scr.* 34, 143.
Mannervik, S. and Cederquist, H. (1983). *Physica Scr.* 27, 175.

Mannervik, S. and Cederquist, H. (1985). *Physica Scr.* 31, 79.
Mannervik, S. and Cederquist, H. (1986). *J. Phys. B: At. Mol. Phys.* 19, L845.
Mannervik, S., Short, R. T., Sonnek, D., Träbert, E., Möller, G., Lodwig, V., Heckmann, P. H., Blanke, J. H., and Brand, K. (1989). *Phys. Rev. A* 39, 3964.
Manson, S. T. (1971). *Phys. Rev. A* 3, 147.
Mariella, R. (1979). *Appl. Phys. Lett.* 35, 580.
Mårtensson, A.-M. and Salomonson, S. (1982). *J. Phys. B: At. Mol. Phys.* 15, 2115.
Mårtensson-Pendrill, A.-M., and Ynnerman, A. (1990). *Physica Scr.* 41, 329.
Martin, G. A. and Wiese, W. L. (1976). *J. Phys. Chem. Ref. Data* 5, 537.
McAlexander, W. I., Abraham, E. R. I., and Hulet, R. G. (1996). *Phys. Rev. A* 54, R5.
McAlexander, W. I., Abraham, E. R. I., Ritchie, N. W. M., Williams, C. J., Stoof, H. T. C., and Hulet, R. G. (1995). *Phys. Rev. A* 51, R871.
McIlrath, T. J. and Lucatorto, T. B. (1977). *Phys. Rev. Lett.* 38, 1390.
McKenzie, D. K. and Drake, G. W. F. (1991). *Phys. Rev. A* 44, R6973. Erratum: *Phys. Rev. A* 48, 4803 (1993).
McKoy, V. (1965). *J. Chem. Phys.* 42, 2959.
Mérawa, M., Rérat, M., and Pouchan, C. (1994). *Phys. Rev. A* 49, 2493.
Meyer, M., Müller, B., Nunnemann, A., Prescher, Th., v. Raven, E., Richter, M., Schmidt, M., Sonntag, B., and Zimmermann, P. (1987). *Phys. Rev. Lett.* 59, 2963.
Molof, R. W., Schwartz, H. L., Miller, T. M., and Bederson, B. (1974). *Phys. Rev. A* 10, 1131.
Morgan, J. D., III. (1989). In M. Defranceschi and J. Delhalle (Eds.). *Numerical determination of the electronic structure of atoms, diatomic and polyatomic molecules* (p. 49). Kluwer (Dordrecht).
Morrison, R. C., Mizell, J. R., Jr., and Day, O. W., Jr. (1996). *Int. J. Quantum Chem.* 57, 355.
Muszyńska, J., Papierowska, D., and Woźnicki, W. (1980). *Chem. Phys. Lett.* 76, 136.
Nagourney, W., Happer, W., and Lurio, A. (1978). *Phys. Rev. A* 17, 1394.
Nesbet, R. K. (1970). *Phys. Rev. A* 2, 661.
Nicolaides, C. A. and Aspromallis, G. (1986). *J. Phys. B: At. Mol. Phys.* 19, L841.
Nicolaides, C. A. and Aspromallis, G. (1988). *Physica Scr.* 38, 55.
Nicolaides, C. A. and Komninos, Y. (1981). *Chem. Phys. Lett.* 80, 463.
Nicolaides, C. A., Komninos, Y., and Beck, D. R. (1981). *Phys. Rev. A* 24, 1103.
Öhrn, Y. and Nordling, J. (1963). *J. Chem. Phys.* 39, 1864.
Orth, H., Ackermann, H., and Otten, E. W. (1975). *Z. Phys. A* 273, 221.
Panigrahy, S. N., Dougherty, R. W., Das, T. P., and Andriessen, J. (1989). *Phys. Rev. A* 40, 1765.
Pauli, G. and Kleindienst, H. (1984). *Theor. Chim. Acta* 64, 481.
Peach, G., Saraph, H. E., and Seaton, M. J. (1988). *J. Phys. B: At. Mol. Opt. Phys.* 21, 3669.
Pegg, D. J., Haselton, H. H., Thoe, R. S., Griffin, P. M., Brown, M. D., and Sellin, I. A. (1975). *Phys. Rev. A* 12, 1330.
Pekeris, C. L. (1958). *Phys. Rev.* 112, 1649.
Pekeris, C. L. (1962). *Phys. Rev.* 126, 143.
Pelzl, P. J. and King, F. W. (1998). *Phys. Rev. E* **57**, xxx, in press.
Perkins, J. F. (1968). *J. Chem. Phys.* 48, 1985.
Perkins, J. F. (1969). *J. Chem. Phys.* 50, 2819.
Perkins, J. F. (1972). *Phys. Rev. A* 5, 514.
Pipin, J. and Bishop, D. M. (1992). *Phys. Rev. A* 45, 2736. Erratum: *Phys. Rev. A* 53, 4614 (1996).
Pipin, J. and Bishop, D. M. (1993). *Phys. Rev. A* 47, R4571.
Pipin, J. and Woźnicki, W. (1983). *Chem Phys. Lett.* 95, 392.
Ponomarenko, D. V. and Shestakov, A. F. (1993). *Chem. Phys. Lett.* 210, 269.
Porras, I. and King, F. W. (1994). *Phys. Rev. A* 49, 1637.
Prasad, S. S. and Stewart, A. L. (1966). *Proc. Phys. Soc.* 87, 159.
Preuss, H. (1961). *Z. Naturf.* 16a, 598.

Radziemski, L. J., Engleman, Jr., R., and Brault, J. W. (1995). *Phys. Rev. A* 52, 4462.
Rassi, D., Pejčev, V., and Ross, K. J. (1977). *J. Phys. B: At. Mol. Phys.* 10, 3535.
Redman, L. T. and Browne, J. C. (1977). *Int. J. Quantum Chem. Symp.* 11, 311.
Reid, C. E. (1972). *Int. J. Quantum Chem.* 6, 793.
Reid, C. E. (1974). *Chem. Phys. Lett.* 26, 243.
Remiddi, E. (1991). *Phys. Rev. A* 44, 5492.
Ritter, G. J. (1965). *Can. J. Phys.* 43, 770.
Roberts, P. J. (1966a). *Proc. Phys. Soc.* 88, 53.
Roberts, P. J. (1966b). *Proc. Phys. Soc.* 89, 789.
Rødbro, M., Bruch, R., and Bisgaard, P. (1979). *J. Phys. B: At. Mol. Phys.* 12, 2413.
Rothery, N. E., Storry, C. H., and Hessels, E. A. (1995). *Phys. Rev. A* 51, 2919.
Russell, D. M. and Greenlee, W. M. (1985). *Phys. Rev. Lett.* 54. 665.
Sack, R. A. (1964). *J. Math. Phys.* 5, 245.
Sansonetti, C. J., Richou, B., Engleman, R., Jr., and Radziemski, L. J. (1995). *Phys. Rev. A* 52, 2682.
Schmid, E. W. and Schwager, J. (1968). *Z. Phys.* 210, 309.
Schmieder, R. W., Lurio, A., and Happer, W. (1971). *Phys. Rev. A* 3, 1209.
Schwartz, C. (1962). *Phys. Rev.* 126, 1015.
Schwartz, C. (1963). In B. Alder, S. Fernbach, and M. Rotenberg (Eds.). *Methods in Computational Physics.* (Vol. 2, p. 241). Academic Press (New York).
Scrinzi, A. (1992). *Phys. Rev. A* 45, 7787.
Shabaev, V. M., Shabaeva, M. B., and Tupitsyn, I. I. (1995). *Phys. Rev. A* 52, 3686.
Shestakov, A. F., Kristenko, S. V., and Vetchinkin, S. I. (1972). *Opt. Spectros.* 33, 223.
Shimizu, F., Shimizu, K., Gomi, Y., and Takuma, H. (1987). *Phys. Rev. A* 35, 3149.
Sims, J. S. and Hagstrom, S. (1971a). *Phys. Rev. A* 4, 908.
Sims, J. S. and Hagstrom, S. A. (1971b). *J. Chem. Phys.* 55, 4699.
Sims, J. S. and Hagstrom, S. A. (1975). *Phys. Rev. A* 11, 418.
Sims, J. S., Hagstrom, S. A., and Rumble, J. R., Jr. (1976a). *Phys. Rev. A* 13, 242.
Sims, J. S., Hagstrom, S. A., and Rumble, J. R., Jr. (1976b). *Phys. Rev. A* 14, 576.
Slater, J. C. (1970). *Int. J. Quantum Chem.* 4, 561.
Smolenskii, E. A. and Zefirov, N. S. (1993). *Phys. Dokl.* 38, 112.
Sonnek, D. and Mannervik, S. (1990). *J. Phys. B: At. Mol. Opt. Phys.* 23, 2451.
Stevens, G. D., Iu, C.-H., Williams, S., Bergeman, T., and Metcalf, H. (1995). *Phys. Rev. A* 51, 2866.
Stevens, G. D., Iu, C.-H., Bergeman, T., Metcalf, H. J., Seipp, I., Taylor, K. T. and Delande, D. (1996). *Phys. Rev. A* 53, 1349.
Stevenson, A. F. (1938). *Phys. Rev.* 53, 199.
Stevenson, A. F. and Crawford, M. F. (1938). *Phys. Rev.* 54, 375.
Storry, C. H., Rothery, N. E., and Hessels, E. A. (1997). *Phys. Rev. A* 55, 128.
Sundholm, D. and Olsen, J. (1990). *Phys. Rev. A* 42, 2614.
Sundholm, D., Pyykkö, P., Laaksonen, L., and Sadlej, A. J. (1984). *Chem. Phys. Lett.* 112, 1.
Switkes, E. (1967). *J. Chem. Phys.* 47, 869.
Szász, L. (1961). *J. Chem. Phys.* 35, 1072.
Temple, G. (1928) *Proc. Roy. Soc.* (London) *A* 119, 276.
Themelis, S. I. and Nicolaides, C. A. (1992). *Phys. Rev. A* 46, R21.
Themelis, S. I. and Nicolaides, C. A. (1995). *Phys. Rev. A* 51, 2801.
Theodosiou, C. E. (1984). *Phys. Rev. A* 30, 2881.
Theodosiou, C. E., Curtis, L. J., and El-Mekki, M. (1991). *Phys. Rev. A* 44, 7144.
Tong, M., Jönsson, P., and Fischer, C. F. (1993). *Physica Scr.* 48, 446.
Vadla, C., Obrebski, A., and Niemax, K. (1987). *Opt. Commun.* 63, 288.
Veseth, L. (1985). *J. Phys. B: At. Mol. Phys.* 18, 3463.
Vinti, J. P. (1932). *Phys. Rev.* 41, 432.

Vinti, J. P. (1940). *Phys. Rev.* 58, 882.
Volz, U. and Schmoranzer, H. (1996). *Physica Scr.* T65, 48.
Wakid, S., Bhatia, A. K., and Temkin, A. (1980). *Phys. Rev. A* 21, 496.
Wang, Z.-W. and Chung, K. T. (1994). *J. Phys. B: At. Mol. Opt. Phys.* 27, 855.
Wang, Z.-W., Zhu, X.-W., and Chung, K. T. (1992a). *Phys. Rev. A* 46, 6914.
Wang, Z.-W., Zhu, X.-W., and Chung, K. T. (1992b). *J. Phys. B: At. Mol. Opt. Phys.* 25, 3915.
Wang, Z.-W., Zhu, X.-W., and Chung, K. T. (1993). *Physica Scr.* 47, 65.
Weinstein, D. H. (1934). *Proc. Natl. Acad. Sci.* (U.S.A.) 20, 529.
Weiss, A. W. (1992). *Can. J. Chem.* 70, 456.
White, R. J. and Stillinger, F. H., Jr. (1970). *J. Chem. Phys.* 52, 5800.
White, R. J. and Stillinger, F. H., Jr. (1971). *Phys. Rev. A* 3, 1521.
Wilson, Jr., E. B. (1965). *J. Chem. Phys.* 43, S172.
Windholz, L. and Umfer, C. (1994). *Z. Phys. D* 29, 121.
Windholz, L., Jäger, H., Musso, M., and Zerza, G. (1990). *Z. Phys. D* 16, 41.
Windholz, L., Musso, M., Zerza, G., and Jäger, H. (1992). *Phys. Rev. A.* 46, 5812.
Wineland, D. J., Bollinger, J. J., and Itano, W. M. (1983). *Phys. Rev. Lett.* 50, 628.
Woźnicki, W., Bylicki, M., Jaskólska, B., and Pipin, J. (1983). *Chem. Phys. Lett.* 95, 609.
Wu, T.-Y. and Shen, S. T. (1944). *Chinese J. Phys.* 5, 150.
Yan, Z.-C. and Drake, G. W. F. (1995a). *Phys. Rev. A* 52, 3711.
Yan, Z.-C. and Drake, G. W. F. (1995b). *Phys. Rev. A* 52, R4316.
Yan, Z.-C. and Drake, G. W. F. (1997a). *J. Phys. B: At. Mol. Opt. Phys.* **30,** 4723.
Yan, Z.-C. and Drake, G. W. F. (1997b). *Phys. Rev. Lett.* **79,** 1646.
Yan, Z.-C., Babb, J. F., Dalgarno, A., and Drake, G. W. F. (1996a). *Phys. Rev. A* 54, 2824.
Yan, Z.-C., McKenzie, D. K., and Drake, G. W. F. (1996b). *Phys. Rev. A* 54, 1322.
Ziem, P., Bruch, R., and Stolterfoht, N. (1975). *J. Phys. B: At. Mol. Phys.* 8, L480.

STORAGE RING LASER SPECTROSCOPY

THOMAS U. KÜHL

Gesellschaft für Schwerionenforschung, Darmstadt,
Germany and Johannes-Gutenberg-Universität, Mainz,
Germany. Presently at University of California, Lawrence
Livermore National Laboratory, Livermore, California

I. Introduction	114
II. Properties of Existing Heavy-Ion Storage Rings	115
III. Kinematic Effects in Storage Rings	117
IV. Laser Experiments in the Electron Cooler	122
A. Laser-Induced Recombination (LIREC) in the Electron Cooler	122
B. Laser-Induced Recombination Using a Single Laser Step	125
1. First LIREC Experiments at the TSR	125
2. LIREC at Reduced Electron Temperatures	127
C. Two-Step Laser-Induced Recombination	128
1. Motivation	128
2. Experiments at the ESR and the TSR	129
3. Doppler-Free Two-Step Laser-Induced Recombination	131
V. Laser Cooling	131
A. Principles of Laser Cooling in the Storage Ring	131
B. Experiments Using Li^+ Ions at TSR and ASTRID	133
1. First Experiments at the TSR	133
2. Experiments on Li^+ Ions at the ASTRID Ring	136
C. Advanced Cooling Schemes	137
1. Experiments Using Be^+ and Mg^+ Ions	137
2. Cooling of the Transverse Velocity Component	138
3. Cooling with Improved Capture Range	138
4. Cooling of Bunched Beams	139
VI. A Test of Special Relativity in the Storage Ring	142
A. Test Schemes for the Theory of Special Relativity	142
B. A Modern Ives and Stilwell Experiment at the TSR	142
1. Analysis of the Transverse Doppler-Effect	142
2. Experimental Setup at the TSR	144
3. Results	145
VII. Quantum Electrodynamics in Strong Fields Probed by Laser Spectroscopy in Highly Charged Ions	146
A. The Ground-State Hyperfine Splitting of Hydrogen-Like Ions	146
B. First Measurement of a HFS Splitting at Optical Frequencies in the ESR	149
1. Experimental Technique	149
2. Result	150
C. Spectroscopy of the Infrared Ground-State Hyperfine Transition in $^{207}Pb^{81+}$	152

1. Motivation	152
2. Experimental Procedure	152
D. QED and Nuclear Contributions to the Hyperfine Splitting	154
VIII. Conclusion and Outlook	155
IX. Acknowledgments	156
X. References	157

I. Introduction

The advent of heavy-ion cooler rings has created a rapidly growing new field of laser spectroscopy. In the past, poor beam quality and short interaction time had restricted the application of lasers at accelerators to exotic single events, and laser spectroscopy appeared as a more or less parasitic user of these installations. The key experimental features of heavy-ion cooler rings in comparison to standard accelerators are the increase in interaction time and the dramatic improvement of the beam quality achieved by electron cooling. There are a number of unique possibilities accessible for laser-based experiments in heavy-ion storage rings. One centers on the physics in the electron cooler where stored ions and the electrons co-propagate with nearly equal velocity, and form an unusual kind of plasma. The analysis of this interaction by laser techniques can help to understand the cooling process in the storage ring and also to examine the physics of highly charged plasmas. The situation in the electron cooler is very peculiar, because the charge state of the ions compares to the hottest available plasma temperatures whereas the relative energy between ions and electrons is lower than 200 meV. The technique used for these studies is laser-induced recombination, the reverse of resonant laser ionization. An interesting process in itself, it provides a selective population of very high-lying Rydberg states, which can be used as a starting point for precision laser spectroscopy in a second step. Another group of experiments seeks to use laser cooling in addition to electron cooling to reduce the thermal energy of the ions in the storage ring below the magnitude of the intrabeam Coulomb fields. If one defines temperatures according to the kinetic energy of the stored ions relative to their mean trajectory, the results of laser cooling of relativistic ions reach levels approaching 1 mK. Therefore, collective effects become important, which may finally lead to a transition to liquid behavior of the beam, or even crystallization. The goal is to study ordering phenomena in the one component plasma formed by the stored beam. Because laser cooling is a non-Liouvillean process, limitations to reach unprecedented space charge density are removed. This might gain practical consequences where high power densities of energetic ions are required, for example, for inertial fusion energy applications. Finally, the exotic properties of the stored ions can be exploited to address fundamental problems of physics. An example of this third group is a test of the special theory of relativity, using the high but very well-controlled velocity of the ions for a test of the special theory of relativity. For the first time it is possible to perform high-precision laser spectroscopy on objects moving considerably faster than the alleged 300 km/s velocity of Earth. In the storage ring, saturation spectroscopy is used to probe the second-order Doppler shift, leading to a new upper limit for deviations from theory. The much higher velocity, compared to conventional experiments, makes this test particularly sensitive to higher order contributions. The full spectrum of possibilities for storage-ring laser spectroscopy is reached at the very high energies available at the ESR, where highly charged ions beyond all previous possibilities are now available for precision laser spectroscopy, and give novel opportunities for the study of quantum electrodynamics (QED) in high magnetic and electric fields. The transition most suited for such research is the ground-state hyperfine structure transition, well known from the 21-cm line in atomic hydrogen. The Z^{3-}-increase leads to a transition energy in the ultraviolet part of the optical spectrum for the case of Bi^{82+}. At the same time, the QED corrections from vacuum polarization and self-energy rise to $2-3\%$ with a large fraction of higher order contributions.

The history of the young field of laser spectroscopy at heavy-ion storage rings exhibits a rapid growth that is still taking place. Following the inauguration of the TSR storage ring at the MPI für Kernphysik in Heidelberg at Christmas in 1987, the first results of spectroscopy on lithium ions were presented at the Workshop of Crystalline Ion Beams in fall 1988 (Wertheim, 1988). In 1989 came the first successes in laser cooling both at the TSR (Schröder *et al.*, 1990) and at the ASTRID storage ring in Aarhus (Hangst *et al.*, 1991). This was followed by the first observation of laser-induced recombination (Schramm *et al.*, 1991) and the first results of the special relativity test (Klein *et al.*, 1993) at the TSR. Laser experiments at the CRYRING in Stockholm started in 1992 (Abrahamson *et al.*, 1993). The application of the new experimental techniques to heavier and more energetic ions at the ESR storage ring at Darmstadt proved to be more complicated. However, with increasing beam current and improved beam quality and definition, the extension to higher charged ions was finally successful, opening the way to the optical spectroscopy of the ground-state hyperfine structure of hydrogen-like Bi^{82+} (Klaft *et al.*, 1994) and Pb^{81+} (Seelig *et al.*, 1997). Continuing progress is documented by a wealth of new laser experiments, and new proposals at all rings including the newly planned ring at Legnaro (Bisoffi *et al.*, 1994).

II. Properties of Existing Heavy-Ion Storage Rings

A listing of essential properties of existing and planned heavy-ion storage rings is given in Table 1. Laser spectroscopy requires high ion-beam intensity, good ion-beam quality, and the availability of particular ion species. As a result, different facilities have proven to be complementary to a large extent. An immediate

TABLE 1
PROPERTIES OF OPERATIONAL AND PROPOSED STORAGE RINGS.

	Low Energy Range			Medium Energy Range	
Name	CRYRING	TSR	ASTRID	ESR	Tarn 2	CRYSTAL
Institute	AFI	MPI	University	GSI	INS	INFN-LNL
Location	Stockholm	Heidelberg	Aarhus	Darmstadt	Tokyo	Legnaro
Circumference	48.6 m	55 m	40 m	108.4 m	78 m	68.8 m
$B \times r$	1.4 Tm	1.4 Tm	1.88 Tm	10 Tm	5.8 Tm	4 Tm
Injector	CRYEBIS + RFQ	Tandem 15 MV + rf	50 kV–200 kV	Synchrotron	Cyclotron	Tandem + ALPI
Ion Mass A	20–208	12–130	4–238	20–238	1–15	1–238
Charge/Mass	0.5–0.3	0.5–0.37	0.5–0.1	0.5–0.39	1–0.2	0.5–0.3
Energy [MeV/u]	9–24	13–27	2–41	3–830	2.5–200	1–45
e-Cooling [MeV/u]	0.1–10	1–30		3–560	2.5–200	1–45
Proposal	1985	1985	1986	1984	1983	
Operation	1991	1988	1991	1990	1988	1992
Reference	Abrahamsson, 1993	Baumann et al., 1988	Stensgaard, 1988	Franzke, 1987	Katayama et al., 1989	Bisoffi et al., 1992

classification is given by the properties of the injector available at the given facility, and by the magnetic field strength of the bending magnets, which limits the kinetic energy of the particles that can be kept in the ring. Storage rings designed for lower energies presently can provide a better beam quality. This is evident in the case of the ASTRID ring, where the injector operates at an accelerating potential of 50–200 kV, which can be controlled very accurately. There is also the advantage that at lower magnetic fields the forces on the stored ions are lower, which offers the very interesting ability to store fragile ions and molecules such as Ca^- (Haugen *et al.*, 1992) and fullerenes (Andersen *et al.*, 1994) in the ASTRID storage ring.

Often the number of ions that can be accumulated in the storage ring is of extreme importance for the feasibility of laser experiments. In most cases so far, much below the limits given by beam instabilities, the injected intensity is a restraining factor. In the medium mass range, the development of more effective electron-cyclotron-resonance (ECR) and electron-beam ion sources (EBIS) can possibly improve this situation. These devices also offer higher charge states without additional stripping. At present as many as 10^{10} ions of low and medium mass can be stored. The interesting cases of highly charged ions can only be provided with sufficient intensities at high-energy machines. The stripping of uranium ions in a fixed metal foil requires a beam energy of 200 MeV per nucleon to reach a 10% efficiency. So far only the SIS (magnetic rigidity 18 Tm)–ESR (magnetic rigidity 9 Tm) facility (see Fig. 1), can meet this requirement. The alternative production scheme by electron bombardment in an electron-beam ion trap (EBIT) presently yields around 1000 ions per filling cycle for hydrogen-like uranium (Marrs *et al.*, 1994), compared to 10^7 particles per pulse at SIS. For heavy ions like bismuth and uranium, beam intensities of 10^8 stored ions are reached (Franzke *et al.*, 1993). Even this value is far below theoretical limits for the beam intensity in the storage ring. A program to increase the current of high-energy ions has been started (GSI, 1995). The plan is to use electron cooling in the main accelerator ring SIS. The increase in phase-space quality should allow a more efficient multiturn injection from the linear accelerator.

An additional benefit of the SIS–ESR facility is the availability of intense beams of unstable nuclei produced by fragmentation of the primary ion beam. At GSI, these secondary ions are separated from the primary beam in flight in the FRS (fragment separator), and the resulting beam of unstable nuclei can be stored in the ESR. This provides an outstanding opportunity, for example, to measure nuclear masses off stability (Irnich *et al.*, 1995).

III. Kinematic Effects in Storage Rings

For detailed understanding of the interaction of laser light with ions moving in a storage ring, it is necessary to give a short description of the ion kinematics in this

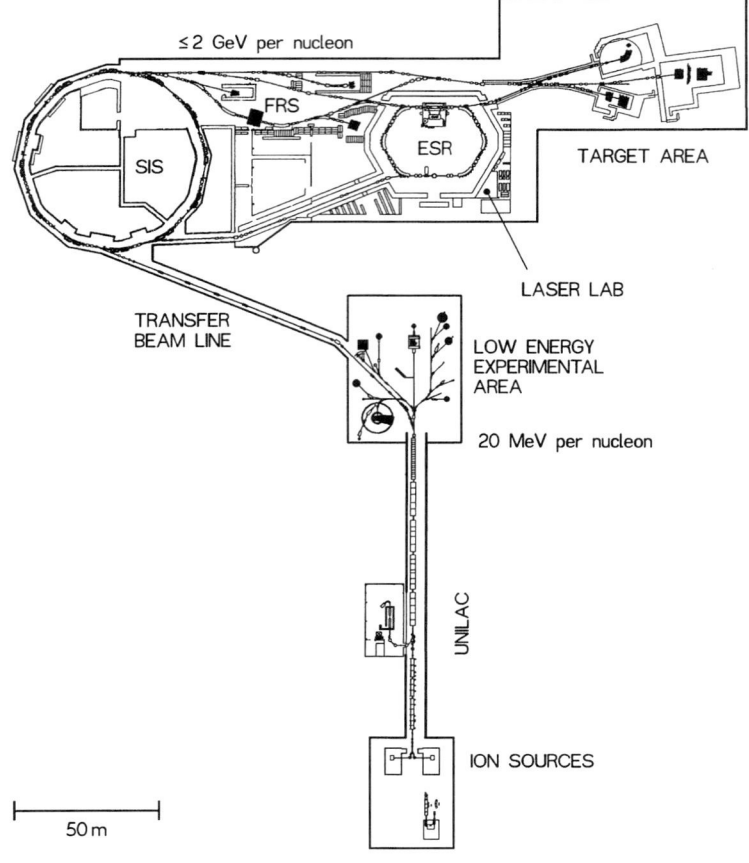

FIG. 1. The SIS–ESR facility at GSI. Heavy ions are pre-accelerated in the UNILAC linear accelerator to energies up to 20 MeV per nucleon. Before injection into the SIS heavy-ion synchrotron, the charge-state of these ions can be increased by passing them through a stripping foil. The final energy is limited by the bending fields of 18 Tm to, for example, 1 GeV per nucleon for uranium. At this energy, naked ions can be prepared by additional stripping. Alternatively, radioactive nuclei are available by fragmentation of the projectiles. All these species can be stored, electron-cooled, and—if necessary—decelerated in the experimental storage ring ESR.

situation. Storage rings contain a number of optical elements that guide and focus the beam. The motion of the ions can be best described in a system of coordinates moving along with the z-axis along the central orbit of the storage ring. Deviations from the central orbit are given by the coordinates $x(s)$ and $y(s)$ for any position s around the ring. No force exists in the direction of the z-axis. The confinement of the ions in the directions perpendicular to this axis by the strong focusing elements of the ring lattice leads to harmonic oscillations of the ions around the mean trajectory. This oscillation can be described by a Hills-type differential equation. For the displacement y from the central orbit in the horizontal plane, this can be expressed as

$$\frac{d^2y}{ds^2} + K(s)y = G(s)\,\Delta p/p_o \tag{1}$$

where $G(s)$, acting only within the dipoles and in the y-direction, is the bending force, $K(s)$ is the focusing strength, and $\Delta p/p_o$ is the relative deviation from the ideal central orbit momentum. A similar relation holds for the vertical plane. At an arbitrary position s along the beam trajectory, the displacement $y(s)$ of an ion with zero momentum deviation $\Delta p = 0$, but an emittance of ϵ is usually expressed through the β-function of the storage ring, which is determined by the lattice layout:

$$y(s) = \sqrt{\epsilon\beta(s)} \cdot \cos(\psi(s) - \delta) \tag{2}$$

The s-dependence includes a variable phase σ and the betatron phase advance

$$\psi(s) = 2\pi \cdot \int_0^s \frac{ds}{\beta(s)} \tag{3}$$

The number of betatron oscillations in one revolution is given by the quantity $Q = \oint ds/\beta(s)$, which is called the tune of the storage ring. The value of Q has to be nonrational in order to avoid resonances with the structure of the storage ring lattice. For a given position in the storage ring, the beam envelope $E(s)$ can be determined as the maximum displacement for an arbitrary value of the betatron phase $\cos(\psi(s) - \delta)$

$$E_y(s) = y_{max}(s) = \sqrt{\epsilon\beta(s)} \tag{4}$$

For an average value of $\beta = 5.1$ m as given for the laser cooling section of the TSR, the envelope of an uncooled beam with an emitance of $\epsilon = 60\pi$mm/mrad would be 3.6 cm, compared to less than 5 mm for a cooled beam of 1πmm/mrad emittance. The transverse velocity at a given point s is given as

$$v_y(s) = \frac{dy}{ds}\cdot\frac{ds}{dt} = -\sqrt{\frac{\epsilon}{\beta(s)}}[(\sin\psi(s) - \delta) - 1/2\beta'(s)\cos(\psi(s) - \delta)]v_0 \tag{5}$$

where v_0 is the mean velocity of the ions, and $\beta'(s)$ the local derivative $d\beta/ds$.

The constant total ion velocity is given by $v_0 = \sqrt{v_x^2 + v_y^2 + v_z^2}$. For collinear laser excitation, as is typical for the experiments described here, the modulation of v_z induced by the transverse oscillations is of interest:

$$v_z = (v_0^2 - v_x^2 - v_y^2) \tag{6}$$

$$\frac{v_z}{v_0} \approx 1 - \left(\frac{1}{2}\right)\frac{v_x^2 + v_y^2}{v_0^2} = 1 - \frac{\partial v_z}{v_0}$$

Averaged over many turns or betatron phases, $\psi(s)$, the maximum value of the longitudinal velocity modulation by a transverse motion, is given by

$$\frac{\partial v_z}{v_0} = \frac{1}{2}\frac{\epsilon}{\beta(s)} \cdot \left(1 + \frac{1}{4}\beta'^2(s)\right) = \frac{1}{2}\frac{E^2/s}{\beta^2(s)} \cdot \left(1 + \frac{1}{4}\beta'^2(s)\right) \tag{7}$$

In the middle of the field-free experimental section we can assume $\beta' = 0$ for both the horizontal and vertical beta functions. Even for the uncooled beam this means a maximum relative transverse velocity v_x/v_0 and v_y/v_0 in the order of a few times 10^{-3}, resulting in a total variation $\partial v_z/v_0 \approx 1.6 \cdot 10^{-5}$. In the ring, ions with momenta differing from p_0 will travel on different paths compared to the central orbit due to the dispersion $\eta(s)$. The length L of the closed orbit for a momentum change of $\partial p = p - p_0 \neq 0$ is changed by $\partial l/L = \alpha_c \cdot \partial p/p_0$, where α_c is the "momentum compaction factor" $\alpha_c = (1/L)\oint G(s)\eta(s)ds$. The revolution frequency as a consequence differs from ω_0 by an amount

$$\frac{\Delta\omega}{\omega_0} = \left(\frac{1}{\gamma^2} - \frac{1}{\alpha_c^2}\right)\frac{\partial p}{p_0} \tag{8}$$

where $\gamma = \sqrt{1 - (v^2/c)}$ is the usual relativistic factor. Often a quantity $\gamma_{TR} = 1/\alpha_c$, the so-called "transition γ," is introduced, which can be identified as the relativistic velocity where the dependence of the revolution frequency from the particle velocity changes sign.

Due to the high beam velocity, relativistic effects are of extreme importance for the light absorption and emission in experiments with high-energy beams. Ions moving at relativistic velocities are described by their 4 vectors (ct, x) and $(E/c, p)$. The Lorentz transformation from a given system Σ to a frame Σ' moving along the z-axis at a velocity $\beta = v_z/c$ for an arbitrary 4-vector $(A, \vec{A})$ is given by

$$A'_0 = \gamma(A_0 - \beta A_z)$$
$$A'_{xy} = A_{xy} \tag{9}$$
$$A'_z = \gamma(A_z - \beta A_0)$$

For laser light impinging onto an absorber moving at relativistic velocity, the resonance frequency is shifted relative to the transition frequency v_0 in the rest frame of the ions according to the relativistic Doppler formula

$$v = v_0[\gamma(1 - \beta \cos \vartheta)]^{-1} \tag{10}$$

where ϑ denotes the angle between the ions and the laser beam ($\vartheta = 0$ for the parallel case). This relation also causes the light emitted from the moving ion to

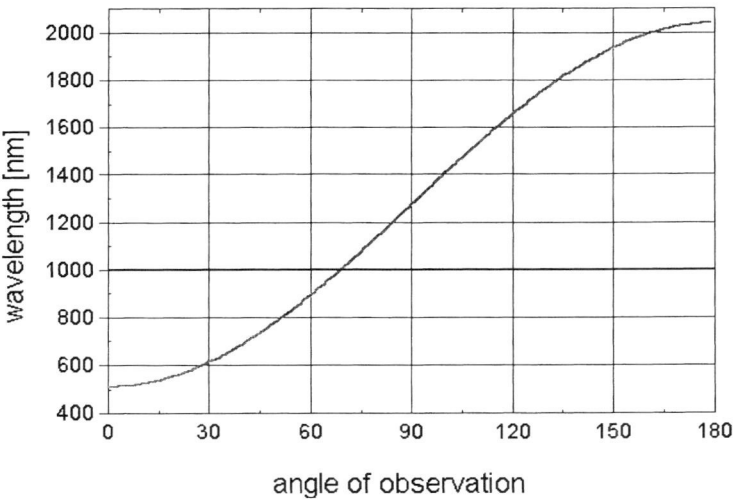

FIG. 2. Dependence of the observed photon wavelength on the emission angle. Light emitted at 1000 nm by ions moving with $v/c = \beta = 0.6$ is observed at widely changed wavelength under different angles of observation.

be observed in the laboratory frame at a strongly shifted frequency. The velocity spread $\delta\beta/\beta$ of the ion beam leads to a Doppler width of

$$\delta_{vDopp} \approx v\gamma^2 \delta\beta \tag{11}$$

The uncertainty of the ion velocity, $\Delta\beta$, enters directly into an uncertainty of the excitation wavelength in the rest frame of the ions, given by

$$\Delta\lambda_0 = \lambda_0 \beta \gamma^2 \frac{\Delta\beta}{\beta} \tag{12}$$

The frequency shift for the fluorescence of light as a function of the angle of detection is depicted in Fig. 2. A large detection solid angle will inevitably result in a large spectral width of the detected light. Equally important for the detection of fluorescence emitted by high-energy ions is the kinematical narrowing of the emission angle toward forward angles. For beam velocities smaller than 10% of the speed of light this is of minor importance. At higher beam velocities, however, the detection efficiency is no longer symmetric. This also holds for the excitation probability. At $\beta = 0.6$ the excitation probability in collinear geometry is a factor of 6 smaller than for anticollinear excitation.

For sub-Doppler resolution one must also take into account the small additional broadening and shift arising from the angle ϑ between laser beam and ion beam in the Doppler formula. At an interaction length of 10 meters and more, angles are easily controlled to better than 1 mrad. This limits a possible shift, which enters by $\Delta_{v\vartheta} = -v\beta^2\vartheta^2$ to $v \cdot 2 \cdot 10^{-7}$ or smaller.

The transverse beam temperature leads to a broadening via the betatron movement of the particles as a result of Eq. (7). For the straight experimental sections this effect can be estimated to be of minor importance for cooled beams. The maximum angle is given from the beam diameter and the length of the straight section, for example, 5 mm and 20 meters or .25 mrad for cooled beams at the ESR.

IV. Laser Experiments in the Electron Cooler

A. Laser-Induced Recombination (LIREC) in the Electron Cooler

Electron cooling of the ion beam is achieved by a collinear superposition of an electron beam copropagating with the ions in the storage ring at nearly the same speed (Budker *et al.*, 1978). The electrons start at a thermal cathode and are recovered after the ion and the electron beams are separated in the collector (Fig. 3). In order to prevent a blow-up of the intense electron beam, it is guided over its complete length by a longitudinal magnetic field. This field also decouples the longitudinal and transverse degrees of freedom of the electron. Due to kinematic narrowing after the electrostatic acceleration of the electrons, the longitudinal beam temperature is much lower than the temperature of the transverse degrees of free-

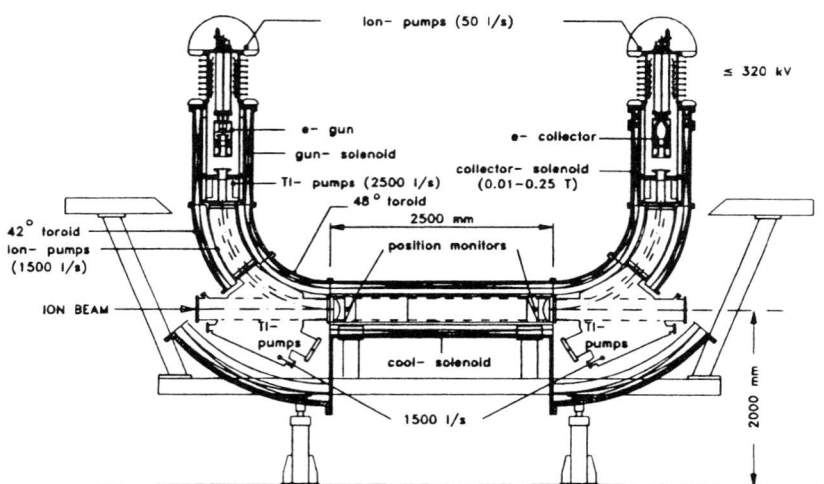

FIG. 3. Schematics of the electron cooler at the ESR storage ring. The electrons are accelerated from the gun by electrostatic fields, merged with the ion beam, and afterwards stopped at their starting potential in the collector.

dom. Typical values are 1–10 meV in the longitudinal direction and 100–500 eV transversally. The interaction between the cold electrons and the ions that provides the cooling force is often described as a scattering process (Poth, 1990).

The force $\vec{F}$ between ions and electrons moving at velocity $\vec{v}_i$ or $\vec{v}_e$, respectively, can be expressed as:

$$\vec{F}(\vec{v}_i) = -4\pi \left(\frac{Ze^2}{4\pi\epsilon_0}\right)^2 \frac{n_e L_c}{m_e} \int f(v_e) \frac{\vec{v}_i - \vec{v}_e}{|\vec{v}_i - \vec{v}_e|^3} d^3 v_e \tag{13}$$

Here, Z and e are the charge of the ion and the electron, n_e and m_e the number and mass of the electrons, L_c the length of the cooler, and ϵ_0 the electric constant.

The situation of highly charged ions embedded in a large number of free electrons can also be looked at as a highly charged plasma. In this view the cooling force can be interpreted as a drag-force between ions and electrons. Once the ions are cooled and, therefore, moving with exactly the same speed as the electrons, this latter picture implies that the ions are surrounded by a charge cloud of electrons. Although the number of particles is much lower than in usual plasmas, recent calculations of the cooling force for highly charged ions indeed suggest that a relatively large number of electrons surrounds the ions in the cooler (Zwicknagel et al., 1996). Study of this dynamic self-screening should give insight into the microscopic behavior of a highly charged plasma.

Binary collisions between the highly charged ions and the electrons are accompanied by spontaneous radiative recombination. An electron is captured into a bound state and a photon is emitted, which carries away the sum of the ionization energy of this state and the kinetic energy of the electron. Analogous to photo ionization, this process can also be stimulated by collision with photons of the correct energy. The idea of using laser-induced recombination in a storage ring was first mentioned in 1979 (Rivlin, 1979) as a way to provide positronium and later also for the production of antihydrogen (Neumann et al., 1983). This was followed by several articles on the theory of this process.

The spontaneous recombination rate depends on the number of the electrons n_e and their energy distribution $f(E)$ according to

$$R^{sp} = n_e \cdot N_i \cdot \sum_{n=1}^{n_{max}} \int_0^\infty f(E') \sigma_n(E') \sqrt{\frac{2E'}{m_e}} dE' \tag{14}$$

N_i is the number of ions, E is the energy of the electrons, and $\sigma_n(E)$ is the energy-dependent recombination cross-section for a given Rydberg state n. Neglecting the Gaunt factor (Omidvar and Guimaraes, 1990), which reduces the cross-section by about 20% for small n, $\sigma_n(E)$ can be approximated as

$$\sigma_n(E) \approx 2.9 \cdot 10^{-21} \cdot \frac{Z^2}{n \cdot E} [eVcm^2] \tag{15}$$

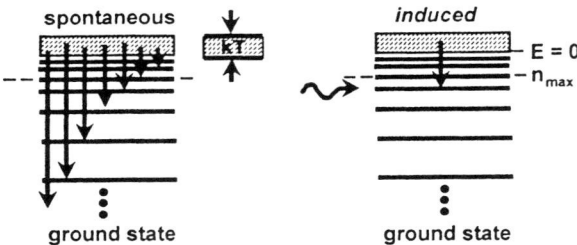

FIG. 4. Schematic picture of the laser-induced recombination process. While spontaneous recombination reaches all bound Rydberg states, laser-induced recombination resonantly populates only a narrow band of states, determined by the laser wavelength and the width of the laser and the electron distribution.

The spontaneous process populates all n-states with preference to the very low-lying main quantum number n. In contrast, the induced process as shown in Fig. 4 predominantly populates a certain quantum number n because the photon energy has to be appropriate to stimulate the transition from the continuum states to the bound states. The integration is now limited to the energy interval given by the energy and bandwidth of the laser and the width of the electron energy distribution:

$$R^{ind} = \frac{Ic^2}{8\pi h\nu^3 \Delta \mathbf{v}} n_e N_i \sum_{n=1}^{n_{max}} \int_{E-h\Delta\nu/2}^{E+h\Delta\nu/2} f(E')\sigma_n(E') \sqrt{\frac{2E'}{m_e}} dE' \quad (16)$$

At light intensities I starting from about 1 MW/cm^2, easily available with pulsed laser systems, the induced recombination into a single Rydberg state will, in resonance, exceed the sum of all spontaneous recombination channels. The increase in recombination probability by the laser process is usually expressed by the gain factor $G_n(E)$, the ratio between induced and spontaneous recombination rate:

$$G_n(E) = \frac{R_n^{ind}}{\sum_{n=1}^{n_{max}} R_n^{spont}}$$

$$\approx \frac{I\lambda^3}{8\pi c} \frac{f(E)}{\sqrt{E}} \cdot \left[\int dE' \frac{F(E')}{\sqrt{E'}} \right]^{-1} \propto \left(\frac{n}{Z}\right)^5 \cdot \frac{I}{Z} \cdot \frac{F(E)}{\sqrt{E}} \quad (17)$$

again neglecting the Gaunt factor. The spectral shape of the induced process reflects strongly the energy distribution of the electrons, which can be probed in this way, but is obviously also sensitive to the influence of external fields on the ionization limit. The possibility of selective population of highly excited Rydberg states gives access to a detailed spectroscopy of these exotic states. Worth mentioning is also the idea to use the inversion created by laser-induced recombination for a short wavelength laser scheme.

B. Laser-Induced Recombination Using a Single Laser Step

1. First LIREC Experiments at the TSR

Laser-induced recombination was demonstrated for the first time in 1990 (Schramm *et al.*, 1991) for protons stored in the TSR in Heidelberg. The experimental scheme is shown in Fig. 5. The experiment employs the merging of three beams: the stored ion beam, the electron beam of the electron cooler, and an intense laser beam. For the detection of charge-changed ions the bending magnet is used as a charge state analyzer. Ions that experience any type of recombination in the electron cooler deviate from the storage orbit due to the increase in magnetic rigidity, and can be registered by a particle detector. The example of protons is the most extreme case, as the recombined hydrogen atoms emerge without any deflection from the ring. The interaction region in the electron cooler has the length of approximately 1.5 m. An electron beam of $\sim$1A current accelerated to approximately 10 keV provides a target thickness of $4.5 \cdot 10^7$ electrons per cm^3. A critical parameter for the observed recombination rate is the temperature of the electron beam. In the original design a transverse temperature corresponding to

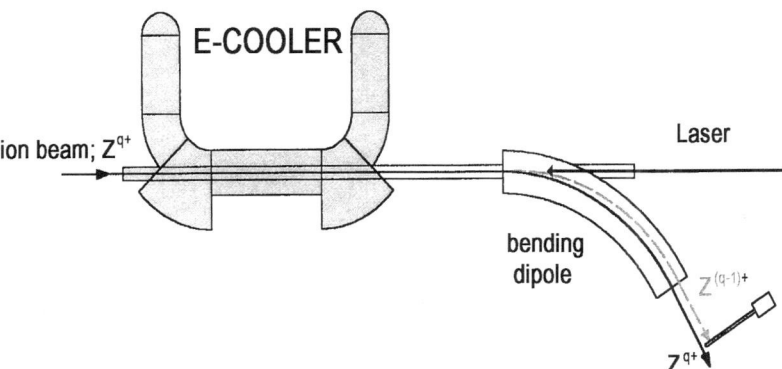

FIG. 5. Schematical view of a laser-induced recombination experiment in the storage ring. Within the electron cooler a laser beam is superimposed collinearly to the ion- and electron trajectory. After capture of an electron, ions will be less strongly deflected in the next bending magnet, and hit a detector.

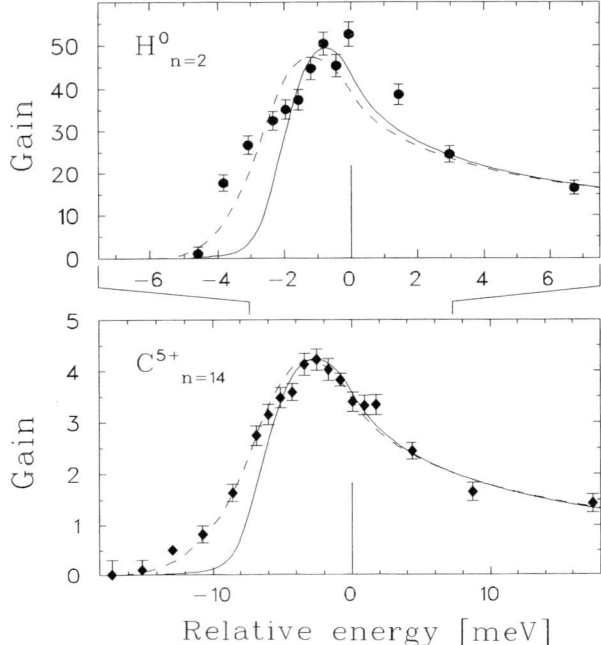

FIG. 6. LIREC enhancement factor as a function of the laser wavelength for protons and C^{6+} ions at the TSR.

approximately 100 meV is reached. The longitudinal temperature is reduced by a factor of 10 due to the kinematic narrowing of the velocity distribution. Under these conditions a spontaneous recombination rate of about 10^5 per second is observed for a proton beam of 1 mA. The induced process can be driven by pulses from an excimer-laser pumped dye-laser system with pulse energies around 10–30 mJ. This laser beam is focused to a beam waist of 5 mm within the electron cooler, providing laser intensities up to 20 MW/cm^2. The gain factor is determined by comparing the count rate at the detector observed outside with that inside a time window coincident with a laser pulse. In hydrogen, recombination induced by 2 eV photons leads to the $n = 2$ levels. Figure 6 shows results from the first laser-induced recombination experiments at the TSR (Wolf et al., 1993). The observed enhancement factor is plotted as the function of the photon energy. The data demonstrate the impressive potential of laser-induced recombination experiments in the storage ring. The gain factor is $G = 40$ for the case of hydrogen and between 4 and 5 for the heavier ions.

The spectral shape can be interpreted in terms of the longitudinal and transverse temperatures of the electron beam. At the low-energy side of the distribution, however, a deviation from the predictions is observed. Seemingly, laser-induced recombination takes place already below the threshold photon energy given by the difference between ionization potential and the binding energy of the recombined Rydberg level. The influence of electric fields in the interaction region, originating mostly from the charge distribution of the electrons and (v × B)-fields caused by inhomogeneity of the toroidal field, is not sufficient to cover this effect. An increased gain factor below the threshold for induced recombination might be explained by processes in the electron ion plasma.

- Three-body collisions might lead to the formation of very highly excited Rydberg states in the first few centimeters of the electron cooler. Because the induced transition rate of the bound-bound transitions is much larger than the induced recombination of free electrons, even a small admixture of prebound electrons could result in a sizable signal at the low-energy end of the spectra.
- A complete mixing of the narrowly spaced high-lying Rydberg states might lower the ionization limit in accordance with the so-called Inglis-Teller criterion found in stellar plasmas.
- Field gradients might be caused by the microscopic charge distribution in the cooler.

2. *LIREC at Reduced Electron Temperatures*

A reduction of the electron beam temperature is desirable in view of the efficiency and temperature limit of electron cooling. At the CRYRING in Stockholm and at the TSR, the transverse temperature in the interaction region now can be reduced by increasing the field of the solenoidal magnet in the electron gun region (Danared *et al.*, 1994) similar to what is done in high-precision electron spectrometers. Lowering the electron temperature in the cooler reduces the mean energy between electrons and ions and, therefore, improves the laser-induced recombination experiments, because (i) the total recombination rate is increased due to the $1/E$-dependence of $\sigma_n(E)$, and (ii) the induced rate is additionally enhanced due to the better overlap of the energy distribution of the free electrons with the laser bandwidth. In addition to a larger total radiative recombination rate, therefore, a larger gain factor is observed. Experiments both at the CRYRING in Stockholm and at the TSR in Heidelberg use this technique. As shown in Fig. 7, the resulting laser-induced spectrum, obtained on Deuterium ions in the CRYRING (Asp *et al.*, 1996), is narrowed but still exhibits a tail on the low-energy side. A detailed study of the change of the spectral shape—which in addition seems to depend on the ion intensity—is still under way.

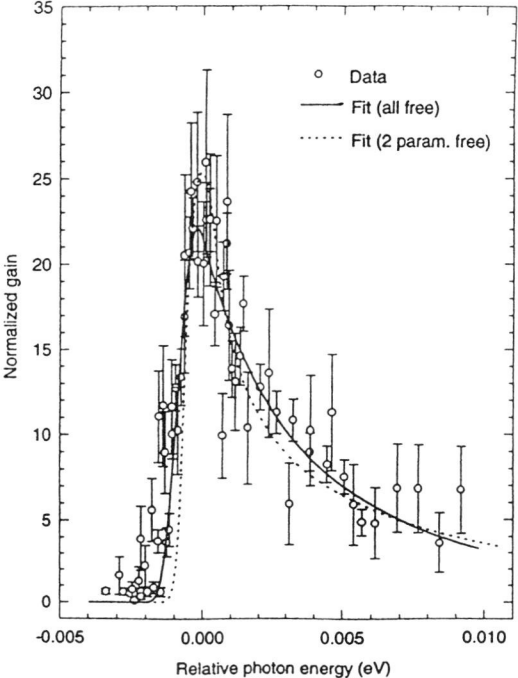

FIG. 7. Laser-induced recombination enhancement factor as a function of wavelength for deuterium ions measured with the improved electron cooler at CRYRING.

C. Two-Step Laser-Induced Recombination

1. Motivation

As shown in the last section, the width of the electron distribution in the continuum is visible in the spectral shape of the induced capture effect. This provides an interesting method for a study of the electron-cooling process. On the other hand, due to the large energy width involved, the high spectral resolution of the laser technique is spoiled. In close analogy to multistep laser ionization, this loss of resolution can be circumvented by splitting the capture process into several steps. In such a scheme the first laser photon populates a specific high-lying Rydberg state. The next step can then fully exploit the spectral resolution given by the laser for a study of the Rydberg states. Obviously, again in analogy to the multistep laser ionization, this process also allows for an increase in the signal size. Using an infrared laser for the first step, the capture efficiency is increased proportional to λ^3 in comparison to a single-step process with shorter wavelengths. This enhancement is especially important for the future application to very highly charged ions because at the beam energies necessary for the complete stripping of heavy ions, the Lorentz field in the bending magnets of the storage ring requires an increasingly large binding energy of the final state. The transfer of population by the second laser step also reduces saturation of the original capture process. The saturated steady-state number of captured electrons changes from

$$N^{1sat} = \frac{w_1}{w_1 + \Gamma_1} N_0 \qquad (18)$$

to

$$N^{2sat} = \frac{w_1}{w_1 + \Gamma_1 + \Gamma_1 w_2/(w_2 + \Gamma_2)} N_0 \qquad (19)$$

where $w_{1,2}$ and $\Gamma_{1,2}$ denote the spontaneous and the induced transition rate for the first and the second state. Because the second transition connects two bound states of the Rydberg ion, a large transition probability can be achieved. The reduction of saturation is, therefore, close to the ratio of the lifetimes of the original and the final Rydberg state, which scales with $1/n^3$. Consequently, gain increases by nearly an order of magnitude for the typical situation where the main quantum number of the final state is about 1/2 of the state reached in the original capture process.

A theoretical framework for the quantitative interpretation of laser-induced electron-ion recombination was presented recently in terms of the density matrix, taking into account stochastic bandwidth and pulse duration of the laser (Schlagheck *et al.*, 1996).

2. Experiments at the ESR and the TSR

A first experiment on two-step recombination at the ESR used bare Ar^{18+} ions stored in the ESR at a velocity of $\beta \approx 0.5$. The experimental set-up is similar to the one used in the TSR (Fig. 5), but two lasers are used to drive the two processes involved. In a first step, electron capture from the continuum into a $n = 83$ Rydberg state is induced by a copropagating Nd:YAG-laser at 1.064 nm.

For the second step, a Ti:Sapphire-laser pumped by a Nd:YAG-laser is used. The laser wavelength tunable from $\lambda = 760-850$ nm corresponds to $\lambda = 438.8-490.7$ nm in the ion rest frame. This allows transitions from Rydberg states with $n = 83$ to $n = 36$ and $n = 37$. The experiments at the ESR employ the field ionization by the Lorentz field in the bending magnet as a means to discriminate between the primary capture process and the secondary stabilization. All ions that underwent a charge change in the electron cooler and maintain hydrogen-like within the bending magnet are registered in a multiwire proportional counting (MWPC) detector mounted within the first dipole magnet of the storage ring behind the electron cooler. Hydrogen-like Ar^{17+} ions in Rydberg states with $n > \sim 50$ are reionized at the entrance into the magnetic field of the dipole. The spontaneous capture rate under the conditions given at the ESR experiment (10^9 stored ions, 500 mA electron current in the cooler) reaches about 50 kHz. The time resolution of the position sensitive MWPC detector of 40 ns is only about a factor of two larger than the interaction time in the electron cooler. The relative timing of the two lasers has to be much better than 2 ns to ensure a maximum overlap of the light pulses within the electron cooler.

In the first experiment (Borneis et al., 1994), the line width was dominated by the laser profile of the high-power Ti:Sapphire laser used for the second laser step. In a following experiment (Borneis et al., 1995) this laser was replaced by a narrow-band dye laser with a line width of 4 GHz, in order to increase the spectral resolution. In this case, resolution is limited by the Doppler width. This is sufficient to resolve fine structure effects. It is also worth mentioning that by laser-induced two-step recombination, laser diagnostics can be performed on stored beams of naked ions. For example, one can determine the velocity of these ions with high precision.

Detailed spectroscopy of fine structure in highly excited Rydberg levels in beryllium-like oxygen ions (Schüssler et al., 1995), as displayed in Fig. 8, was achieved recently at the TSR. The experimental technique is similar to the one described for the ESR measurements. An excimer pumped dye laser drives the second step after the initial capture with a Nd:YAG laser. At the lower beam velocities at the TSR, field ionization in the bending fields is less important than at the ESR. A signal is also detected when only the Nd:YAG laser is interacting with the ions and the electrons. Detection has therefore to rely on the increase in signal due to the reduction in saturation for the first step. Alternatively, reionization by an additional, delayed laser pulse is used to discriminate the low-lying final states.

The detailed spectroscopy of the fine structure in few-electron systems is an

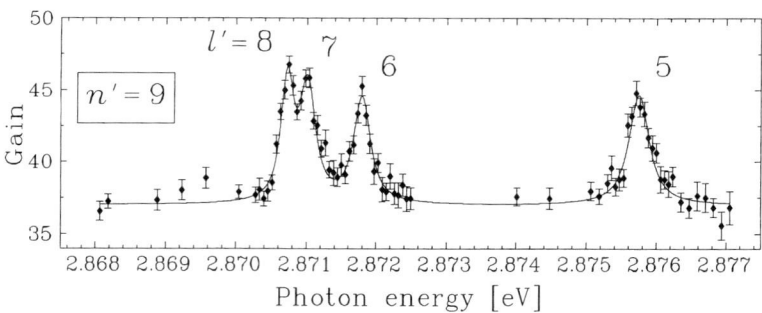

FIG. 8. Fine-structure spectrum of beryllium-like oxygen ions recorded by two-step laser-induced recombination at the TSR.

important test case for atomic physics calculations, probing long-range correlations between the electrons. The understanding of many-body effects is also useful for the analysis of the QED effects in very heavy few-electron systems.

3. Doppler-Free Two-Step Laser-Induced Recombination

Laser-induced recombination can address a number of questions about the condition of the plasma in the electron cooler. The still unclear shape of the energy dependence below the ionization threshold might be of immediate interest for the clarification of open details in the laser cooling process. At high electron densities and highly charged ions, small shifts are predicted to result from dynamic screening in the cooler plasma (Zwicknagel *et al.*, 1996). Using lasers with small bandwidths, the resolution of bound-bound Rydberg transitions in these experiments is usually limited by the Doppler-width of the ion beam, which is about 20 GHz. Further improvement of the resolution is possible by introducing the Doppler-free two-photon technique. If the bound-bound transition is induced by coherently adding the photons of two counterpropagating laser beams, the residual Doppler width can be minimized.

A first Doppler-free two-step LIREC experiment was performed at the ESR. An $^{40}Ar^{18+}$ beam at an energy of 64 MeV/u was used, driving the enhanced electron capture with a pulsed Nd:YAG laser at 1064 nm. The wavelength of this laser beam, copropagating with the ion beam, was shifted to 1540 nm in the moving ion system, populating mainly the level $n = 74$. The Doppler-free two-photon transition was performed by copropagating 355 nm photons from a frequency-tripled Nd:YAG laser and counterpropagating photons at 746 nm from a flashlamp-pumped Ti:Sapphire laser. The two lasers had intensities of 1.2 and 0.6 MW/cm^2, respectively, in the ion rest frame. Their bandwidth was narrowed by seeding with narrow-band cw-lasers. The transition probability for the two-photon process was sufficiently enhanced because the energy of the frequency-tripled Nd:YAG laser, in the ion rest frame, was near the transition $n_1 = 74$ to $n_2 = 37$. The pulsed Ti:Sapphire laser was running at a repetition rate of 16 2/3 Hz, while the two YAG lasers were firing with 50 Hz. This allowed subtracting the background of one-step and, if present, the Doppler-broadened two-step electron capture induced by the two Nd:YAG lasers from the narrow signal of two-step two-photon recombination. Further experiments with this method are in preparation.

V. Laser Cooling

A. Principles of Laser Cooling in the Storage Ring

Improvements in the use of light to control the motion of charged and neutral particles have stimulated the development of elaborate cooling and trapping techniques. Usually, particles are decelerated and kept at the slowest possible velocity. Temperatures in the μK-range have been reported (JOSA, 1989). At the high velocities in the storage ring, laser cooling minimizes the velocity components relative to the mean movement through space of the beam particles. In close analogy to the situation of ions and atoms at low velocities, the absorption of photons causes an acceleration of the ions in the direction of the laser beam. Except for the fact of relativistic transformation just discussed, the geometry of the subsequent emission of photons will be isotropic. This leads to the spontaneous cooling force

$$\vec{F}_{sp} = \frac{\hbar \vec{k} \Gamma}{2} \cdot \frac{S}{1 + S + 4((\Delta - \vec{k}\vec{v})\Gamma)} \qquad (20)$$

In this equation, Γ is the natural line width of the transition, S is the saturation parameter, and Δ is the detuning of the laser relative to the resonance frequency in the rest frame. With the resonance behavior related to the natural line width, this cooling force allows a narrow limit for the achievable temperature T, $k_B T = \hbar\Gamma/2$, called the Doppler limit (Minogin and Lethokov, 1981). k_B denotes the Boltzmann factor. To control the velocity of fast-moving ions a unidirectional cooling force, leading only to a deceleration, is not sufficient (Javanainen et al., 1985). There must be a second, counteracting force to keep the ions at the desired velocity. This can be achieved by different mechanisms: In a pure laser scheme, a second laser in opposite geometry is used. Any other kind of acceleration can also serve this purpose, because resonance dependence in one direction is sufficient. To supply a well controlled and continuously acting force, an induction accelerator was used at the TSR (Ellert et al., 1992). The drawback of this method is the restriction in time given by the required steady increase of the magnetic field. Another possibility is to combine laser cooling and electron cooling. In such a scheme, the velocity of the electrons has to be detuned from the ion velocity in order to provide a force counteracting to the light force. The spontaneous cooling force is strong enough to change the beam velocity equivalent to electrostatic acceleration in a field of several kV. The balance between light force and electron cooling force can be used to measure the electron cooling force as a function of the velocity difference between ions and electrons. Acceleration by high-frequency fields, typically used in ion accelerators, requires a bunching of the beam. This broadens the velocity distribution, and therefore deteriorates the situation for the laser process. As shown recently (Hangst et al., 1995) this is more than compensated by the fact that the ions are kept in the bunch and can undergo many subsequent cooling cycles, and therefore bunched cooling might even be applicable for the cooling of very dense high-energy beams (Habs et al., 1995).

Besides the obvious importance of obtaining even higher brightness than by electron cooling, the interest in laser cooling of stored ion beams has been spurred by the observation of crystallization in ion traps (Birkl et al., 1992). A summary of activities toward crystallization in a storage ring was given by Habs and Grimm (Habs et al., 1995). Collective behavior of the particle beam should appear as soon as the thermal energy is lower than the Coulomb energy amongst the particles of the beam. This ratio is usually expressed as the factor

$$\Gamma_{ord} = \frac{E_{Coul}}{k_B T}$$

In fact, beam ordering has been observed at the NAP-M proton storage ring in Novosibirsk as a consequence of electron cooling (Dementev et al., 1979). Close to the resonance of an allowed transition, the light force is orders of magnitude larger than the force acting in the electron cooler. The Doppler-limited temperature—if achieved in transverse as well as longitudinal direction—corresponds to $\Gamma \approx 10000$ for example, for a beam of 10^7 Be^+ ions at the TSR. This is about 50 times the critical value $\Gamma_{crit} = 170$ required for the formation of crystals. Three-dimensional ordering in a storage ring constitutes a very fundamental process, comparable to the formation of atomic clusters, as external fields are only restricting the beam in two dimensions.

For particles at rest, crystallization was observed in a circular quadrupole ring (Birkl et al., 1992). Theoretical studies demonstrated that at least for low particle numbers, where the crystallization would be a one-dimensional string, ordering should be possible despite the different heating processes, also for fast particles in a storage ring (Schiffer et al., 1986; Wertheim, 1988). Recent calculations of the heating due to the shear forces, however, indicate that a very specific design of the storage ring and the applied laser cooling geometry will be necessary to achieve such crystalline structures (Toepffer, 1997). A number of new proposals for such storage rings have been made, ranging from a table-top RFQ ring (Habs et al., 1995) to the CRYSTAL storage rings for relativistic particles (Bisoffi et al., 1994). Ordered particles moving at relativistic speed might be technically useful as a source of continuously tunable radiation by means of coherent scattering. Once crystallized, the beam might even be accelerated to higher energies (Wei J et al., 1994), enabling unprecedented brightness and energy densities.

B. EXPERIMENTS USING Li^+ IONS AT TSR AND ASTRID

1. First Experiments at the TSR

First experiments on laser cooling in a storage ring were performed at the TSR (Schröder et al., 1990). Li^+ ions were chosen because of a transition within the metastable triplet states, well suited for laser excitation and laser cooling by standard cw dye-lasers. The metastable $2\ ^3S_1$ level is the lowest excited state of

He-like lithium. At an excitation energy of 59 eV, this level has a transition lifetime of 50 seconds as an electric dipole decay to the ground state is prohibited both for parity and spin reasons. A strong allowed transition ($\tau = 43$ ns, $\lambda = 548.5$ nm) connects this metastable state with the $2\,^3P$-fine structure multiplet. The decay of the $2\,^3P_{1,2,0}$ sublevels leads almost exclusively back to the $2\,^3S_1$ state. In the case of ^7Li ($I = 3/2$) all these levels exhibit hyperfine structure splitting. According to the selection rules of electric dipole transitions, the $2\,^3P_2$ ($F = 7/2$) sublevel can only be excited starting from the $2\,^3S_1$ ($F = 5/2$) hyperfine level and can also only decay by this channel. This forms a two-level system that, because it has no obvious pumping channels to other levels, is almost ideal for laser cooling. The saturation intensity for this transition is 8.8 mW/cm^2. At the TSR, $^7Li^+$ ions are produced from $Li\,H^-$ ions in a gas target in the tandem accelerator of the MPI Heidelberg and then are injected into the storage ring. The metastable state is strongly populated in the break-up reaction in the tandem accelerator and more than 10% of the ions in the storage ring are in this excited state. In a single injection cycle without beam accumulation, 10^7–10^8 particles are injected into the storage ring at an energy of 13.3 MeV corresponding to a speed $\beta = v/c = 0.064$. For the ring vacuum of approximately 10^{-8} Pa, storage times are close to 50 seconds, in the order of the lifetime of the metastable state. The storage time is mainly limited by collision with molecules that result in ionization. The accelerator and the TSR allow one to choose the ion velocity over a wide range up to nearly 5% of the speed of light. By proper choice of the ion velocity, the resulting Doppler shift can be used to shift the excitation wavelength in collinear geometry to 514 nm, the green emission line of the argon ion laser. At this same velocity the excitation wavelength in antiparallel geometry is shifted to 584.8 nm. This provides a very efficient laser cooling scheme using a standard single-mode dye laser working at rhodamin 6 G and a fixed frequency single-mode Ar^+-ion laser. A schematic diagram of the laser cooling experiment is shown in Fig. 9.

Laser cooling is achieved in the following way: The ion velocity is chosen slightly higher than required for resonance with the Ar^+-ion laser. As a result, the laser can excite only ions in the low-energy tail of the velocity distribution. The counter-propagating dye laser is scanned starting from a wavelength longer than required for resonance. Tuning this laser toward shorter wavelength resonance will first occur only with beam ions in the fast tail of the velocity distribution. Because the ions are decelerated when undergoing excitation with the dye laser, the wavelength scan leads to an accumulation of ions at a lower energy. This detuning is stopped at the wavelength where the ions decelerated by the dye laser come into resonance also with the argon ion laser. Thereby, a narrow velocity distribution with well-defined energy is achieved as a stationary state. A further detuning of the dye laser leads to a strong fluorescence signal when both lasers interact with the cooled ions. The velocity distribution can be probed by blocking

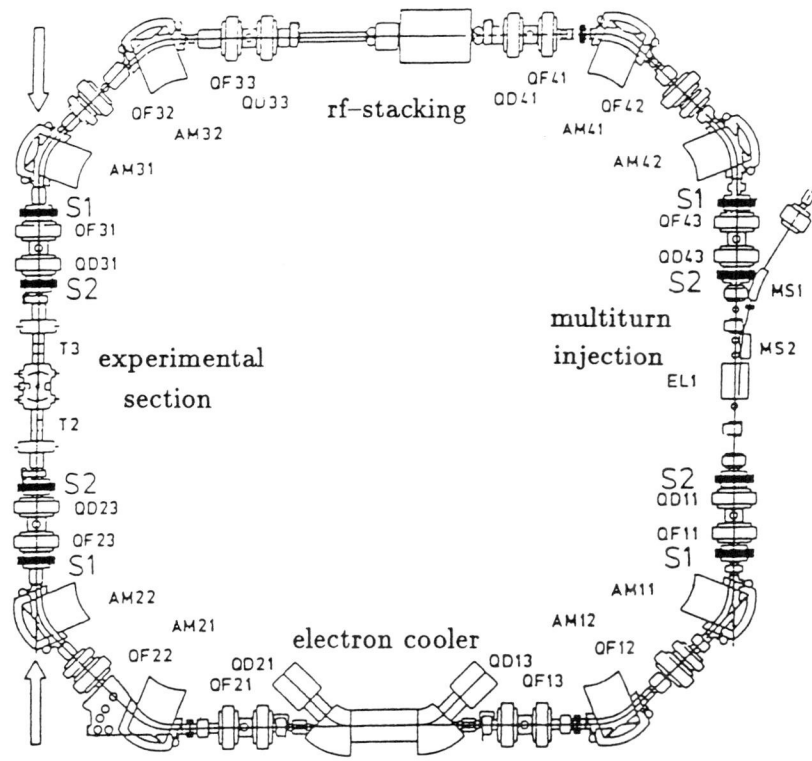

FIG. 9. Schematics of the laser cooling experiment at the TSR. Instead of using two counterpropagating lasers, a stable cooling velocity can also be achieved with one laser and a counteracting accelerating force.

the fixed frequency laser and observing the fluorescence signal obtained by tuning the dye laser from lower to higher frequencies. Compared with the fluorescence signal observed before the cooling process, the signal is strongly enhanced due to the reduced Doppler width and the hitherto better spectral overlap with the narrow laser bandwidth profile. Results are shown in Fig. 10.

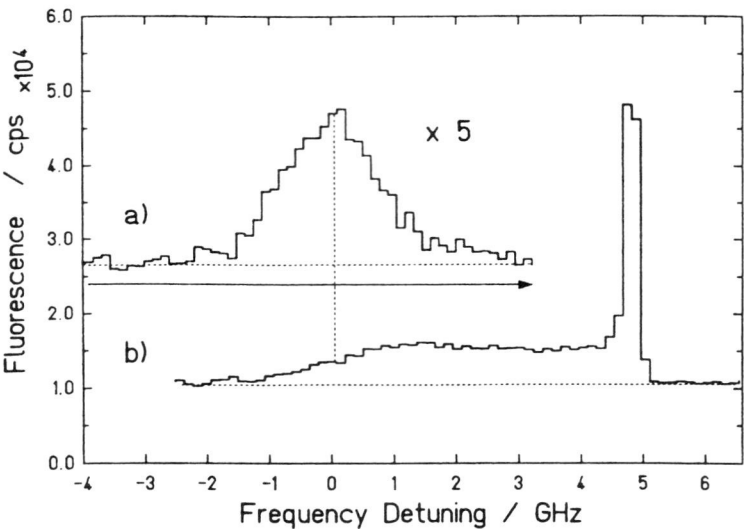

FIG. 10. Laser cooling of $^7Li^+$ ions at the TSR verified by the Doppler width of laser-induced fluorescence spectra: (a) before cooling, (b) after cooling. As a result of the cooling process, the ions are decelerated, leading to the displacement of the narrow signal line produced by the cooled ions. The broad background is due to ions lost out of the capture range of the cooling process.

2. Experiments on Li^+ Ions at the ASTRID Ring

The ASTRID storage ring in Aarhus pursues a double purpose. The ring is designed both for the generation of synchrotron radiation from stored electrons as well as for the storage of heavy ions. The heavy-ion injection system consists of an electrostatic accelerator based on the usual mass separator design and is capable of handling acceleration voltages up to 500 kV. This design makes the system suitable for the study of atomic physics and molecular physics at low charge states. Because the injection system produces a small energy spread, the design does not so far include an electron cooler; instead an intensive program of laser cooling was undertaken from the beginning.

The excellent beam quality in the ASTRID ring was expected to lead to substantially lower final temperatures of the laser cooled beams, in comparison to the higher energy rings. The experimental set-up for laser cooling is similar to the one at the TSR. At the lower beam velocity the Doppler-shift is not large enough to reach resonance with the Ar^+ laser. Therefore, two single-mode dye lasers are used. The final beam temperature after laser cooling was reported to be 1 mK, more than a factor of 10 better than in the early TSR experiments (Hangst et al., 1991). This low temperature is most probably a consequence of a very low density of metastable Li^+ ions in the ASTRID injection procedure.

C. ADVANCED COOLING SCHEMES

1. Experiments Using Be^+ and Mg^+ Ions

In order to increase the percentage of ions in the beam that are affected by the laser cooling, it is essential to use ground-state transitions. Suitable transitions in singly-charged ions are only found in the ultraviolet part of the spectrum. This usually necessitates the complication of frequency doubling. Due to the large Doppler shift, the TSR offers a relatively easy solution for laser cooling of Be^+ ions (Petrich et al., 1994). At rest the transition wavelength between the $^2S_{1/2}$ ground state and the $^2P_{3/2}$ level is 313 nm. At the TSR this transition can be Doppler-shifted to allow excitation using the 300.2 nm radiation of a single-mode Ar^+-ion laser. In comparison to the Li^+ case, the shorter transition lifetime of 8.7 ns leads to a larger laser cooling force. Due to the hyperfine splitting of 1.25 GHz in the ground-state, however, Be^+ does not have a two-level system suitable for laser cooling. To avoid optical pumping, two argon-ion lasers have to be used to induce simultaneously the transition from $F = 1$ to $F' = 2$, and $F = 2$ to $F' = 3$. The 1.25 GHz frequency difference of the two lasers, corresponding to the HFS splitting of the ground-state, has to be kept constant by active stabilization. Instead of a counterpropagating second laser beam, the ions in the ring are accelerated by means of an induction accelerator. In this device the accelerating force is caused by the electric field induced inside a magnet at a change of the magnetic field. The induction accelerator (INDAC) (Ellert et al., 1992) installed at the TSR supplies an accelerating force of 1 meV/m. At a constant rate of change of the magnetic field, the acceleration is constant and does not degrade the beam quality. The principle of laser cooling using this device is similar to the laser scheme. The ion velocity is chosen to produce a Doppler-shifted transition wavelength slightly off from the laser wavelength. Then the induction accelerator is used to bring the ions into resonance. The ions are kept at constant velocity by the laser cooling force, counteracting the acceleration by the INDAC. Changing of the acceleration force of the INDAC allows a measurement of the achievable cooling force (Petrich et al., 1994). To avoid a reduction of the effective cooling force by spatial fluctuations, the laser beams were actively stabilized. An important difference compared to the early experiments with Li^+ lies in the fact that a short cooling phase with the electron cooler was introduced before the laser cooling. In these experiments temperatures of less than 5 mK were obtained at about 10^6 particles in the ring.

In experiments with different INDAC settings the maximum achieved cooling force was determined to be 7.9 eV/m, about an order of magnitude below the theoretical value. Both this reduced cooling force and the final temperature limit, each with a strong dependence on the particle number, can be attributed to the influence of the heat transfer between longitudinal and transverse motion caused by intrabeam scattering. A realistic model shows that this is the result from binary collisions between the ions. In a number of these collisions the change in velocity is sufficient to detune the ions from resonance via the Doppler effect, and thereby reduces the cooling efficiency. At the ASTRID storage ring, similar experiments have been performed on Mg^+ ions (Hangst et al., 1991) using an intense frequency-doubled dye laser. The atomic level scheme of Mg^+ provides a two-level system starting from the ground state, ideal for laser cooling. Laser cooling of Mg^+ ions at beam velocities close to 10% of the speed of light, requiring a higher bending field, are planned for the near future at the ESR. There the objective is to demonstrate the preparation of ultradense energetic beams (Habs and Grimm, 1995).

2. Cooling of the Transverse Velocity Component

A further improvement of laser cooling will necessarily require cooling of the transverse velocity component. Only if both the transverse and the longitudinal motion are cooled, can a suppression of the effects of intrabeam scattering be reached. Several cooling schemes have been discussed. Presently it is unclear how direct two-dimensional cooling can be achieved in a fast moving beam. One possibility may be to use the magneto-optic force (Grimm et al., 1990). As a side-effect of the intrabeam scattering, however, a pure longitudinal cooling force should finally result also in reduction of the transverse velocity. This was established by recent experiments at the TSR, which showed a reduction of the beam size after laser cooling (Miesner, 1996).

As shown in Fig. 11, laser cooling restitutes the transverse beam temperature to values better than achieved by electron-cooling within a few seconds after the laser is switched on.

3. Cooling with Improved Capture Range

In order to employ a transfer of cooling between the longitudinal and transverse direction, an extended cooling time has to be achieved. For this the loss of ions out of the capture range of laser cooling has to be prevented. Two schemes of improved capture range have been tested so far. In one case, the Doppler effect is used to cover a wider velocity range (Wanner et al., 1994) by the introduction of an electrostatic potential along the cooling section. It is possible to design the field increase to induce a rapid adiabatic excitation. This is a well-known technique in magnetic resonance spectroscopy. Usually, the applied frequency is changed rapidly. In order to use the Doppler shift for this process, the ion velocity has to be changed rapidly over a short distance. In this way a condition similar to a μ-pulse excitation can be realized for ions starting with different velocities. First results prove the principle of this method.

A more straightforward solution is to apply laser light with an artificially broadened emission spectrum (Calabrese et al., 1995). Successful experiments recently

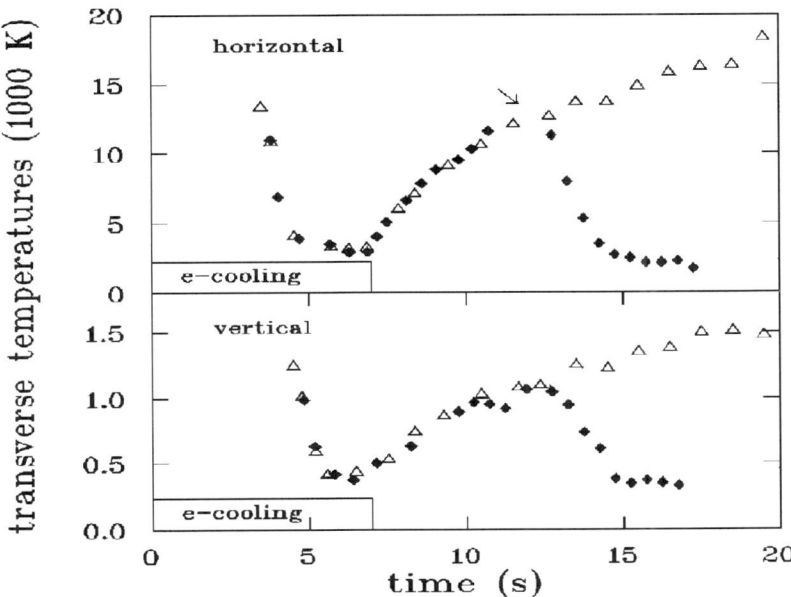

FIG. 11. Transverse cooling of a beam of Be^+ ions after longitudinal laser cooling in the TSR. The transverse temperature is directly monitored via observation of the beam diameter. After switching off the initially applied electron cooling, the beam size increases. Laser cooling, started as indicated by the arrow, is demonstrated to reduce the size of the beam effectively.

demonstrated the applicability of this method (Engel *et al.*, 1997, Calabrese *et al.*, 1997).

4. Cooling of Bunched Beams

Significant experimental work has been reported on laser cooling of bunched beams (Hangst *et al.*, 1995; Miesner *et al.*, 1996). Although the additional confinement in the longitudinal direction makes this approach less attractive for fundamental crystallization studies, it is of great interest for the preparation of intense cold beams for other experiments. For instance, laser cooling of bunched beams might be an important technique for achieving inertial fusion with energetic heavy-ion beams.

In a storage ring bunching of the beam can be provided by a single radio-frequency cavity. The frequency applied to the cavity is chosen to match the mean revolution frequency of the ions in the ring. Ions out of phase with this frequency are either accelerated or decelerated. This accelerating potential provides a quasi-harmonic restoring force toward the center of the bunch, similar to the situation

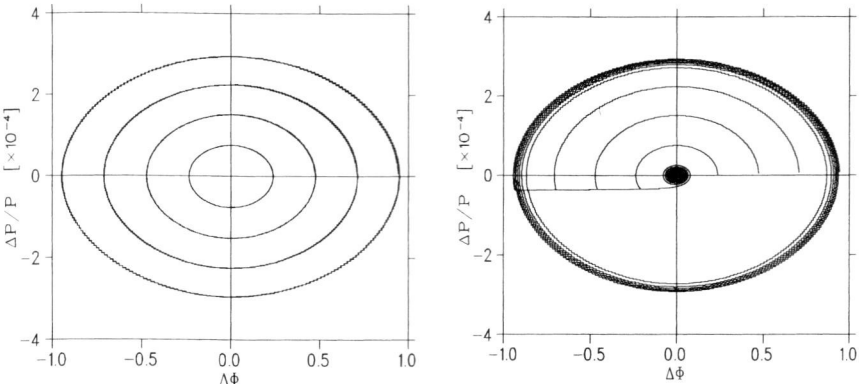

FIG. 12. Schematic representation of the oscillatory motion of ions in a bunch circulating in a storage ring. Calculated phase space trajectories of the ions in the RF buckets with (right side) and without (left side) laser cooling. The vertical scale gives the relative velocity, the horizontal scale the relative spatial phase. Trajectories of ions with four different starting conditions, beginning in the right upper quadrant, are shown. The laser tuning is chosen for excitation at slightly negative relative velocity. The spontaneous light force of the laser leads to damping of the oscillatory ion movement depending on the amplitude of the ion motion.

of a Paul-trap circulating around the ring. The ions within the bunch oscillate in velocity and in phase relative to the cavity potential as depicted in Fig. 12. As in a normal ion-trap experiment, these oscillations can be damped by laser cooling: Absorption of photons from a collinear laser beam can be used to selectively accelerate ions that deviate from the mean velocity of the bunch. Tuning the laser from the wing of the Doppler profile toward the center—or application of a corresponding shift of the cavity frequency—will eventually damp the relative ion movement. Because loss of particles is prevented by the radio-frequency potential, and ions getting out of the capture range of the laser cooling force due to collisions can be recollected by subsequent cooling cycles, an efficient compression of the bunch is possible with just a single laser for cooling. The first experiment using laser cooling of a bunched beam was achieved at the ASTRID storage ring (Hangst *et al.*, 1996). Figure 13 shows the effect of laser cooling to bunched Li^+ ions at the TSR. In the 3-dimensional spectrum the fluorescence intensity is plotted with regard to the position within the bunch, and, in foreground–background direction, the relative velocity of the excited ions. This spectrum was recorded by time-resolved fluorescence spectroscopy of the cooled ions (Grieser *et al.*, 1996).

The bucket length in this experiment was 5.5 m, that is, a relative distance of 0.86 m corresponds to a relative spatial phase $\Delta\Phi = 1$. In accordance with the

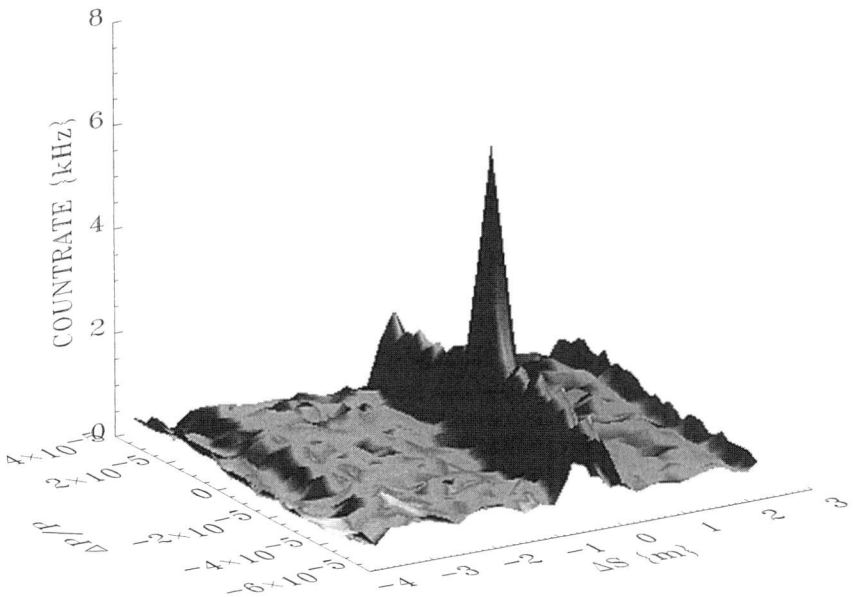

13. Narrowing of the bunch length of Li^+ ions in the TSR after laser cooling. The spatial and momentum distribution of the ions in the bunch is directly monitored via the time resolve laser-induced fluorescence.

simulation in Fig. 12, laser cooling leads to a compression of the ions in the center of the bunch (0 m). Independently from the detuning of the probe laser, the ions can be localized by their fluorescence signal, because they are excited by the Ar^+ laser used for the laser cooling. The measurement exhibits a bunch length $l = 0.53(7)$ m (FWHM). The additional excitation by the scanning dye laser probes the velocity distribution of the ions. A steep and narrow velocity distribution is found for the laser-cooled ions in the bunch center.

More recent experiments at TSR demonstrate the potential of this method to reach an ordering parameter Γ_{ord} larger than 1 for about 10^6 particles in a bunch of less than 50 centimeter length (Grimm et al., 1995). Temperatures corresponding to a transition into a highly viscous state seem to be in reach soon. It is particularly hopeful that there is evidence that intrabeam heating mechanisms will be strongly reduced as soon as ordering parameters of 2–5 are achieved. Other experiments used the advantages of the bunched cooling to selectively prepare a beam of Li^+ ions in the metastable state (Grieser et al., 1996). In principle this cooling scheme should also apply to very intense beams, and a proposal has been presented to use the ESR to study this possibility to intensities up to 10^{11} Mg^+ ions as a test case for cooling energetic Hg^+ ions.

VI. A Test of Special Relativity in the Storage Ring

A. Test Schemes for the Theory of Special Relativity

The mathematical framework of special relativity developed in response to the Michelson-Morley experiment (Michelson and Morley, 1887) around the turn of our century has itself become the subject of improved experiments. The most precise modern version of the Michelson-Morley experiment was published in 1979 (Brillet and Hall, 1979), providing an upper limit of $5 \cdot 10^{-9}$ for any anisotropy of the speed of light. A second, important test for the validity of the theory of special relativity is the Kennedy and Thorndike experiment (Kennedy and Thorndike, 1932). This experiment is testing the velocity dependence of an interferometrically defined length in an unsymmetric interferometer. The most recent value for this experiment (Hils and Hall, 1990) states an upper limit of $7 \cdot 10^{-5}$.

An important consequence of the theory of special relativity is the dilatation of time in a moving system. In a spectroscopic experiment, the quantity can be tested by a precise measurement of the transverse Doppler effect. A first experimental proof had been achieved by Ives and Stilwell (Ives and Stilwell, 1938) in experiments on a fast-moving hydrogen beam. In the special situation at the storage ring one can do this experiment at relativistic velocities. This provides a large increase in the measured effect in a much cleaner experimental situation. For comparison, it should be kept in mind that Earth cannot be regarded as a reference system at rest. The anisotropy of background radiation from outer space is usually explained by assuming a velocity of 350 km/s relative to some interstellar frame of reference.

B. A Modern Ives and Stilwell Experiment at the TSR

1. Analysis of the Transverse Doppler-Effect

A storage ring makes it possible to perform precision laser spectroscopy on a wide variety of atomic species that can otherwise not be obtained. In particular, storage rings can provide ions moving at relativistic speed in a beam of superb quality. At the TSR storage ring in Heidelberg this has been used for an improved test of SRT by performing laser spectroscopy of a precisely known transition. The ions are treated as moving clocks with optical frequencies v_0, traveling in a well-controlled trajectory. Within the framework of special relativity, using the Lorentz transformation as given in Section III.10 for two reference frames moving at a relative speed $\vec{w}$, the resonance frequencies observed in parallel and antiparallel geometry are

$$v_p = v_0\left(\gamma\left(1 - \left|\frac{\vec{w}}{c}\right|\right)\right)^{-1} \quad \text{and} \quad v_a = v_0\left(\gamma\left(1 + \left|\frac{\vec{w}}{c}\right|\right)\right)^{-1} \quad (21)$$

where $\gamma = \gamma_{SRT} = (1 - (\vec{w}/c)^2)^{-1}$ is valid. To account for possible deviations from SRT, the following formulation can be found in the literature (Will, 1992):

$$T = a^{-1}(t - \epsilon \cdot x) \quad (22)$$

for the time coordinate, and

$$X = d^{-1}x - \frac{d^{-1}}{b^{-1}}\frac{\left(\frac{\vec{w}}{c} \cdot x\right)\frac{\vec{w}}{c}}{\frac{\vec{w}^2}{c}} + \frac{\vec{w}}{c} \cdot T \quad (23)$$

for the length coordinates. The Lorentz transformation is a special case of this procedure where

$$a^{-1} = b = \sqrt{1 - \frac{\vec{w}^2}{c}}, \quad d = 1, \quad \text{and} \quad \epsilon = \frac{\vec{w}}{c} \quad (23)$$

Using an expansion of the parameter a as

$$a(\vec{w}) = 1 + \left(\alpha - \frac{1}{2}\right)\left(\frac{\vec{w}}{c}\right)^2 + \left(\alpha_2 - \frac{1}{8}\right)\left(\frac{\vec{w}}{c}\right)^4 + \cdots \quad (24)$$

the relation holds:

$$\gamma = \gamma_{SRT}\left(1 + \alpha\left(\frac{\vec{w}}{c}\right)^2 + \alpha_2\left(\frac{\vec{w}}{c}\right)^4 + \cdots\right) \quad (25)$$

Besides these absolute deviations, an Ives and Stilwell experiment would also be sensitive to an additional anisotropy

$$A = \alpha\frac{\vec{w}}{c}\cos(\Omega) \quad (26)$$

where Ω denotes the angle of the reference frame of the experiment and a movement relative to some hypothetical "universal" frame of reference. A candidate for a universal frame of reference might be deduced from the anisotropy of the 3K background radiation. For this case an experiment fixed to Earth would be exposed to modulations of such a deviation by the sidereal motion. In recent years a number of high-resolution laser spectroscopy versions of an Ives and Stilwell experiment have been reported and analyzed that set limits on α. Three of them

(Kaivaola et al., 1985; Riis et al., 1985; McGowan et al., 1993) have been performed as two-photon spectroscopy on a neon atomic beam with $\beta = 0.0036$ and report as their best result an absolute limit $\alpha \leq 1.6 \cdot 10^{-6}$ (McGowan et al., 1993) and a limit on sidereal influence (according to Eq. 26) of $\alpha \leq 1.4 \cdot 10^{-6}$ (Riis et al., 1985). Because the anisotropy of the cosmic background radiation implies a velocity of the earth's frame of reference of 350 km/s, it is desirable to perform these experiments at higher velocities as at LAMPF (McArthur et al., 1986), where an experiment has been performed on an accelerated beam of H^- at $\beta = 0.84$. This work achieved a spectral resolution of $2.7 \cdot 10^{-4}$ and obtained an upper limit of $1.3 \cdot 10^{-4}$ for α_2 due to the large Doppler shift.

2. Experimental Setup at the TSR

The experiments at the TSR storage ring were performed on singly charged $^7Li^+$ ions accelerated to 13.3 MeV, corresponding to a velocity of $\beta = 0.064$. The experimental scheme, described in detail in (Grieser et al., 1994a), was mostly identical to the one used for laser cooling shown in Fig. 9 For the experiments, 10^6 $^7Li^+$ ions were stored in the ring. The triplet system of helium-like $^7Li^+$ exhibits a well resolved and precisely known fine and hyperfine structure multiplet. For the stringent test of relativity, high-resolution saturation spectroscopy was performed in the Λ-system formed by the transitions $^3S_1\{F = 3/2\} => ^3P_2\{F' = 5/2\}$ and $^3S_1\{F = 5/2\} => ^3P_2\{F' = 5/2\}$.

The experiment used a single-mode Ar^+ laser stabilized on a $^{127}I_2$ saturation resonance—a recommended frequency standard (CIPM, 1984)—and a tunable dye laser modified for better frequency stability. The frequency of the scanning laser was determined by comparison with simultaneously recorded $^{127}I_2$ resonances of the $R(99)$ 15–1 transition (Grieser et al., 1994b). In order to assure a well-defined collinearity of the two laser beams, they were delivered through the same polarization preserving single-mode fiber. This bichromatic beam, entering collinear with the ion beam, was focused on a plane mirror behind the experimental section to produce the counterpropagating geometry. An angular accuracy better than 40 μrad was achieved by adjusting the retro-reflected laser beam for maximum transmission back into the fiber. The optical resonances were detected in fluorescence by two photomultipliers equipped with filters to suppress stray light. The alignment of the laser beams and the ion beam was controlled relative to four capacitive position pick-ups mounted within the experimental section of the TSR. Before being subjected to laser spectroscopy, the injected ion beam was first electron-cooled for 7 seconds. During this time the laser beams were blocked mechanically in order to avoid optical pumping while the longitudinal velocity changed during the electron cooling process. The spectrum displayed in trace 1 of Fig. 14 is the Doppler-broadened fluorescence of the $F = 5/2 => F' = 7/2$ transition, and is produced by sending only the beam from the dye laser to the experiment. The HWHM corresponds to a velocity spread of $\delta v/v = 3 \cdot 10^{-5}$. The satu-

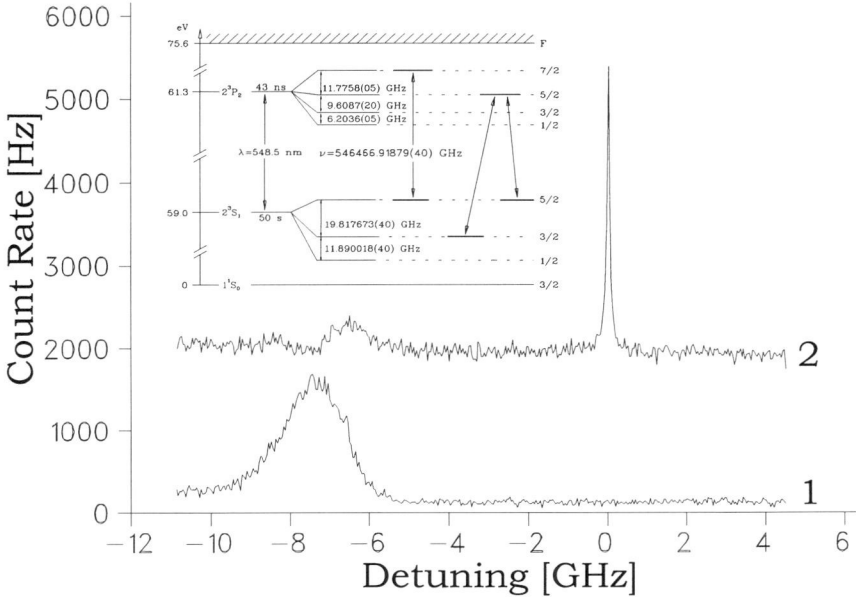

FIG. 14. Doppler-free spectroscopy of $^7Li^+$ at the TSR. The essential part of the level scheme is shown on top. The Doppler-broadened signal from the $F = 5/2 \Rightarrow F' = 7/2$ transition excited by the dye-laser shows a dip caused by the simultaneous excitation by the fixed-frequency Ar^+ laser. The narrow Λ-resonance occurs, when the dye-laser is tuned to the $F = 3/2 \Rightarrow F' = 5/2$ transition.

ration spectroscopy signal is obtained by keeping the Ar^+ laser at resonance with the $F = 5/2 \Rightarrow F' = 5/2$ transition. Depletion of the $F = 5/2$ level by optical pumping is evident when the population is probed by the dye-laser scan. Excitation of the $F = 3/2 \Rightarrow F' = 5/2$ transition yields a strong fluorescence, as the two lasers now simultaneously excite the Λ-system.

3. Results

The measured frequency is finally quoted as $v_{exp} = 512667592.4(3.1)$ MHz. The effect of AC Stark shift of the Λ-resonance was investigated by measurements at different ion velocities where the fixed frequency Ar^+ laser is not resonant at the center of the Doppler profile of the velocity distribution. An influence smaller than 1 MHz was deduced. The shift due to photon recoil can be experimentally established by changing the role of the two lasers. A maximal shift of 0.7 MHz was observed. The line width of the resonance of 60 MHz is twice the expected line width. This line width is probably caused by irregularities in the phase front of the laser beams. Given this fact, it leads to a systematic shift of 2.7 MHz, and this is the largest contribution to the experimental uncertainties, including the statistical quality of the data.

The prediction of special relativity can be calculated by taking the product of the frequencies of the $F = 3/2 => F' = 5/2$ and the $F = 5/2 => F' = 5/2$ transition in $^7Li^+$ at rest (Kowalski et al., 1983) divided by the Doppler-shifted value given by the frequency of the Ar^+ laser. This leads to v_{SRT} = 512667588.2(7) MHz. The frequency difference between these values of 4.2(3.2) MHz is—requiring 2σ confidence—still consistent with zero and confirms the validity of special relativity. This corresponds to an upper limit for the deviation from the expectation of special relativity of $\delta\alpha \leq 8 \cdot 10^{-7}$ (Grieser et al., 1994a).

The TSR result represents a considerable improvement in comparison to experiments at lower velocity. A further increase in sensitivity to deviations from the predictions of special relativity is desirable and possible with the advantages of storage-ring laser spectroscopy. Considering a precise velocity control by laser cooling of the $^7Li^+$ beam and an increased quality of the laser wave-fronts, the final accuracy should be determined by the angular setting of laser and ion beam, which has been quoted as smaller than 0.29 MHz. Recent experiments show that precision spectroscopy after laser cooling is also possible in a bunched beam. This provides the fascinating possibility of working with a very small number of ions because the time window, where the ions pass the detector, can be exactly defined by the bunch structure, allowing for a very sensitive lock-in detection scheme. Moreover, laser manipulation can even remove the unwanted ground-state ions (Grieser et al., 1996). Because only a small number of metastable Li^+ ions would be required, this opens a way to improve the measurement further in an experiment with faster ions at the ESR. Such an experiment might be performed at 1/3 of the speed of light, where the co- and counterpropagating lasers would have a frequency ratio of two. Higher velocity would make the experiment a particularly sensitive test of any possible contribution from α_2.

VII. Quantum Electrodynamics in Strong Fields Probed by Laser Spectroscopy in Highly Charged Ions

A. THE GROUND-STATE HYPERFINE SPLITTING OF HYDROGEN-LIKE IONS

In neutral and low-charge atoms, the predictions of QED are in perfect agreement with experiment. However, the mathematical formulation used for these calculations is a perturbative technique using an expansion in $Z\alpha$. Radiative QED effects in hydrogenic ions, on the other hand, increase by a scaling factor of Z^4, and the size of these higher-order contributions makes these systems exceptionally interesting for testing QED. This opportunity has given rise to renewed experimental effort to investigate the Lamb shift in heavy ions (Mokler et al., 1997). Completely new and complementary is the possibility to test QED in the combination of strong electric and magnetic fields by precision laser spectroscopy of the ground-state hyperfine splitting in high-Z hydrogenic systems. In this case, instead of the propagator of the free electron, the modified propagator in the field of the nucleus has to be introduced in the calculation of the magnetic interaction. In this situation, both the $Z\alpha$ perturbative treatment and the usual renormalization procedure break down.

Experimental results on the hyperfine structure of highly charged systems, therefore, represent a very direct way to check the predictions of QED in a completely new regime. The magnetic interaction of the $1s$ electron in the nucleus causes a hyperfine splitting in the ground state of hydrogen and hydrogen-like ions of nuclei possessing a nuclear spin I of at least one-half. The 1.4 GHz splitting frequency in hydrogen, corresponding to a transition wavelength of 21 cm, has been extensively studied by precision microwave spectroscopy, and in hydrogen masers the splitting frequency has been determined to an accuracy of $7 \cdot 10^{-13}$ (Essen et al., 1971). A comparison of this splitting energy to the energy difference in more highly charged systems could provide important experimental and theoretical improvements for our understanding of effects due to nuclear structure and quantum electrodynamics (QED).

Because only a single electron is involved, the calculation of the hyperfine splitting from the Dirac equation is straightforward, even including the effect of the extended nucleus (Finkbeiner et al., 1993; Schneider et al., 1993). The ground-state hyperfine splitting of hydrogen-like ions is conveniently written as a product of the nonrelativistic solution multiplied by correction factors

$$\Delta E(\mu) = \frac{4}{3}\alpha(\alpha Z)^3 \frac{\mu}{\mu_N} \frac{m}{m_p} \frac{2I+1}{2I} mc^2 \cdot \{A(\alpha Z)(1-\delta)(1-\epsilon) + x_{rad}\} \quad (27)$$

Here α is the fine structure constant; Z is the nuclear charge; m is the electron mass and m_p the proton mass; μ is the nuclear magnetic moment; μ_N is the nuclear magneton; and I is the nuclear spin. The relativistic correction $A(\alpha Z)$ is obtained from exact solution of the Dirac equation with a Coulomb potential. The factor $(1-\delta)$ corrects for the finite spatial distribution of the nuclear charge (Breit-Schawlow correction); $(1-\epsilon)$ corrects for the finite spatial distribution of the nuclear magnetization (the Bohr-Weisskopf effect) and x_{rad} is the QED correction. The nuclear charge radii of most stable isotopes are known well enough to yield a precision of better than 10^{-4} in the size of the hyperfine structure contribution. The authors of (Schneider et al., 1993) could also verify that different parameterizations of the nuclear charge distribution produce even smaller changes in the calculated splitting. The expectation of QED effects of the order of a percent for highly charged ions, combined with high-resolution laser spectroscopy, implies the possibility of studies with a precision comparable to Lamb shift measurements.

Until recently, however, it has not been possible to extend measurements of the $1s$-splitting beyond $^3He^+$ (Schuessler et al., 1969). There are two reasons for this.

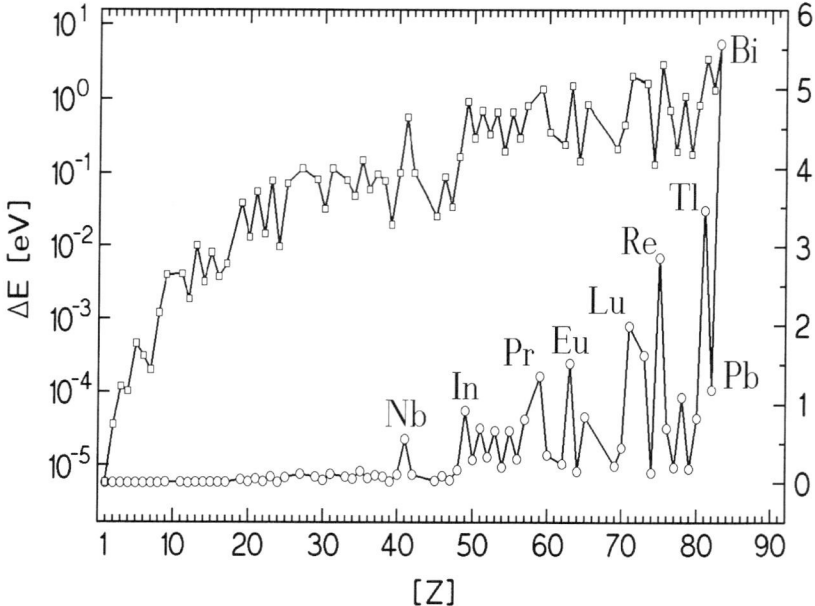

FIG. 15. Transition wavelengths for the ground-state HFS transition of selected heavy hydrogen-like ions. Values are displayed in linear (round symbols, right axis) and logarithmic (rectangular symbols, left axis) scale.

First, the splitting energies increase proportional to Z^3 and, therefore, transitions in higher charged ions are inaccessible to microwave spectroscopy. Second, for more highly charged ions, where the transition energy approaches the optical part of the electromagnetic spectrum, it has not been possible until now to produce and maintain samples of highly charged hydrogen-like ions. For the highest Z atoms, the increase of the splitting with Z^3 shifts the HFS transition into the optical region of the spectrum and makes it accessible to precision laser spectroscopy (Kühl, 1986). Transition energies for highly charged ions are shown in Fig. 15. An exceptional candidate is hydrogen-like bismuth, the heaviest stable nuclide existing in nature: The combination of the high nuclear charge $Z = 83$, a magnetic moment of 4.1106(2) nm, and a nuclear spin of $I = 9/2$ results in a transition wavelength shorter than 250 nm. It is also experimentally important that the transition lifetime is also shortened according to the E^3-dependence of the transition probability. For Bi^{83+} the calculation (Finkbeiner et al., 1993) predicts 0.41 ms, compared to 10^7 years (!) for the same transition in hydrogen.

At the SIS–ESR complex it is possible to produce, store, and cool more than 10^8 highly charged ions of any element in the ring. These may have any charge state up to and including bare uranium. In this situation it is possible for the first time to perform precision laser spectroscopy on high-Z few-electron atoms.

B. First Measurement of a HFS Splitting at Optical Frequencies in the ESR

1. Experimental Technique

The ground-state HFS splitting of hydrogen-like $^{209}Bi^{82+}$ was measured at the ESR heavy-ion cooler-storage ring at GSI (Klaft *et al.*, 1994). Highly charged ions were produced with an efficiency of about 30% by in-flight stripping of lower-charged bismuth ions accelerated to 220 MeV/nucleon in the heavy-ion synchrotron SIS. Hydrogen-like ions were then selected from the resulting charge state distribution and injected into the ESR storage ring. The high initial velocity spread of the ions was significantly reduced by electron cooling (Franzke *et al.*, 1993). Because the beam lifetime in the storage ring is longer than one hour, it was possible to increase the beam intensity in the storage ring by accumulation of a number of cycles of the SIS. Up to $1.8 \cdot 10^8$ ions of $^{209}Bi^{82+}$ were stored at a velocity $\beta = 0.58666(11)$ c.

The experimental arrangement is shown in Fig. 16. The ion beam was overlapped in a collinear, antiparallel ($\theta = 180°$) geometry with 10 mJ laser pulses of 30 ns duration from a dye laser pumped by an excimer laser. The alignment and overlap of the ion beam (8-mm diameter) and the laser beam (10-mm diameter) in a field-free straight of the storage ring was controlled to an accuracy of 0.33 mrad by pairs of mechanical scrapers separated by 6 meters. The Doppler effect shifted the wavelength for laser excitation in the given geometry from $\lambda_0 = 250$ nm in the rest frame of the ions to $\lambda = 489$ nm in the laboratory frame at the given β. This large shift makes it extremely important to determine the ion velocity accurately. The velocity was deduced from the acceleration voltage of the electron cooler, which can be determined with an absolute accuracy of 10 V, including the effects of space charge potentials. Due to the cooling process the ion velocity can be assumed to be identical to the velocity of the electrons. Two-step laser-induced recombination measurements in Ar^{18+} (see Section V) verified the accuracy of this determination (Borneis *et al.*, 1995). Schottky analysis of the signal induced in capacitative probes showed that electron cooling reduced the relative spread in velocity to $\delta\beta/\beta = 7 \cdot 10^{-5}$ at the given beam intensity. This corresponds to a Doppler width of 40 GHz in the laboratory frame. Due to its low transition probability, the M1-decay of the $F = 5$ sublevel takes place along the entire ESR orbit of 108.25 m. To access a maximum solid angle, a 60-cm long-light collection system, consisting of elliptical and cylindrical mirrors, was installed in the ultrahigh vacuum beam tube of the storage ring. This focuses $5 \cdot 10^{-4}$ of the total emission in the ring through sapphire windows onto three photomultipliers outside of the

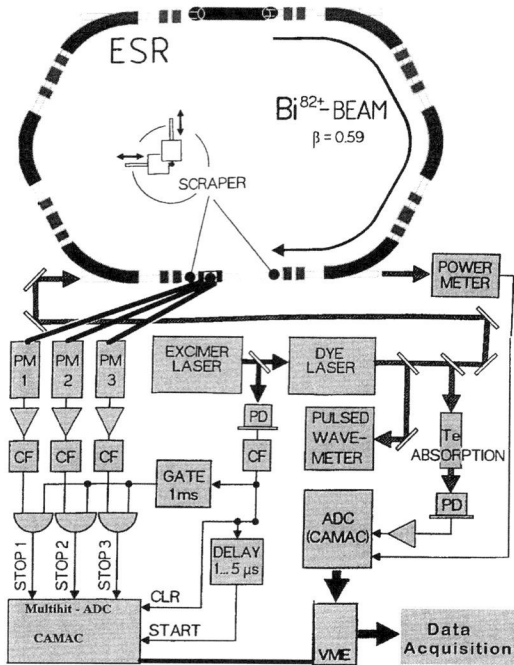

FIG. 16. Experimental arrangement: Laser pulses excite hydrogen-like Bi^{82+} ions, circulating in the ESR with an energy of 220 MeV/nucleon, from the $F = 4$ to the $F = 5$ hyperfine structure level in the ground state. A mirror system reflects fluorescence light into three photomultipliers. Time-resolved detection serves for the reduction of background and a measurement of the transition lifetime.

vacuum. The Doppler effect causes the spectral distribution of the emitted light to range from 128 nm to 490 nm. Photons were accepted between the transmission cut-off of the sapphire windows at about 200 nm and the edge of a color-glass filter at 450 nm. This filter is necessary to reject scattered laser light. The detected fluorescence exhibits a time structure because only a portion of the ions in the ring can be excited by each laser pulse. As a result, the light is modulated corresponding to the revolution period of 615 ns of the ions in the ring. This pattern is maintained throughout the decay time because of the low velocity spread of the stored ions and can, therefore, be used for the discrimination of background.

2. Result

The ground-state HFS resonance in $^{209}Bi^{82+}$ was found at a laser wavelength of $\lambda = 477.794(4)$ nm and is shown in Fig. 17a. The quoted error is 10% of the Doppler width, which was 40 GHz as expected. The laboratory value λ

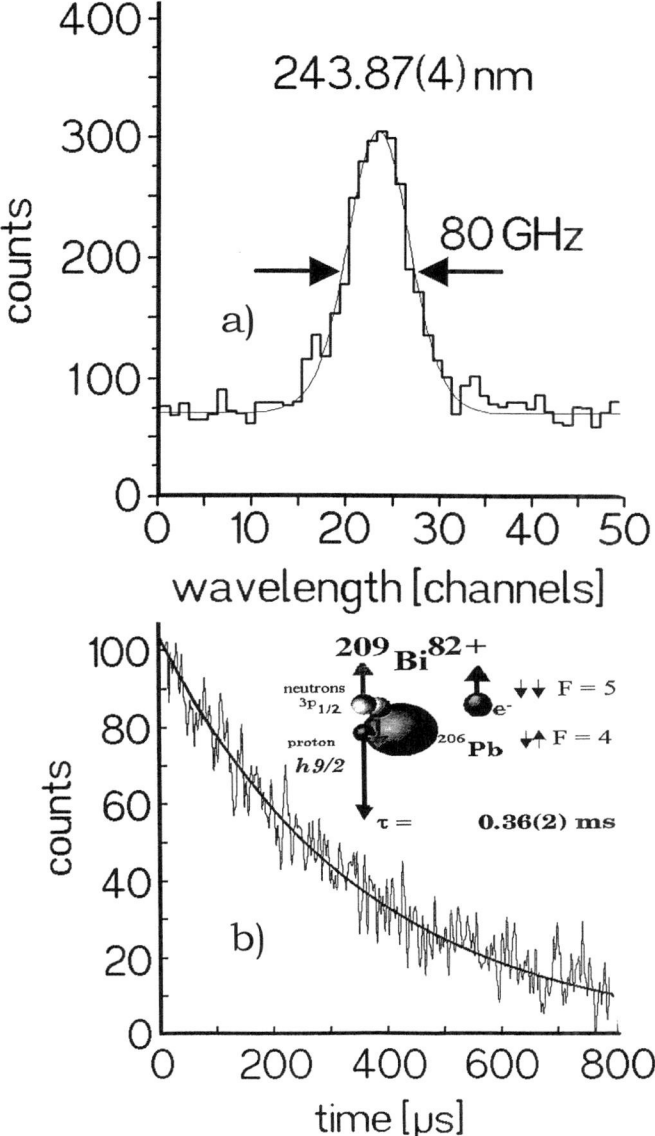

FIG. 17. (a) Resonance signal of the M1-transition in the ground state of $^{209}Bi^{82+}$. The Doppler width of the line is determined by the velocity spread of the stored ions. (b) Lifetime measurement of the $F = 5$ ground-state hyperfine level of $^{209}Bi^{82+}$. The fluorescence intensity is plotted as a function of the time delay with respect to the pulsed excitation. The insert sketches the main nuclear configuration contributing to the hyperfine splitting.

corresponds to $\lambda_0 = 243.87(4)$ nm in the rest frame of the ions. The error is dominated by the $\delta\beta/\beta \approx 10^{-4}$ uncertainty of the ion velocity in the ring entering directly through the Doppler shift (see Eq. 10).

By improved determination of the effective cooler potential, the error margin was later reduced by a factor of two (Borneis et al, 1994).

The effects of divergence and relative angle of the laser and ion beams are negligible compared to the uncertainty in velocity. The fluorescence decay curve acquired after excitation by the laser pulse at resonance is shown in Fig. 17b and implies a lifetime of $\tau_0 = 0.351(16)$ ms for an ion at rest.

C. Spectroscopy of the Infrared Ground-State Hyperfine Transition in $^{207}Pb^{81+}$

1. Motivation

The measurements of the ground-state hyperfine splitting of $^{209}Bi^{82+}$ at GSI, and more recent measurements in $^{165}Ho^{66+}$ $^{185,186}Re^{86+}$ (Crespo et al., 1995) at the Super EBIT electron beam ion trap at LLNL, stimulated a large number of theoretical calculations of the wavelengths of these transitions (Finkbeiner et al., 1993; Shabaev, 1994; Schneider et al., 1993–1995; Shabaev et al., 1995–1997; Labzowsky et al., 1995; Tomaselli et al., 1995; Bastug et al., 1996; Blundell et al., 1997). Satisfactory agreement between theory and experiment, however, was not reached. The calculations for bismuth yield a value 1 nm or $5 \cdot 10^{-3}$ larger than the measured value. On the basis of the precision assigned to the corrections this discrepancy is significant, but corrections for the nuclear effects vary considerably depending upon how much the nuclear core is assumed to be polarized.

In order to disentangle the QED corrections and the contributions due to the nucleus, a measurement of the ground-state hyperfine splitting in $^{207}Pb^{81+}$ was performed at the ESR. This candidate is exceptional as the nuclear structure is particularly simple. The lead nucleus is well described by the single-particle model, and the effects of core polarization are expected to be small. Its nuclear magnetic moment is given very exactly by that of the $p_{1/2}$ neutron hole in the doubly magic nucleus ^{208}Pb (82 protons and 126 neutrons). Furthermore, the nuclear radius of ^{207}Pb has been precisely measured by electron scattering and muonic x-ray studies, which permits accurate calculation of effects due to the finite size of the nucleus.

2. Experimental Procedure

The infrared wavelength of the ground-state hyperfine transition in ^{207}Pb transition creates difficulties that can be nicely handled in the storage ring. Although the wavelength in the rest frame of the ions is outside the range of efficient photo

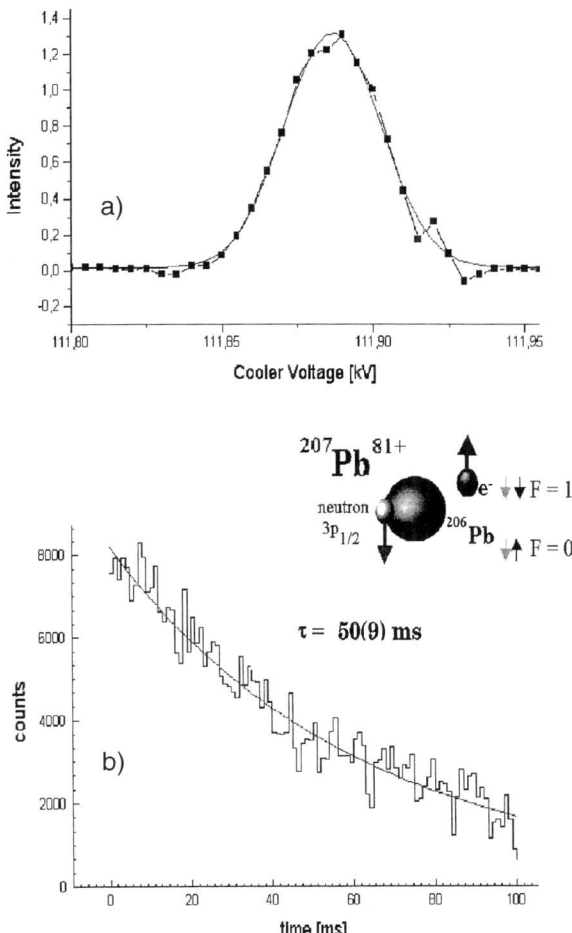

FIG. 18. Results of the measurement on $^{207}Pb^{81+}$: (a) Laser induced fluorescence spectrum. (b) Decay-time measurement.

cathode materials, standard near-IR photomultipliers sensitive up to about 800 nm can be used because the wavelength of the fluorescence light from ions circulating in the ESR is Doppler-shifted by nearly a factor of two by the ions' large velocity. The storage ring is also essential because of the long lifetime of the infrared transition of about 50 ms, which makes the fluorescence intensity rather low. With a storage ring, however, rather accurate measurements of the lifetime of the upper hyperfine level are possible as the beam has a storage time in the ring, limited mostly by electron capture in the electron cooler, of about 20 minutes. The long lifetime of the beam also means that de-excitation by collisions, which could reduce the detectable fluorescence, must be very low. Consequently a measurement of the transition lifetime will not suffer from additional loss processes. To improve the collection of fluorescence photons the array of mirrors was slightly tilted with respect to the beam direction to enhance the detection of light emitted into forward angles. The storage ring also allowed to use a fixed frequency Nd : YAG laser, doubled to 532.222(5) nm (pulse energy 250 mJ), to search for the resonance. Tuning was done by changing the beam velocity to vary the Doppler shift (Fig. 18).

A significant advance in technique, indispensable for the spectroscopy of the extremely long-lived infrared transition, was achieved by making use of bunching the circulating ions. Applying a radio-frequency acceleration voltage with an amplitude of 20 V compressed the circulating ions into two bunches. Each bunch, taking 60 nano-seconds to pass an observation point, was about 11 meters long, about 10% of the ring circumference. By using the bunched beam, both the efficiency with which ions are excited by the laser pulses and the background rejection were improved. In a coasting beam only a small fraction of ions can be excited by the pulsed laser. These excited ions would redistribute over the full circumference after about 1 ms, thus in addition preventing a selective detection. As a result the new method achieves an order of magnitude improvement of the signal-to-background ratio. The higher rate of intrabeam collisions resulting from the higher density of bunched ions increases their velocity spread only slightly compared to that of the coasting beam. In this experiment the fractional spread ranged from $\delta\beta/\beta = 1 \cdot 10^{-4}$ to $\delta\beta/\beta = 3 \cdot 10^{-4}$, depending on the number of ions in the ring.

D. QED AND NUCLEAR CONTRIBUTIONS TO THE HYPERFINE SPLITTING

To separate out the QED effects in Eq. 27, one must evaluate the amount of the HFS splitting without QED. This requires a combination of the prediction of the Dirac-Fock calculations (Finkbeiner *et al.*, 1993; Schneider *et al.*, 1993) with information on the nuclear magnetization distribution, which leads to the $(1 - \epsilon)$ Bohr-Weisskopf correction. For lead and bismuth the size of this correction amounts to 2% of the hyperfine splitting. Accurate assessment of the Bohr-Weisskopf contribution is closely related to the problem of nuclear magnetism itself. Note that for ^{209}Bi the shell-model prediction for the magnetic moment differs from the experimental value by nearly a factor of two. Recently, reasonable success has been obtained to reproduce the magnetic moments of heavy nuclides, including bismuth, very well (Tomaselli *et al.*, 1994). Because the charge distribution obtained in these calculations is also in agreement with the measured charge radius, it seems reasonable to use this model to determine the magnetic moment distribution in the nucleus.

Use of the magnetic moment distribution given by this procedure in combination with the Dirac-Fock calculations results in a theoretical transition wavelength

TABLE 2
CONTRIBUTIONS TO THE HYPERFINE SPLITTING OF THE GROUND STATE
OF HYDROGEN-LIKE LEAD AND BISMUTH.

	$^{209}Bi^{82+}$	$^{207}Pb^{81+}$
RMS radius	5.519 fm	5.497 fm
Magn. moment (corr.)	4.1106	0.58219
Point nucleus (Dirac)	212.320 nm	880.017 nm
Incl. Breit-Schawlow	238.791(50) nm	989.66(1) nm
Incl. Bohr-Weisskopf	243.91(38) nm	1019.1(1.6) nm
Vacuum Polarization	−1.64 nm	−6.83 nm
Self Energy	+2.86 nm	+11.9 nm
Total QED	+1.22 nm	+5.08 nm
Theory incl. QED	245.13(58) nm	1024.2(2.0) nm
Experimental Result	243.87(2) nm	1019.5(2) nm

of $\lambda = 243.91(38)$ nm for bismuth and 1019.1(1.6 nm) for lead, which leaves little room for any contribution from QED. The quoted uncertainty of the theoretical value in Table 2 mainly reflects the difficulty of verifying the nuclear model: Comparison with the experimental result from muonic bismuth (Rüetschi et al., 1984) is limited by the experimental accuracy of 10% reached for the hyperfine splitting in the muonic bismuth. Calculations of the vacuum polarization and self-energy contributions were recently accomplished (Persson et al., 1996; Blundell et al., 1996), predicting a shift of +1.22 nm in bismuth, and accordingly 5.08 nm in lead. The accuracy of these calculations is limited mainly by the size estimate of untreated higher order corrections, and the different approaches agree very well.

At present, an obvious discrepancy remains between the theoretical predictions and the experimental results. Because the new laser approach can be applied to a family of test cases, it can be used to provide a wider experimental basis for the separation of nuclear and QED effects, which is necessary to fully exploit the new experimental possibility for a test of QED calculations. The signal-to-noise ratio obtained for the resonance line clearly demonstrates the potential for extending this technique to other, experimentally more demanding candidates. A possibility of eliminating the nuclear parameters for the test of QED will be given by a measurement of the 2s-hyperfine splitting in lithium-like $^{209}Bi^{80+}$ (Schneider et al., 1995; Shabaev et al., 1996). The wavelength of this transition is close to 1500 nm (Beiersdorfer et al., 1998), still accessible in the storage ring using the bunched beam method.

VIII. Conclusion and Outlook

In the short time period since the inauguration of the first heavy-ion storage ring at the TSR in Heidelberg, a large number of laser spectroscopy experiments have been performed at such facilities. The experiments reviewed in this article have demonstrated that the combination of storage ring and laser technology allows for a complete new class of data experiments. At the present stage it does not seem overoptimistic to foresee a further increase in the number and quality of experiments. Each group of research discussed has the potential to develop into a field of its own.

Laser-induced recombination has already become a tool for precision spectroscopy of highly excited Rydberg states in highly charged ions and for the study of plasma physics in the electron–ion interaction in the cooler.

Laser cooling of stored ions will certainly be pursued not only as a subject in itself but also as a way to provide the ultimate beam quality for precision experiments, and possibly also dense and brilliant beams for inertial fusion concepts. Laser cooling can provide extremely cold beams moving at relativistic velocities. The final temperatures reached in laser cooling experiments already correspond to the regime in which collective movement is dominant in the longitudinal direction of the beam. First results in cooling of the transverse direction have been reported. Cooling of bunched beams will allow detailed studies of the intrabeam heating process in the additional confinement given by the bunching field. Tests of relativity at higher b will be able to achieve a larger sensitivity to higher order deviations from the theory of special relativity. The existence of minute anisotropies due to the stellar constellations can be examined.

The test of QED in the presence of the high magnetic hyperfine field has reached a point where the final clarification by separation of nuclear and atomic properties seems to be at hand. In the future, whatever the outcome should be, laser spectroscopy could be used also to explore nuclear effects by comparative studies of neighboring elements and, possibly, radioactive isotopes.

There are a number of cases where laser spectroscopy has just started. As an example, one should think of spectroscopy of molecules in the storage ring and a continuation of the laser Lamb shift excitations at medium mass ions under the improved conditions given at the storage rings. In the future, storage rings, beams at even higher velocities, and of radioactive ions, will become available. This will open up the possibility to directly excite allowed groundstate transitions in highly charged ions. Even nuclear tansitions will come into reach, due to the large Doppler shift.

IX. Acknowledgments

For this report different kinds of important contributions must be acknowledged:

- Encouragement and advice that initiated and sustained the effort to collect existing information into a review article. Here I would like to name espe-

cially E. W. Otten and G. Huber, University of Mainz and H.-J. Kluge, University of Heidelberg/GSI.
- Help from colleagues to support information on their ongoing research. These are the members of the groups around ASTRID, CRYRING, TSR, and the Legnaro Laboratory, especially J. Hangst and H. Shen, Aarhus; H. Gao, M. Larsson, and R. Schuch, Stockholm; R. Grimm, D. Habs, H. J. Miesner, U. Schramm, T. Schüssler, D. Schwalm, and A. Wolf, Heidelberg; R. Calabrese, Ferrara; L. Moi, Siena; B. Fricke, Kassel; S. Schneider, Frankfurt; G. Soff, Dresden; E. Kankeleit and M. Tomaselli, Darmstadt; G. Huber and R. Grieser, Mainz.
- Scientific and technical contributions by all collaborators to those results originating from or including the GSI laser group, in particular R. Grieser, I. Klaft, R. Klein, P. Merz, G. Huber, Mainz; M. Grieser, D. Habs, H. J. Miesner, W. Petrich, U. Schramm, T. Schüssler, D. Schwalm, B. Wanner, A. Wolf, Heidelberg, and St. Becker, S. Borneis, C. Bruske, A. Dax, S. Dutta, T. Engel, B. Franzke, C. Holbrow, O. Klepper, D. Marx, R. Neumann, S. Schröder, P. Seelig, M. Steck, GSI.

In particular I would like to thank C. Holbrow for valuable advice and revisions.

X. References

Abrahamsson, K., Andler, G., Bagge, L., Beebe, E., Carle, P., Danared, H., Egnell, S., Ehrnsten, K., Engstrom, M., Herrlander, C. J., Hilke, J., Jeansson, J., Kallberg, A., Leontein, S., Liljeby, L., Nilsson, A., Paal, A., Rensfelt, K.-G., Rosengard, U., Simonsson, A., Soltan, A., Starker, J., and af Ugglas, M. (1993). *Nucl. Instr. Meth.* B79, 269.
Andersen, L. H., Brink, C., Haugen, H. K., Hvelplund, P., Yu, D. H., Hertel, N., and Moller, S. P. (1994). *Chem. Phys. Lett.* 217, 204.
Asp, S., Schuch, R., DeWitt, D. R., Biedermann, C., Gao, H., Zong, W., Andler, G., and Justiniano, E. (1996). *Nucl. Instr. Meth.* B117, 31.
Bastug, T., Fricke, B., Finkbeiner, M., and Johnson, W. R. (1996). *Z. Phys.* D37, 281.
Baumann, P., Blum, M., Friedrich, A., Geyer, C., Grieser, M., Holzer, B., Jaeschke, E., Krämer, D., Martin, C., Atl, K., Mayer, R., Ott, W., Povh, B., Repnow, R., Steck, M., Steffens, E., and Arnold, W. (1988). *Nucl. Instr. Methods Phys. Res. Sect.* A268, 531.
Beierdorfer, P., Osterheld, A. L., Scofield, J. H., Crespo Lopez-Urrutia, J. R., and Widman, K. (1998). *Phys. Rev. Lett.* 80, pp. 3022-5.
Birkl, G., Kassner, S., Walther, H. (1992). *Physikal. Blätter;* and Waki, I., Kassner, S., Birkl, G., and Walther, H. (1992). *Phys. Rev. Lett.* 68, 2007.
Bisoffi, G., Calabrese, R., Ciullo, G., Dainelli, A., Guidi, V., Gustafsson, S., Labrador, A., Lamanna, G., Moisio, M. F., Petrucci, F., Pisent, A., Ruggiero, A., Stagno, V., Tecchio, L., Variale, V., and Yang, B. (1994). CRYSTAL Feasibility Study, LNL-INFN Report 80/94, Legnaro.
Blundell, S. A., Cheng, K. T., and Sapirstein, J. (1997). *Phys. Rev.* A55, 1857.
Borneis, S., Bosch, F., Engel, T., Klaft, I., Klepper, O., Kühl, T., Marx, D., Moshammer, R., Neumann, R., Schröder, S., Seelig, P., and Völker, L. (1994). *Phys. Rev. Lett.* 72, 207.

Borneis, S., Becker, St., Engel, T., Klaft, I., Klepper, O., Kohl, A., Kühl, T., Marx, D., Meier, K., Neumann, R., Schmitt, F., Seelig, P., and Völker, L. (1995). *AIP Conference Proceedings* 329, RIS-94, H.-J. Kluge, J. E. Parks, K. Wendt, Eds. (New York), 150.

Brillet, A., and Hall, J. L. (1979). *Phys. Rev. Lett.* 42, 549.

Budker, G. I., and Skrinsky, A. N. (1978). *Us. Fiz. Nauk* 124, 561; *Sov. Phys.* 21, 277.

Calabrese, R., Guidi, V., Lenisa, P., Grimm, R., Miesner, H. J., Mariotti, E., and Moi, L. (1996). *Hyper. Interact.* 99, 259.

Comité International des Poids et Mesures (CIPM). (1984). Documents concerning the New Definition of the Meter, *Metrologia* 19, 163, 81 session (1992).

Crespo Lopez-Urrutia, J. R., Beiersdorfer, P., Savin, D. W., and Widman, K. (1997). *AIP Conference Proceedings* (No. 392, Pt. 1). Application of Accelerators in Research and Industry, Fourteenth *International Conference*, Denton, TX, 6–9 November 1996, AIP, p. 87–8.

Danared, H., Andler, G., Bagge, L., Herrlander, C. J., Hilke, J., Jeansson, J., Kallberg, A., Nilsson, A., Paal, A., Rensfelt, K.-G., Rosengard, U., Starker, J., and Ugglas, M. A. (1994). *Phys. Rev. Lett.* 72, 3775.

Dementev, E. N., Dikanskii, N. S., Medvedko, A. S., Parkhomchuk, V. V., and Pestrikov, D. V. (1980). *Sov. Phys. Tech. Phys.* 25, 1001.

Ellert, C., Habs, D., Jaeschke, E., Kanbara, T., Music, M., Schwalm, D., Sigray, P., and Wolf, A. (1992). *Nucl. Instrum. Methods* A314, 399.

Engel, T., Wuertz, M., Borneis, S., Becker, St., Klaft, I., Kohl, A., Kuehl, T., Laeri, F., Marx, D., Meier, K., Neumann, R., Schmidt, F., Seelig, P., and Völker, L. (1997). *Hyper. Interact.* 108, 251.

Essen, L., Donaldson, R. W., Bangham, M. J., and Hope, E. G. (1971). *Nature* 229.

Finkbeiner, M., Fricke, B., and Kühl, T. (1993). *Phys. Lett.* A176, 113.

Franzke, B. (1987). *Nucl. Inst. Meth. Phys. Res., Sect.* B24, 18.

Franzke, B., Beckert, K., Bosch, F., Eickhoff, H., Franczak, B., Gruber, A., Klepper, O., Nolden, F., Raabe, P., Reich, H., Spädtke, P., Steck, M., and Struckmeier, J. (Edited by S. T. Corneliussen). (1993). *Proceedings of International Conference on Particle Accelerators* (Washington, DC) 17–20 May 1993, IEEE (New York, NY), 1645.

Grieser, R., Klein, R., Huber, G., Dickopf, S., Klaft, I., Knobloch, P., Merz, P., Albrecht, F., Grieser, M., Habs, D., Schwalm, D., and Kühl, T. (1994a). *Appl. Phys.* B59, 127.

Grieser, R., Dickopf, S., Huber, G., Klein, R., Merz, P., Bönsch, G., Niclolaus, A., and Schnatz, H. (1994b). *Z. Phys.* A348, 147.

Grieser, R., Merz, P., Huber, G., Schmidt, M., Sebastian, V., Then, V., Grieser, M., Habs, D., Schwalm, D., and Kühl, T. (1996). *Hyper. Interact.* 99, 145.

Grimm, R., Ovchinnikov, Yu. B., Sidorov, A. I., and Letokhov, V. S. (1991). *Opt. Comm.* 84, 18.

GSI. (1995). Beam Intensity Upgrade of the GSI Accelerator Facility, GSI Report 95-05, ISSN 0171 4546.

Habs, D., Huber, G., Kühl, T., and Schramm, U. (1995). ESR Proposal.

Habs, D., and Grimm, R. (1995). *Ann. Rev. Nucl. and Particle Sci.* 45, 391–428.

Hangst, J. S., Kristensen, M., Nielsen, J. S., Poulsen, O., Schiffer, J. P., and Shi, P. (1991). *Phys. Rev. Lett.* 67, 1238.

Hangst, J., Nielsen, J. S., Poulsen, O., Shi, P., and Schiffer, J. P. (1995). *Phys. Rev. Lett.* 74, 4432.

Haugen, H. K., Andersen, L. H., Andersen, T., Balling, P., Hertel, N., Hvelplund, P., and Moller, S. P. (1992) *Phys. Rev.* A46, R1.

Hils, D., and Hall, J. L. (1990). *Phys. Rev. Lett.* 64, 1697.

Irnich, H., Geissel, H., Nolden, F., Beckert, K., Bosch, F., Eickhoff, H., Franzke, B., Fujita, Y., Hausmann, M., Jung, H. C., Klepper, O., Kozhuharov, C., Kraus, G., Magel, A., Münzenberg, G., Nickel, F., Radon, T., Reich, H., Schlitt, B., Schwab, W., Steck, M., Summerer, K., Suzuki, T., and Wollnik, H. (1995). *Phys. Rev. Lett.* 75, 4182.

Ives, H. E., and Stilwell, G. R. (1938). *J. Opt. Soc. Am.* 28, 215.

Javanainen, J., Kaivola, M., Nielsen, U., Poulsen, O., and Riis, E. (1985). *J. Opt. Soc. Am.* B 2, 1768.

JOSA. (1989). Special Issue on Laser Cooling. *J. Opt. Soc. Am.* B6.
Kaivaola, M., Poulsen, O., Riis, E., and Lee, S. A. (1985). *Phys. Rev.* 54, 255.
Katayama, T., Chida, K., Honma, T., Hattori, T., Katsuki, K., Mizobuchi, A., Nakai, M., Noda, A., Noda, K. Sekiguchi, M., Soga, F., Tanabe, T., Ueda, N., Watanabe, S., Watanabe, T., and Yoshizawa, M. (Editors, F. Bennett, J. Kopta). (1989). *Proceedings of the 1989 IEEE Particle Accelerator Conference: Accelerator Science and Technology* (Cat. No. 89CH2669–0), (Chicago, IL) 20–23 March 1989, IEEE (New York, NY), 37.
Kennedy, R. J., and Thorndike, E. M. (1932). *Phys. Rev.* 42, 200.
Klaft, I., Borneis, S., Engel, T., Fricke, B., Huber, G., Kühl, T., Marx, D., Neumann, R., Seelig, P., Schröder, S., and Völker, L. (1994). *Phys. Rev. Lett.* 73, 2425.
Klein, R., Grieser, R., Hoog, I., Huber, G., Klaft, I., Merz, P., Kühl, T., Schröder, S., Grieser, M., Habs, D., Petrich, W., and Schwalm, D. (1992). *Z. Phys.* 342A, 455.
Kowalski, J., Neumann, R., Noehte, S., Suhr, H., zu Putlitz, G., and Herman, R. (1983). *Z. Phys.* A313, 147.
Kühl, T. (1988). *Physica Scripta* T22, 144.
Labzowsky, L. N., Johnson, W. R., Schneider, S. M., and Soff, G. (1995). *Phys. Rev.* A51, 4597.
Marrs, R. E., Elliott, S. R., and Knapp, D. A. (1994). *Phys. Rev. Lett.* 72, 4082.
McArthur, D. W., Butterfly, K. B., Clark, D. A., Donahue, J. B., and Gram, P. A. M. (1986). *Phys. Rev. Lett.* 56, 282.
McGowan, R. W., Giltner, D. M., Sternberg, S. J., and Lee, S. A. (1993). *Phys. Rev. Lett.* 70, 251.
Michelson, A. A., and Morley, E. W. (1887). *Am. J. Sci.* 34, 333.
Milsner, H.-J., Grimm, R., Grieser, M., Habs, D., Schwalm, D., Wanner, B., and Wolf, A. (1996). *Phys. Rev. Lett.* 77, 623.
Minogin, V. G., and Letokhov, V. S. (1981). *Phys. Rep.* 73, pp. 1–65.
Mohler, P. H., and Stöhlker, Th. (1996). *Adv. At. Mol. Opt. Phys.* 37, 297.
Neumann, R., Poth, H., Winnacker, A., and Wolf, A. (1983). *Z. Physik* A313, 253.
Omidvar, K., and Guimaraes, P. T. (1990). *Astrophys. J. Suppl. Ser* 73, 555.
Persson, H., Schneider, S. M., Greiner, W., Soff, G., and Lindgren, I. (1996). *Phys. Rev. Lett.* 76, 1433.
Petrich, W., Grieser, M., Grimm, R., Gruber, A., Habs, D., Miesner, H.-J., Schwalm, D., Wanner, B., Wernoe, H., Wolf, A., Grieser, R., Huber, G., Klein, R., Kühl, T., Neumann, R., and Schröder, S. (1993). *Phys. Rev.* A48, 2127.
Poth, H. (1990). *Phys. Rep.* 196, 135.
Riis, E., Andersen, L.-U., Bjerre, N., Poulsen, O., Lee, S. A., and Hall, J. C. (1985). *Phys. Rev. Lett.* 60, 81.
Rivlin, L. A. (1979). *Sov. J. Qu. E.* 9, 353.
Rüetschi, A., Schellenberg, L., Phan, T. Q., Piller, G., Schaller, L. A., and Schneuwly, H. (1984). *Nucl. Phys.* A422, 461.
Schiffer, J. P., and Poulsen, O. (1986). *Europhysics Letters* 1, 55.
Schlagheck, P., Maragakis, P., and Lambropoulos, P. (1996). *Z. Physik* D37, 19.
Schneider, S. M., Schaffner, J., Greiner, W., and Soff, G. (1993). *J. Phys.* B26, L529 and L581.
Schneider, S. M., Greiner, W., and Soff, G. (1994). *Phys. Rev.* A50, 118.
Schneider, S. M., and Stöcker, H. (1995). Private communication.
Schramm, U., Berger, J., Grieser, M., Habs, D., Jaeschke, E., Kilgus, G., Schwalm, D., Wolf, A., Neumann, R., and Schuch, R. (1991). *Phys. Rev. Lett.* 67, 22.
Schramm, U., Schüssler, T., Habs, D., Schwalm, D., and Wolf, A. (1996). *Hyper. Interact.* 99, 309.
Schramm, U., Schüssler, T., Habs, D., Schwalm, D., and Wolf, A. (1995). *Nucl. Instr. Meth.* B98, 309.
Schröder, S., Klein, R., Boos, N., Gerhard, M., Grieser, R., Huber, G., Karafillidis, A., Krieg, M., Schmidt, N., Kühl, T., Neumann, R., Balykin, V., Grieser, M., Habs, D., Jaeschke, E., Krämer, D., Kristensen, M., Music, M., Petrich, W., Schwalm, D., Sigray, P., Steck, M., Wanner, B., and Wolf, A. (1990). *Phys. Rev. Lett.* 64, 2901.
Schuessler, H. A., Fortson, E. N., and Dehmelt, H. G. (1969). *Phys. Rev.* 187, 5.

Schüssler, T., Schramm, U., Ruter, T., Broude, C., Grieser, M., Habs, D., Schwalm, D., Wolf, A. (1995). *Phys. Rev. Lett.* 75, 802.

Seelig, P., Borneis, S., Dax, A., Engel, T., Faber, S., Gerlach, M., Holbrow, C., Huber, G., Kühl, T., Marx, D., Meier, K., Merz, P., Quindt, W., Schmitt, F., Tomaselli, M., Volker, L., Winter, H., Würtz, M., Beckert, K., Franzke, B., Nolden, F., Reich, H., Steck, M., and Winkler, T. (1998). To be published.

Seurer, M., and Toepffer, C. (1997). *Hyper. Interact.* 108, 333.

Shabaev, V. M. (1994). *J. Phys.* B27, 5825.

Shabaev, V. M., Shabaeva, M. B., and Tupitsyn, I. I. (1995). Hyperfine structure of hydrogenlike and lithiumlike atoms. *Phys. Rev.* A52, 3686.

Shabaev, V. M., and Yerokhin, V. A. (1996). *Pis'ma Zh. Eksp. Teor. Fiz.* 63, 309; (JETP Letters 63), 316.

Shabaev, V. M., Tomaselli, M., Kühl, T., Artemyev, A. N., and Yerokhin, V. A. (1997). *Phys. Rev.* A 56, 252.

Stensgaard, R. (1988). *Physica Scripta* T22, 315.

Tomaselli, M., Schneider, S. M., Kankeleit, E., and Kühl, T. (1995). *Nucl. Phys.* C51, 2989.

Wanner, B., Grieser, M., Grimm, R., Gruber, A., Habs, D., Miesner, H.-J., Nielsen, J. S., Petrich, W., Schwalm, D., and Wernoe, H. (Ed. J. Bosser) (1994). *Proc. Workshop on Beam Cooling and Related Topics,* Montreux, Switzerland (4–8 Oct. 1993). CERN Report 94-03, 354.

Wei, J., Li, X.-P., and Sessler, A. M. (1994). *Phys. Rev. Lett.* 73, 3089.

Wertheim. (1988). *Proceedings of the Workshop on Crystalline Ion Beams.* Wertheim, Germany, October 4–7; GSI-Report 89-10 (1989).

Will, C. M. (1993). *Theory and Experiment in Gravitational Physics,* Cambridge University Press New York.

Wolf, A., Habs, D., Lampert, A., Neumann, R., Schramm, U., Schussler, T., and Schwalm, D. (1993). *Proc. Thirteenth International Conference on Atomic Physics,* Munich, 3–7 August 1992, *AIP Conference Proceedings* 275, 228.

Zwicknagel, T., Toepffer, C., and Reinhard, P. G. (1996). *Contr. 1st Euroconference on Atomic Physics with Stored Highly Charged Hyper. Interact.* 99, 286.

LASER COOLING OF SOLIDS

CARL E. MUNGAN

*Department of Physics, The University of West Florida,
Pensacola, Florida*

TIMOTHY R. GOSNELL

*Condensed Matter and Thermal Physics Group,
Los Alamos National Laboratory,
Los Alamos, New Mexico*

I. Introduction	161
A. Overview of Basic Concepts	161
B. Summary of Experimental Results	163
C. Comparison to Laser Doppler Cooling	164
II. Historical Review of the Thermodynamics of Fluorescence Cooling	165
A. Viability of Anti-Stokes Cooling Processes	165
B. Thermodynamics of Electroluminescence	167
C. Thermodynamics of Photoluminescence	170
III. Working Substances for Fluorescence Cooling	175
A. Gases	175
B. Organic Dye Solutions	178
C. Semiconductors	181
D. Ruby	184
E. Rare-Earth Ions in Solids	186
IV. Laser Cooling of Ytterbium-Doped ZBLANP Glass	188
A. Photothermal Deflection Spectroscopy	191
B. Bulk Cooling Experiments	198
V. Fundamental Limits	204
A. Thermodynamics	204
1. General Refrigerator	205
2. A Three-Level Optical Refrigerator	206
3. A Simpler Approach to the Three-Level Optical Refrigerator	208
4. A Model Three-Level Optical Refrigerator	210
B. Minimum Temperature	215
C. Maximum Cooling Power and Optical Refrigerators	217
VI. Conclusions and Prospects	221
VII. Acknowledgments	223
VIII. Notes	224
IX. References	224

I. Introduction

A. Overview of Basic Concepts

Laser cooling of a solid may occur when the average energy of the photons emitted by the solid is larger than the energy of the ones it absorbs. More formally, the

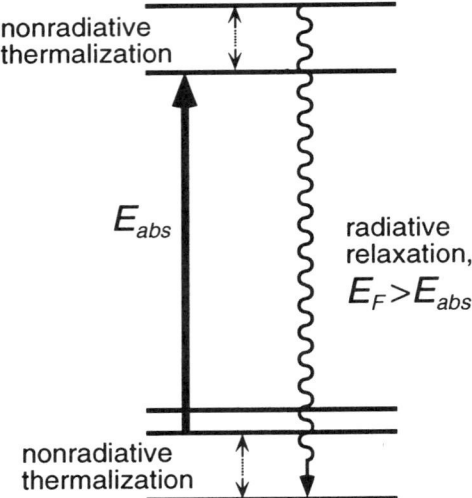

FIG. 1. Generic energy-level diagram for an impurity ion in a solid host. The arrows indicate pump and relaxation processes leading to anti-Stokes fluorescence cooling of the system.

anti-Stokes emission, which occurs at frequencies larger than that of the pump laser, must dominate the *Stokes emission* that occurs at smaller frequencies. A crucial additional requirement is that the nonradiative decay rates of the laser-pumped states be negligible in comparison to their radiative decay rates. For the sake of specificity, these states will be taken to be those of a set of isolated ions embedded within an insulating host, although in general they could equally well be those of gas-, liquid-, or solid-phase neutral atoms or molecules, or even those of the energy bands of intrinsic semiconductors. A typical system will consist of a ground-state manifold and an excited-state manifold well separated from one another, with at least one of these manifolds split into two or more levels. (See Fig. 1.) For a host temperature of T, a cooling cycle begins by optical pumping of the thermally populated high-lying levels of the ground-state manifold to low-lying states of the excited-state manifold. Next, both the excited- and ground-state manifolds rethermalize, typically through the net absorption of host-lattice phonons by the ions. Complete *intra*manifold thermalization is often assured because this process typically occurs on a picosecond time scale (Miniscalco, 1993), at least for intramanifold energy splittings on the order of $k_B T$, whereas the time scales for *inter*manifold radiative decay will range from nanoseconds for strongly allowed electronic transitions to as long as milliseconds for vibrational or nominally forbidden electronic transitions (Miniscalco, 1993). To conclude the cooling cycle, the excited manifold radiatively relaxes back to the ground manifold, after which both manifolds will most likely undergo additional rethermalization.

In practice, however, nonradiative processes that deexcite the active ions across the intermanifold energy gap compete with the radiative decay. Although direct multiphonon relaxation between the two manifolds will be strongly suppressed for an energy gap ~10 times larger than the host's maximum phonon frequency, more problematic will be nonradiative quenching of the excited active ions by low concentrations of unwanted bulk or surface impurities. The net effect of both processes is most simply quantified in terms of the fluorescence quantum efficiency, η_Q, defined as the ratio of the average number of emitted photons per pump photon absorbed.

In terms of the quantum efficiency, fluorescence cooling of the sample can be understood more quantitatively in the following way. Let E_F be the average energy of an emitted photon, with a corresponding wavelength of $\lambda_F = hc/E_F$. Similarly, let E_{abs} be the energy of a pump photon of wavelength λ. An empirical rule of thumb, one that has come to be known as Vavilov's Law,[1] states that λ_F is independent of λ, or very nearly so (Vavilov, 1945; Zander and Drexhage, 1995). In that case, the average cooling energy per cycled photon is $E_{cool} = \eta_Q(E_F - E_{abs}) - (1 - \eta_Q)E_{abs}$. Multiplying by the cycling rate to give powers rather than energies, one obtains an expression for the cooling efficiency, η, defined as the ratio of the cooling power to the absorbed laser power. (Technically, this should be called the relative cooling efficiency, in contrast to the absolute cooling efficiency, defined as the ratio of the cooling power to the *incident* laser power. These two efficiencies are, however, simply related via the absorptance of the sample.) The result is $\eta = (\lambda - \lambda_F^*)/\lambda_F^*$, where $\lambda_F^* \equiv \lambda_F/\eta_Q$. Knowledge of the fluorescence spectrum therefore suffices to determine η under the best-case scenario of $\eta_Q = 1$. The reason anti-Stokes fluorescence cooling is so difficult to obtain in practice is implicit in the last two expressions: Because $\eta > 0$ implies $\eta_Q > \lambda_F/\lambda \approx 1 - k_B T/E_{abs}$ and because E_{abs} will typically be at least 20 $k_B T$ in order to minimize multiphonon relaxation, η_Q must be close to unity. The high emission quantum efficiencies of most solid-state laser materials therefore make them natural candidates for anti-Stokes fluorescence coolers. In this context, one can view these materials as optically pumped lasers running in reverse.

B. Summary of Experimental Results

The first observation of net fluorescence cooling of a solid was reported in 1995 (Epstein *et al.*, 1995a). Trivalent ytterbium ions doped into a heavy-metal fluoride glass were pumped in the near-infrared at ~1 μm; in the initial experiments, a temperature drop of 0.3 K below room temperature was measured for a 43-mm³ sample in the shape of a rectangular parallelepiped. In later experiments, a 16-K drop was reported for a 250-μm-diameter optical fiber (Mungan *et al.*, 1997b). Spectroscopic measurements have since indicated that relative cooling efficiencies similar to those observed in these room-temperature experiments should be obtained even at liquid-nitrogen temperatures (Mungan et al., 1997a; Mungan et al., 1997c).

There have been two other published accounts of attempts to laser-cool solids. In both experiments, however, parasitic heating processes dominated the cooling effect and net refrigeration was not observed. In 1968, a large crystal of Nd^{3+}: YAG was inserted into a 1.064-μm laser cavity containing an independent, flash-lamp-pumped crystal of Nd^{3+}:YAG (Kushida and Geusic, 1968). Trace amounts of impurities such as Dy^{3+}, which are difficult to chemically separate from Nd^{3+}, were thought to be responsible for net heating of the sample. More recently, there has also been encouraging progress toward the goal of optically cooling a thin wafer of *GaAs* (Gauck et al., 1997). Unfortunately, the large index of refraction of this material helps trap the emitted radiation within the sample, thus increasing the probability of reabsorption and subsequent heat-generating nonradiative decay. However, strong ongoing interest in industrial opto-electronic applications of direct-gap semiconductors should lead to further progress in this approach to laser cooling.

Outside the realm of solid materials, anti-Stokes cooling of molecules in gases and liquids has also been observed. In 1981, low-pressure carbon dioxide was laser cooled by 1 K starting from 600 K (Djeu and Whitney, 1981). The experiment consisted in optical pumping with a 10.6-μm CO_2 laser between the $v_1 = 1$ symmetric-stretching vibration and the higher-frequency $v_3 = 1$ antisymmetric stretch. Because radiative relaxation of the antisymmetric mode to the vibrational ground state was highly favored, thermal repopulation of the depleted fundamental symmetric mode, assisted by a buffer gas of *Xe*, cooled the CO_2-*Xe* mixture.

More recently, two different European groups have optically cooled laser dyes dissolved in ethanol at room temperature. Zander and Drexhage (1995) employed a photothermal lensing technique to observe a transition from heating to cooling in a dye-laser-pumped region interior to a larger volume of a rhodamine 6G solution. They did not, however, attempt to measure overall, net cooling of the entire sample. This claim has since been made for solvated rhodamine 101 by Clark and Rumbles (1996). However, there may have been less question (Mungan and Gosnell, 1996) about their results had these workers begun with photothermal measurements and more detailed spectroscopic studies such as those of Zander and Drexhage.

Anti-Stokes laser cooling of materials is thus an active field of experimental study at present. The topic likewise continues to provide fertile ground for theorists interested in the interactions between light and matter, as shall be discussed at length in Section II.C.

C. COMPARISON TO LASER DOPPLER COOLING

Anti-Stokes laser cooling of solids, liquids, and gases is closely related to the technique of laser Doppler cooling of free atoms (Chu, 1991; Cohen-Tannoudji and Phillips, 1990; Wineland and Itano, 1987), for the practical development and extensions of which the 1997 Nobel prize in physics was awarded to S. Chu, C. Cohen-Tannoudji, and W. Phillips. This process is also crucial in the cooling of dilute gases down to the Bose-Einstein condensation temperature (Anderson *et al.*, 1995). The idea behind the technique was first proposed for neutral atoms by Hänsch and Schawlow (1975) and can be understood as arising from the radiation pressure of pairs of counterpropagating laser beams directed along three mutually perpendicular axes. Translational cooling occurs when the optical frequency is slightly detuned toward the low-frequency wing of an appropriate atomic-absorption line, wherein the pump light will be Doppler-shifted into resonance only for those atoms that are moving toward a particular laser source, thus slowing them down like ping-pong balls striking an oncoming bowling ball.

It is easier to understand the conceptual similarity between the anti-Stokes and Doppler cooling techniques by considering the latter from an energy rather than a momentum point of view. The atoms absorb low-energy photons and then—on average—isotropically re-radiate photons, so that the emitted light is not Doppler-shifted and hence is of higher mean frequency. That is, the mean fluorescence is up-shifted relative to the absorbed light, with the difference representing the amount of heat carried away from the atoms.

Thus both Doppler and anti-Stokes cooling involve the emission of photons of higher mean energy than those absorbed. On the other hand, certain differences are also evident: The first technique involves translational cooling of noninteracting two-level atoms, whereas the second method works by cooling the internal degrees of freedom, at least two of which are coupled to the surrounding medium by thermalizing collisions. Furthermore, anti-Stokes cooling is a much more efficient process: even restricting attention to gases, it has been estimated that Doppler cooling is about six orders of magnitude less efficient (Djeu and Whitney, 1981). Of course, the intended applications of the two techniques are quite different.

II. Historical Review of the Thermodynamics of Fluorescence Cooling

A. Viability of Anti-Stokes Cooling Processes

Early in this century, Raman (1928) in India discovered the spectroscopic effect named after him, namely, that if a beam of monochromatic light passes through a medium, the scattered light will contain lines having lower frequencies (called Stokes lines) and higher frequencies (called anti-Stokes lines) than the frequency of the incident beam. This nomenclature for the set of lower-frequency lines reflects an outdated principle known as Stokes' rule or law, which states that a substance cannot emit light of wavelength shorter than that of the exciting radiation (Stokes, 1852). Exceptions to this rule (Wood, 1928), however, were already well known by the time of Raman's experiments and had accordingly come to be known as anti-Stokes emission. Although it was understood that the excess energy necessary for such anti-Stokes radiation arose from thermal population of low-frequency sideband levels, such as the rotational states within a vibrational manifold, a controversy nevertheless erupted regarding whether a system can emit light that *on average* is of higher frequency than the incident radiation. In this situation, the "energy yield" or "luminescence efficiency" is said to exceed unity (Chukova, 1969a, 1969b, 1971, 1976; Stepanov and Gribkovskii, 1968). The contrary view was that the second law of thermodynamics requires that any anti-Stokes fluorescence be accompanied by enough entropy-compensating Stokes emission to ensure that the luminescence efficiency would be always less than one. This question was first clearly discussed by Pringsheim (1929), who adopted the former position, arguing that a continual cooling of a gas by anti-Stokes scattering is thermodynamically allowable because the system is not closed—as for any refrigerator, an outside agency supplies work to the system.

Pringsheim's position was subsequently opposed in print by Vavilov, leading to an intriguing quartet of short papers appearing at the end of World War II. Vavilov (1945) argued that an excitation–fluorescence cycle is reversible and hence an energy yield greater than unity would be equivalent to the (unallowable) complete transformation of heat into work. Introducing the concept of the average emission frequency, he pointed out that all spectroscopic data available at that time, including measurements on dye solutions (Jablonski, 1933) and on uranium glass, indicated that the intensity of the luminescence fell to zero as the excitation frequency decreased toward this average value. Pringsheim (1946) responded that the optical cycle cannot be reversible because the monochromatic and unidirectional incident beam is converted into isotropic, broad-bandwidth radiation with a corresponding increase in its entropy. Putting it another way, he argued that irradiation of a low-temperature sample by a source of high "excitation temperature" entails a thermodynamically irreversible energy flow. Pringsheim included in his response a brief discussion of the fundamental conditions necessary for cooling, namely, a "quantum efficiency" or "quantum yield" (to be distinguished from the energy yield) as near unity as possible and thermal equilibration of the excited state prior to emission. (Ironically, Pringsheim weakens his response by suggesting that the explanation for the refuting evidence of the spectroscopic data just mentioned is that the quantum yield in many systems decreases when their absorption bands are pumped in the wings. In the condensed phase, for example, this was attributed to the increased electrostatic perturbation experienced by the outlying centers giving rise to these wings rather than to spurious impurity effects.) Vavilov (1946) immediately rebutted by considering a reversible cycle built around an ideal optical cavity into which the anti-Stokes cooling sample is placed along with a Stokes heating sample. In his analysis of this system, the net effect of the cycle was the (impossible) work-free transfer of heat from a cooler to a hotter reservoir. Furthermore,

Vavilov criticized Pringsheim's suggestion that the loss of directionality of the scattered light is an indication of irreversibility by citing an apparatus—dating back to 1743—wherein an arrangement of convex lenses and plane mirrors surrounding the sample could be used to re-collimate the fluorescence. He also suggested, with the example of a gaseous cooling sample, that there may be a contradiction between the requirement that the density be kept low enough to prevent excited-state quenching through nonradiative collisional deexcitation while simultaneously maintaining thermal equilibrium between the pumped and the fluorescing excited states. Presciently, he ends his article by calling for experimental measurements of the luminescence efficiency of solid phosphors pumped at long wavelengths.

Finally, Vavilov's concerns prompted a paper from Landau (1946) in which he presented for the first time a sound thermodynamic argument proving that photoluminescence energy yields can in fact exceed unity. To calculate the radiation entropy, he applied Bose statistics to the photon gas, integrating over the spectral bandwidth and the solid angle. After introducing the concept of an "effective temperature" or "brightness temperature" of the source, defined as the temperature of blackbody radiation whose spectral intensity matches that of the pump radiation at its peak wavelength, the final result was that the energy yield exceeds unity by an amount proportional to the ratio of the sample temperature and the effective temperature. Both Vavilov and Landau dismissed this excess as being insignificant, even though it clearly can be of the order of several percent or more. Nonetheless, the essential physical viability of the concept had thus been established.

B. THERMODYNAMICS OF ELECTROLUMINESCENCE

In the early 1950s, the phenomenon of electroluminescence of semiconductor diodes was discovered (Haynes and Briggs, 1952; Lehovec et al., 1951; Newman, 1953). It was quickly noticed that the energies of the photons emitted at the shortest wavelengths exceeded the applied electrical energy per injected electron; Lehovec et al. (1953) deduced that the difference corresponds to the withdrawal of internal energy from the semiconductor lattice. The possibility of using this effect for cooling purposes was briefly discussed by Tauc (1957). He derived an expression for the cooling power that reduces to $(E_g/eV - 1)iV$ when certain transport contributions and Joule heating are neglected. Here i is the electrical current, V is the forward bias across the diode, and E_g is the energy of the bandgap, across which the luminescence is assumed to occur with unit quantum efficiency.

This concept was taken seriously enough in 1956 to merit application for a U.S. patent, which was granted three years later (Bradley, 1959). The patented device consists of a series of electrically pumped semiconductor cells (*Si, Ge, CdSe, InSb, AlSb, CdTe, CdS, GaSb,* and *GaP* with carefully controlled impurity concentrations were specifically cited as possible compositions) laid out in a flat grid over which a thin layer of a transparent, heat-exchanging liquid such as CCl_4 would circulate. The resistivity and geometry of the cells would be tailored to reduce ohmic losses and total internal reflection, respectively. Advantages of this cooler would include high efficiency and the absence of moving mechanical parts; its intended applications were gas liquefaction and cooling of infrared detectors.

This initial interest in electroluminescence cooling developed independently of the earlier work on photoluminescence cooling summarized previously in Section II.A. From a theoretical point of view, however, the two processes are similar, differing only in the method of excitation, provided that secondary issues such as electrical Joule heating are ignored. Weinstein (1960), working at General Electric's lighting division, appears to have been the first to recognize this concordance and address the general thermodynamic issue of converting heat into light. He was interested in calculating the "technical efficiency" of the process, defined as the ratio of the emitted luminescence power to the rate at which work is input, and consequently equal to unity plus a cooling efficiency. A "flux temperature" (to use the term adopted by later workers), T_{F_h}, of the fluorescence was defined as the ratio of the emitted power to the net rate at which the light carries entropy away from the sample. In general, this entropy flux and hence T_{F_h} depend on the thermal radiation incident on the sample from the environment, a quantity that depends in turn on the ambient temperature T, taken to be the same as that of the sample. In this way, T_{F_h} correctly reduces to T in the limit where the rate of input work falls to zero. It follows simply from the second law of thermodynamics that the technical efficiency must be smaller than $T_{F_h}/(T_{F_h} - T)$, which is, incidentally, the same as the reciprocal of the maximum efficiency of a solar cell when T_{F_h} refers to the input light.

Weinstein's calculation of the radiation energy and entropy essentially follows the derivation of Landau (1946), with the resulting flux densities for these quantities reducing to the familiar blackbody expressions when the output radiation field is in thermodynamic equilibrium with the surroundings. However, for situations of practical interest, such as emission within a narrow band, the luminescence power is very large compared to the ambient thermal emission within the same spectral bandwidth. Assuming that the fluorescence is isotropic, T_{F_h} then becomes approximately equal to the "brightness temperature" of the fluorescence T_{B_h}, defined as the temperature at which a blackbody would have the same intensity as at the spectral peak of the emission. Note that T_{F_h} is infinite if the emitted radiation is either monochromatic or strictly unidirectional. Because the coolest blackbody that significantly emits in the visible range has a temperature of about 800 K, the maximum theoretical cooling efficiency of a visibly fluorescing sample at room temperature is about $300/(800 - 300) = 0.6$. However, this impressive thermodynamic result is of little practical utility because it assumes that the energy levels of the sample can be arranged in any way one pleases, while at the same time preserving unit radiative quantum efficiency.

In a brief letter published in 1965, Gerthsen and Kauer (1965) applied a physical model of an electroluminescent diode to the calculation of the technical efficiency. They showed that for unit quantum efficiency Weinstein's expression, $T_{F_h}/(T_{F_h} - T)$, reduces to E_g/eV when the rate of radiative recombination is explicitly modeled in terms of the product of the electron and hole concentrations. This result was in agreement with that of Tauc for an electroluminescent semiconductor diode; accordingly, Gerthsen and Kauer viewed the diode as a heat pump that transfers energy from a low-temperature reservoir (the lattice) to a high-temperature reservoir (the radiation field).

Moving closer to an actual experiment, Landsberg and Evans (1968) reexamined Weinstein's treatment of the thermodynamics of an electroluminescent diode and noted that one can calculate the rate of excess-entropy generation within the diode in terms of experimentally observable quantities. This entropy production depends on the sample temperature, forward-bias voltage, mean emitted-photon energy, quantum efficiency, and the brightness temperature of the fluorescence. The latter quantity, in turn, can be calculated by assuming uniform, isotropic emission of light with a known spectral profile and depends additionally on the electric current and the surface area of the diode's emitting region. For a room-temperature *GaAs* diode, the experimental data at the time indicated a technical efficiency of less than 6%, far below the cooling threshold of 100%, implying a large rate of entropy generation within the diode amounting to 30–50 k_B per injected electron. More encouragingly, however, the technical efficiency for a typical set of experimental conditions would rise to 120% if all nonradiative processes could be eliminated. After reevaluating previous measurements, Landsberg and Evans suggested that reduced sample temperatures and careful epitaxial growth might result in the necessary increased quantum efficiencies.

More recently, Berdahl (1985) has treated the problem of electroluminescence cooling by semiconductor diodes. He defined a coefficient of performance (COP) as the ratio of the cooling power produced to the rate of external electrical work required. In the forward-biased, light-emitting mode, the Carnot value of the COP is $T/(T_R - T)$, where T is the temperature of the diode and T_R is the temperature of the diode's radiative environment. For operation at 0° C in room-temperature surroundings, the COP is therefore about 10. Assuming values of unity for the emissivity, refractive index, and quantum efficiency of a diode in vacuum, a cooling rate of 650 W per square meter of junction area was deduced for an (unrealistic) energy gap of 3–5 $k_B T$ and an actual COP of 1.25. Cooling would still be obtained for significantly lower quantum efficiencies (due, for example, to Auger recombination), albeit with a reduced COP, but this may be of little practical consequence given that Joule heating and other parasitic effects had not been taken into account.

In his paper, Berdahl offered some explicit advice for the experimentalist. In order to defeat the limitations of refractive indices greater than one, he suggested frustrating the total internal reflection occurring at the semiconductor surface by bringing an external absorber within a couple of microns of the diode. In practice, however, avoiding accidental thermal conduct between the two optical elements would probably prove difficult. Berdahl's most innovative suggestion, however, was to run the diode in its reverse-biased mode, in which case it cools a neighboring object by absorbing its thermal emission, a phenomenon he termed "negative luminescence." The effect of parasitic losses in the diode can then be reduced by mounting it on a separate heat sink. To our knowledge, however, no one appears to have followed up on this idea.

C. THERMODYNAMICS OF PHOTOLUMINESCENCE

The concept of anti-Stokes photoluminescence cooling did not begin to arouse serious practical interest until the advent of the laser, which could serve as the requisite source of high-intensity, narrow-bandwidth radiation. The story begins in the late 1950s, when Scovil and Schulz-DuBois (1959) at Bell Laboratories realized that a maser run backwards would act like a refrigerator. They considered a system with three levels, labeled 0 through 2 (Fig. 2), where states 1 and 2 are in thermal equilibrium with a cold reservoir of temperature T_c; these two states are separated by a transition frequency of ν_{12}. In turn, the 0 and 2 states are coupled to a hot reservoir at temperature T_h and are separated by a transition frequency of ν_{02}. Scovil and Schulz-Dubois then defined the efficiency of maser action as $\eta_M = \nu_{01}/\nu_{02}$ (i.e., the maser output energy divided by the input pump energy, assuming unit quantum efficiency), where T_c is taken to be the sample temperature. For example, T_c might be the temperature of a crystalline lattice to which is coupled spin levels of maser-active rare-earth ions, while T_h would be the temperature of a microwave gas discharge used to pump the system. From knowledge of the Boltz-

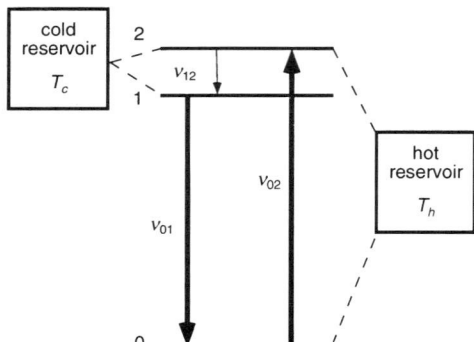

FIG. 2. A three-level maser viewed as a heat engine. Reversing the direction of operation yields a microwave-pumped refrigerator.

mann population factors, one then obtains inversion on the 0–1 transition provided $\eta_M \leq \eta_C$, where the Carnot efficiency is $\eta_C = (T_h - T_c)/T_h$. Because the maser can be regarded as a heat engine, as depicted in Fig. 2, this result is a simple expression of the second law of thermodynamics. Reversing the cycle then cools the crystal and an analogous analysis leads to $\eta \leq COP_C$. Here the cooling efficiency is simply $\eta = \nu_{12}/\nu_{01}$, the cooling energy divided by the maser energy, whereas the Carnot COP of the refrigerator is $COP_C = T_c/(T_h - T_c)$.

Mazurenko (1965a, 1965b) has applied thermodynamics to the irreversible generation of stimulated radiation in a laser. He used Prigogine's (1954) "local" formulation of the second law to consider the rate of entropy change for a set of particle oscillators in contact with both a nonequilibrium radiation field and a thermal reservoir, the latter taken to be the host medium. Mazurenko then derived an inequality for the laser efficiency η_L, defined as the ratio of the laser output power to the absorbed optical pump power. Landsberg and Evans (1968) have since generalized this result to obtain $\eta_L \leq (1 - T/T_{F_p})/(1 - T/T_{F_L})$, where T is the temperature of the thermal reservoir, and T_{F_p} and T_{F_L} are the flux temperatures of the pump and laser radiation, respectively, defined as the ratio of the corresponding energy and entropy fluxes. The limiting value is the ratio of two Carnot efficiencies because we can interpret a laser in terms of a combined forward- and reverse-running pair of heat engines, as depicted in Fig. 3, with the output work of the forward-running engine driving the operation of the reverse engine. Both share a common cold reservoir at temperature T, but the first engine's hot reservoir has a flux temperature of T_{F_p}, while the second's is of temperature T_{F_L}. Because an ideal laser carries energy but no entropy, $T_{F_L} = \infty$ in this case, and thus $\eta_L \leq (T_{F_p} - T)/T_{F_p}$. Running a laser backwards as an optical cooler, what was formerly the pump radiation will now be identified as the output fluorescence with an affiliated flux temperature of T_{F_h}; the technical efficiency will then be given by the reciprocal of the laser efficiency η_L. Hence the technical efficiency must be less than $T_{F_h}/(T_{F_h} - T)$, which agrees with the inequality derived by Weinstein in the case of electroluminescence cooling.

Kafri and Levine (1974) have also emphasized entropy considerations in lasing and cooling cycles, expressing the change in the entropy of the medium during thermal relaxation as $\Delta S_c \geq \Delta S_p + \Delta S_h$, where ΔS_p and ΔS_h are the medium's entropy changes during absorption of the pump and during the subsequent optical emission, respectively. For laser cooling, $\Delta S_p \approx 0$ and $\Delta S_c < 0$, so that coherent emission (for which $\Delta S_h > 0$) is not possible. If the cooling system consists of three energy levels 0 through 2, with the pumping on transition 1–2 and the fluorescence on 2–0, then the ratio of the populations of levels 1 and 0 is $n_1/n_0 = \exp(\Delta S_c/k_B)$ if these two levels are in thermal equilibrium with each other. Thus the refrigeration process depletes ("cools") the population of level 1. Kafri and Levine suggested that a possible practical application of this scheme would be to the increase of gain in a laser whose lower level is 1 (and whose upper level is presumably substantially above 2).

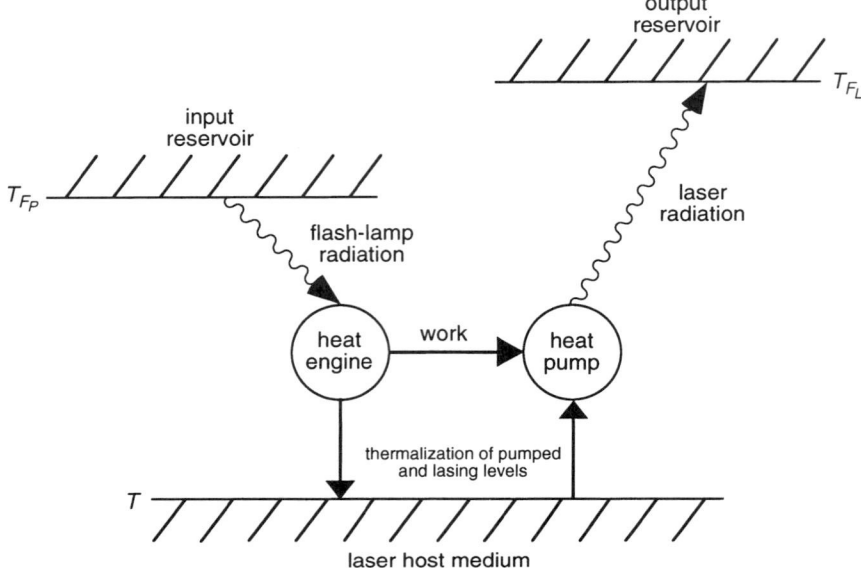

FIG. 3. Double-heat-engine picture for analyzing the efficiency of laser operation. The optical pump is taken to be a flash lamp, for the sake of specificity. The input and output radiation fields in this model are taken to constitute heat exchanges with thermal reservoirs having the indicated temperatures. Note that $T_{F_L} \to \infty$ if the laser radiation is ideal, as indicated in the figure by the higher level for the output reservoir.

For work performed up to 1980, a unified review of the thermodynamics of systems that convert light into heat or work, or vice-versa, was prepared by Landsberg and Tonge (1980). In complete generality, they considered an energy converter to be a box, into which energy and entropy is entering at certain rates from a pumping system and is flowing out at another pair of rates to a sink. Also contributing to energy and entropy flow to and from the box are terms for the conventional transfer of heat to a thermal reservoir and the delivery of work. Adding one final term to quantify the irreversible generation of entropy within the box (such as would arise from nonradiative relaxation of optically pumped centers, or Joule heating during electrical excitation), a pair of energy- and entropy-balance equations for steady-state or cyclic operation of the converter are readily written down. Applied to a laser cooler, the two balance equations imply $\eta \leq T/(T_{F_h} - T)$ once again, where η is the "first-law" cooling efficiency (ratio of the rate of heat withdrawn from the thermal reservoir at temperature T to the input optical pump power) and where as before the fluorescence flux temperature T_{F_h} is defined as the ratio of the emitted rates of energy and entropy.

An important clarifying point that Landsberg and Tonge emphasize in their review—one not fully appreciated by previous workers—is the difference between flux temperatures, which are not "absolute thermodynamic temperatures" (i.e., partial derivatives of energy with respect to entropy at constant volume), and brightness temperatures, which are. In any case, the right-hand side of their inequality is effectively a Carnot efficiency, whose calculation requires a determination of the entropy carried away by the *nonequilibrium* emitted radiation field. Landsberg and Tonge argue that this entropy is given by the usual *equilibrium* expression, namely, an integral of the logarithm of photon occupation numbers over all radiation modes contained within the spectral bandwidth, range of solid angles, and polarization directions of the emission. However, the fluorescence-energy flux density can also be written as an integral over these same photon occupation numbers. Hence, given a knowledge of the fluorescence spectrum, the entropy rate can be related to the energy rate, so that T_{F_h} is ultimately expressible purely in terms of the emission intensity. [This analysis implicitly assumes that T_{F_h} is much larger than the ambient temperature of the surroundings—see Weinstein (1960), whose results Landsberg and Tonge are essentially rederiving, for a discussion of this point.] As a simple example, Landsberg and Tonge considered the case where the fluorescence spectrum is constant over a narrow band of frequencies and zero elsewhere; they graphed values of the fluorescence flux temperature versus the emission center frequency for various choices of the energy flux density per unit bandwidth. They also quoted expressions applicable to the case of a Gaussian spectrum, which may be adaptable to some of the experimental results discussed later in Section III.

Recent experimental successes in cooling glasses, semiconductors, and liquids have led to a resurgence of theoretical interest in laser cooling of condensed matter. Oraevsky (1996) and other Russian workers (Rivlin and Zadernovsky, 1997; Zadernovskii and Rivlin, 1996) have considered the processes that limit the lowest attainable temperature and the rate of cooling in laser-excited semiconductors. They all calculated the absorption coefficient from semi-empirical equations describing electron-hole creation, excitonic effects, and intraband absorption by free carriers. Including corrections for impurity-related and Auger recombination of the charge carriers, Oraevsky found that it should be possible to cool *GaAs* from 300 to 10 K by using a laser intensity near the saturation level, that is, about 2000 W/cm^2 at room temperature and 300 W/cm^2 at the lowest temperatures. These calculations neglected the trapping of emitted photons by total internal reflection and further assumed that any surface states had been passivated by growing *GaInP*$_2$ on both sides of the *GaAs* layer. For simplicity, Oraevsky's model sample was taken to be optically thin, planar, and uniformly irradiated.

On the other hand, Zadernovskii and Rivlin focused on balancing the rate of cooling due to radiative recombination against the blackbody heat load from the room-temperature surroundings. They estimated a low-temperature limit of 3 K for a pump-laser intensity of 12 W/cm^2. In their most recent article, they plotted the expected final sample temperature versus the pump-photon flux density for various values of the laser linewidth and of the ambient temperature. They explicitly checked that the phonon-relaxation time of the charge carriers remains at least two orders of magnitude shorter than the radiative-decay time at all temperatures of interest, so that the electrons and holes would be sure to thermalize with the lattice prior to recombination. Thus these theoretical investigators continue to be optimistic about the prospects for laser cooling of semiconductors, although their calculations do not help to resolve the experimental difficulties to be discussed in Section III.C.

Andrianov and Samartsev (1997a) have briefly reviewed the experimental results on laser cooling of ZBLANP glass doped with Yb^{3+} (Epstein et al., 1995a) and of rhodamine 101 in liquid ethanol (Clark and Rumbles, 1996), again with an eye toward estimating the theoretical limits on the cooling rates and on the coldest temperatures that can be attained. The optical Bloch equations for the populations of the relevant levels were solved, allowing in principle for both spontaneous and stimulated Stokes and anti-Stokes transitions, in addition to direct resonance fluorescence. The experimental systems were divided into two components—the set of active ions or molecules and the embedding medium—and each assigned separate heat capacities. With the help of photon-echo data previously obtained on materials similar to the laser-cooling systems analyzed in their paper, Andrianov and Samartsev deduced that under the experimental conditions of interest, the temperature difference between the ytterbium ions and the glass host is only 30 μK, whereas between the rhodamine 101 molecules and the ethanol solvent the temperature difference is about 3 mK. Unfortunately, too many unknown parameters exist in their resulting theoretical equations to make quantitative tests or predictions.

Finally, a few authors have begun to address the possibility of exploiting coherent effects in laser cooling. Kosloff and coworkers (Bartana et al., 1997; Geva and Kosloff, 1996) have derived a generalized quantum master equation that predicts the existence of a "refrigeration window" for a three-level system coupled to hot and cold quantum heat baths. The system is optically pumped at extremely high intensities; therefore the pump is modeled as a classical electromagnetic field that performs work on the system. Operation within the refrigeration window allows heat to be extracted out of the cold bath until its temperature reaches absolute zero, thus overcoming the weak-field saturation limit on the cooling as first calculated by Scovil and Schulz-DuBois (1959). Nevertheless, the results are in compliance with the three laws of thermodynamics. In particular, the cooling rate drops to zero at 0 K, in agreement with the third law, although it is admitted that the issue of thermalization time scales probably becomes significant before this ultimate limit is reached.

In other work involving optical coherence, Andrianov and Samartsev (1996a, 1997b) have considered exploiting induced superradiance in a two-level system as a cooling technique. In one scheme, taking Frenkel excitons in a pure crystal as the two-level systems, the required population inversion would be created using an intense, nonresonant pulse. Lloyd (1997a, 1997b) has discussed the concept of a coherently pumped, quantum optical refrigerator. He showed that if the Rabi frequency for the laser-driven atoms is larger than the radiative relaxation rate of the excited state, then a cyclical process using two π-pulses of the same frequency results in cooling with significantly higher efficiency than what is achieved with incoherent pumping. In particular, the cooling rate need not approach zero as the Carnot limit is approached. Lloyd has also considered what he termed a quantum-mechanical "Maxwell's demon." In an explicit example, the idea is applied to a two-level system, conveniently visualized as a magnetic spin. First, an incident pulse acquires information about the state of the spin: if it is in the thermally excited high-energy level, a second pulse, of area π, coherently extracts its energy. Such information measurement can be performed using a technique known as spin-coherence double resonance, for example. In this thought experiment, decoherence due to spin dephasing introduces thermodynamic inefficiencies in accordance with the second law.

An important advantage of coherent cooling processes is that the cycling time can be reduced, thus increasing the cooling power. However, high-power pulsed excitation is required by such schemes. So far, no experimental attempt to achieve coherent, solid-state refrigeration has been performed, but this is clearly a promising arena for future work.

III. Working Substances for Fluorescence Cooling

A. GASES

In his pioneering article, Pringsheim (1929) discussed the possibility of cooling Na vapor by anti-Stokes processes. The idea was to pump the D_1 line ($1\,^2S_{1/2} \to 2\,^2P_{1/2}$) at 5896 Å using suitably filtered D_1 light obtained from an independent sodium source. The gas pressure of the cooling sample would be kept low enough so that collisional deexcitations would occur only rarely and therefore the atomic relaxation would be primarily radiative. At the same time, the Na pressure would be high enough to permit at least partial thermalization among the pair of higher-lying $2\,^2P_{1/2}$ and $2\,^2P_{3/2}$ levels. To reduce the heat load, the gas would be kept in a transparent, thermally insulating dewar. Finally, emission on the D_2 line ($2\,^2P_{3/2} \to 1\,^2S_{1/2}$) at 5890 Å would cool the gas.

Twenty-one years later, the French researcher Kastler (1950) expanded upon Pringsheim's idea, which he called an *effet lumino-frigorique* ("photo-refrigerating effect"). As applied to the cooling of sodium vapor, achieving the dual requirements of minimized collisional quenching of the optically pumped level and maximized thermalization within the narrow $2\ ^2P$ manifold is accomplished by introducing an inert buffer gas, such as helium or argon, into the sample cell. Suggested pressures were 1–10 μtorr for the alkali vapor and ~0.1 torr for the buffer. At moderate optical intensities it would be possible to excite every Na atom about 10,000 times per second. Because the 17 cm^{-1} frequency splitting between the D_1 and D_2 lines corresponds to a temperature difference of 24 K, a peak cooling rate of over 10 K/s was predicted. Kastler never attempted this experiment himself, and in fact concludes his abstract with a sentence that we translate as, "Even if one succeeds in realizing the necessary experimental conditions for radiative cooling, this effect is likely to remain a scientific curiosity rather than a practical technique for obtaining low temperatures."

Pringsheim (1946), in his response to Vavilov's criticisms (see Section II.A), proposed that diatomic gases such as I_2 could be radiatively cooled via the vibrational sidebands of their electronic transitions. For example, the pump light could be filtered so as to populate an excited electronic manifold with only those molecules initially lying in the thermally populated, fifth vibrational level ($v'' = 4$) of the electronic ground state. These molecules may then relax by emitting anti-Stokes radiation corresponding to transitions to the $v'' = 3, 2, 1$, or 0 levels of the ground state. The relative intensities of these various lines depend only on the Franck-Condon factors and can be larger than the Stokes or the Rayleigh line strengths in an appropriate system. Because cooling is no longer dependent on collision-mediated thermalization in the excited-state manifold, an advantage of this idea is that the sample pressure can now be made as low as one pleases in order to eliminate collisional quenching (at the expense of a reduced cycling rate).

A similar idea consists in pumping the rotational sidebands of the vibrational levels of a heteronuclear diatomic gas such as carbon monoxide (Djeu, 1978). By pumping the R branch of the rovibrational spectrum (e.g., $v = 0, J = 2'' \to v = 1, J = 1'$), the rate of P-branch relaxation ($v = 1, J = 1' \to v = 0, J = 0''$) can be enhanced, which cools the gas as the ground rotational manifold rethermalizes. After accounting for the branching ratio for radiative relaxation via the P and R branches, and provided that nonradiative vibrational–translational deexcitation is negligible, the average cooling energy is found to be $2BJ'$ per cycle, where B is the rotational constant of the molecule. Evidently it is an advantage in this experiment to pump the highest J' levels possible, provided that the intrarotational thermalization times remain short in comparison to the cycling time.

Yet another possibility would be to excite a high vibrational overtone (say $v = 3$) of a gaseous CO molecule that subsequently relaxes through a sequence of $\Delta v = 1$ radiationless energy-transfer exchanges with neighboring, unexcited CO molecules (Treanor et al., 1967; Yardley, 1971). Because of vibrational anharmonicity, the $3 \to 2$ and the $2 \to 1$ transition energies are smaller than the $0 \to 1$ energy and hence the exchanges are endothermic, again cooling the gas. In similar fashion, one might be able to achieve cooling by pumping large Δv transitions of solid-state systems such as CN^- doped in alkali halides (Epstein *et al.,* 1995b). For that matter, one could mix two different but nearly resonant species in the sample and pump the lower-energy system. The rate of phonon-assisted energy transfer to the second species is enhanced by insisting that this latter species' concentration be much higher than that of the first. In essence, the combination of the two species yields a two-level upper manifold in which the higher-energy state is highly degenerate, thus favoring radiative relaxation out of this state and cooling of the lattice. All of these cases require a high radiative efficiency for the final $1 \rightarrow 0$ decay and the absence of trapping centers such as heavy isotopes of CO and CN^- or other contaminating impurities. Also, the efficiencies of these schemes depend on large coupling strengths for the operative radiationless transitions relative to other possible decay channels and on sufficient absorption strengths of the optically pumped transitions. As far as CO and CN^- are concerned, many of these ideas are therefore probably not viable in practice. Nevertheless, they provide helpful starting points for thinking about other possible fluorescence-cooling mechanisms.

A final scheme for anti-Stokes cooling of gases has been actually demonstrated experimentally and was in fact the first observation of non-Doppler fluorescence cooling of any material system. In the early 1980s, Djeu and Whitney (1981) at the Naval Research Laboratory successfully cooled CO_2 gas by 1 K along the path of a 1-cm-diameter pump beam passing through a sample cylinder whose walls were maintained at 600 K. The 10.6-μm (100) $\rightarrow$ (001) vibrational combination transition was pumped by using a 300-W CO_2 laser running on the $P(20)$ line. Cooling resulted from 4.3-μm anti-Stokes emission of the (001) antisymmetric-stretching mode to the (000) vibrational ground state, a process highly favored owing to the 99.8% radiative branching ratio for this transition. Thermal repopulation of the laser-depleted (100) symmetric-stretching mode was assisted by three factors: a near resonance of this state with the first overtone of the (010) bending mode, the 600-K ambient temperature, and mixing of the CO_2 sample at a partial pressure of 64 mtorr with Xe at a partial pressure of just under 0.2 torr. Xenon was chosen as the buffer gas because of its low thermal conductivity and because it only weakly deactivates the CO_2 (001) state; its partial pressure in the sample cell was determined by trial and error to optimize the cooling rate. In the absence of the Xe buffer, the carbon dioxide pressure was set just at the onset of collision-induced relaxation of the (001) state. The resulting CO_2 density, in turn, determined the choice of the cell diameter, namely 12.7 cm, in order to minimize self-absorption of the 4.3-μm radiation. The inner walls of the cell were painted flat black to prevent reflections of the fluorescence back into the gas. The temperature changes were determined by measuring the axial-pressure changes with a capacitance manometer; although disagreeing in absolute magnitude with theoretical calculations, the overall shape of the measured temperature dependence on the Xe partial pressure was verified, thus supporting the cooling results.

B. Organic Dye Solutions

The question of whether anti-Stokes emission from fluorescent dyes dissolved in liquid solvents could yield a cooling effect was first asked by Vavilov (1945), which reflected spectroscopists' great interest in such systems over the preceding two decades (Jablonski, 1933; Wood, 1928). As late as 1970, this question was still being answered in the negative, based on a theoretical analysis of experimental results for a variety of molecules including rhodamine-B and fluorescein in ethanol (Ketskeméty and Farkas, 1970). Two years later, however, Erickson (1972) performed a careful set of measurements of anti-Stokes emission by rhodamine 6G in both ethanol and glycerol solvents and found that the luminescence quantum efficiencies remain independent of the wavelength of excitation, even for pump frequencies 2500 cm^{-1} smaller than the 0–0 transition frequency. Specifically, yellow emission out of the first-excited singlet S_1 manifold was observed following 632.8-nm *HeNe* excitation starting from high-lying rovibrational levels in the singlet-ground-state S_0 manifold. The long-wavelength absorption was exponential in $h\nu/k_B T$ and could be simply modeled by assuming a Boltzmann population distribution over a set of equally spaced levels in the ground electronic manifold. All transitions to the excited state originating in these levels were taken to occur with the same cross section. From his data, however, Erickson estimated an absolute fluorescence quantum efficiency for rhodamine 6G in ethanol of only 0.88. As a possible explanation for this low value, he noted a broad luminescence peak at ~665 nm that did not fit his model, nor could the peak be attributed to triplet states, dimers, or photobleaching products, and thus was ultimately attributed to fluorescent impurities that must have quenched the rhodamine emission.

Simultaneously, Chang *et al.* (1972) were making similar measurements on rhodamine-B in methanol and in polyurethane thin films. Assuming the absorption coefficient of the host to be negligible, they deduced that the ratio of the cooling power to the absorbed laser power (i.e., the cooling efficiency) is given by $\lambda \eta_Q / \lambda_F - 1$, where λ is the pump wavelength, η_Q is the fluorescence quantum efficiency, and λ_F is the mean fluorescence wavelength. For 632.8-nm excitation of rhodamine-B in methanol, this group deduced that a minimum value for the quantum efficiency of 0.94 is required to obtain cooling. Chang *et al.*'s conference abstract ended with the sentence, "This cooling effect is being investigated," yet no further results of this research were published and the field of laser cooling of condensed materials remained dormant for the next 20 years.

It was only in the early 1990s that Drexhage's research group in Siegen, Germany succeeded in laser cooling a solvated organic dye (Zander, 1991; Zander and Drexhage, 1995). They found it necessary to modify Chang *et al.*'s expression for the cooling efficiency by subtracting the quantity A_{nf}/A_{dye}, where A_{nf} is the spurious absorptance due to both nonfluorescent impurities and the solvent itself, and

A_{dye} is the absorptance of the dye molecules alone. In the wavelength region of interest, the intrinsic ethanol absorptance was independently measured and found to be significantly lower for the monodeuterated form of the solvent than for the nondeuterated form. In turn, Erickson's model for the long-wavelength dye absorption was fit to A_{dye}, thus revealing a constant residual background absorptance of 2.3×10^{-4} (for a 1-cm path length) in this wavelength region. This excess absorption was attributed to impurities present in the solute but, impressively, was fully eliminated when the starting dye was purified by column chromatography. The mean fluorescence wavelength λ_F was calculated as $[\int F_\lambda \, d\lambda][\int F_\lambda \, d\lambda/\lambda]^{-1}$, where F_λ measures the emission spectrum, although this expression is not rigorously correct unless F_λ is the spectral intensity normalized by the photon energy—see Eq. (2) of Section IV.A. In any case, λ_F was thus found to be 563 nm for a 10^{-5}-M solution of rhodamine 6G at 293 K. Combining their spectroscopic results, Zander and Drexhage concluded that cooling of the dye solution would occur, at an optimal pump wavelength of 579 nm, if the fluorescence quantum efficiency exceeded a threshold value of 0.984.

A photothermal lensing technique was then used to measure the actual value of η_Q. In this experiment, the output beam from a krypton-ion-pumped dye laser was brought to a focus about one Rayleigh length in front of the sample. Within the sample itself, this pump beam's radial intensity distribution generated a temperature gradient in the solution that was associated, in turn, with a gradient in the liquid's refractive index. Because the index decreased with increasing temperature, a weak diverging lens formed in the liquid at pump wavelengths that heated the sample. Defocusing was then observed for a weak but stable *HeNe* laser probe beam that overlapped the pump beam. After blocking the unabsorbed pump radiation with a bandpass filter, the degree of defocusing could be detected by measuring the probe beam power transmitted through a pinhole centered on the *HeNe* optical axis.

A key advantage of this photothermal lensing technique is that an analytic expression for the time dependence of the signal can be obtained. It is a function of just two parameters, t_c and θ, where t_c is a characteristic thermal diffusion time that depends on the diameter of the pump beam and on the density, specific heat, and thermal conductivity Λ of the solution. The quantity θ measures the rate of heat deposition and depends on the difference between the absorbed and emitted power, the laser wavelength, the gradient in the refractive index, and Λ. Thus, in principle, the data could be fit with no free parameters to yield the fluorescence quantum efficiency. In practice, the model's dependence on the inexactly known thermo-optic coefficients was eliminated by comparison with the signal obtained from a reference sample composed of a nonfluorescent compound dissolved in the same solvent. The results were $\eta_Q = 0.980$ for an air-equilibrated solution of rhodamine 6G-perchlorate in C_2H_5OD, a value that rose to 0.990 after deaeration of the sample with a nitrogen gas stream. Thus, by all indications, the sample would cool if pumped in the wavelength range of 570–585 nm. Zander and Drexhage indeed observed a switch in the sign of the thermal lens from diverging to converging as the pump wavelength was tuned beyond 570 nm, conclusively demonstrating cooling of the internal pumped volume of the sample. They measured a peak absolute cooling efficiency P_{cool}/P_{laser} of 3.2×10^{-5} at a pump wavelength of 575 nm, corresponding to a cooling power of 1 μW and to a relative cooling efficiency P_{cool}/P_{abs} of 1.1%.

More recently, Clark and Rumbles (1996) have also observed laser cooling of a rhodamine dye dissolved in alcohol. Specifically, rhodamine 101 was chosen, as it exhibits little triplet-state crossover or two-photon absorption. Their sample consisted of a 0.3-mL volume of 10^{-4}-M rhodamine 101 in acidified ethanol, which was sealed into a cylindrical, fused-silica tube following degassing of the solution by freezing and vacuum pumping. The tube was suspended in an evacuated cryostat, the temperature of which was initially stabilized, but it is not clear whether the cryostat temperature was subsequently monitored for drifts over the multihour course of each cooling run. The sample was optically pumped in the 580–680-nm range with up to 350 mW from an argon-ion-pumped cw dye laser. At 15-minute intervals, the dye laser was blocked and the fluorescence intensity measured at 620 nm using a monochromator. (This fluorescence was excited by an auxiliary 1-mW *HeNe* laser beam focused into the pumped volume of the liquid.) The sample temperatures could then be deduced from these intensity values via a previously determined calibration curve obtained by measurement of the emission intensity for known temperatures in the range of 150–300 K. The dye solution was observed to heat when pumped at 583 and 605 nm and to cool when pumped at 620 and 634 nm, as expected because the former wavelengths correspond to photon energies larger than the average emitted photon energy whereas the latter wavelengths correspond to smaller energies. In particular, a maximum temperature drop of 3 K below ambient (290 K) was measured after pumping the sample at 634 nm for 4 hours with 350 mW of laser power.

Several criticisms, however, have been directed toward Clark and Rumbles' results (Mungan and Gosnell, 1996). An emissivity near unity is estimated for an ethanol–silica combination in the range of wavelengths over which radiative emission is significant for a room-temperature blackbody. This implies that the sample was subject to a blackbody thermal load that was larger than the actual laser-induced cooling power, a problem that is not considered in the original paper. Second, Clark and Rumbles observed a linear change of the sample temperature even after 4 hours exposure to the pump laser, at which point no further measurements were made. However, theoretically an exponential dependence of temperature on pumping time is expected; their sample's heat capacity of 3 J/K would have entailed a time constant of only 40 min. Hence the temperature should have leveled off within 2 hours. Finally, the original experiments did not include a reference run performed with the dye laser tuned to the null wavelength where the sample should neither heat nor cool. Such a run would have provided a good measure of the overall thermal drifts of the system and a powerful test of the underlying theory.

These criticisms are partially resolved in a reply by Rumbles and Clark (1996). A 4-hour reference run of the type just described showed that no net heating or cooling occurred during the run. However, temperature fluctuations in the latter measurement spanned a range of 3 K, leaving unaddressed the issue of thermal drifts of the system. The question of the radiative heat load is not discussed in Rumbles and Clark's reply except for an unsubstantiated comment that the radiative coupling between the sample and surroundings must be "less than optimum." More positively, a new run showed the return of the sample temperature to the ambient value following laser cooling at a pump wavelength of 635 nm. This result is encouraging, but again the detailed shape of the curve at long times is unclear. One would expect the time dependence of the sample temperature during both the cooling and the reequilibration periods to follow an exponential form with the same time constant. This prediction deserves testing.

In any case, it may prove difficult to scale up the dye results to obtain larger temperature changes. Clark and Rumbles themselves note that increasing the solute concentration leads to the formation of nonfluorescent molecular aggregates such as dimers, as well as to an efficiency-reducing increase in the mean emission wavelength due to reabsorption of the emitted radiation. Already, the peak emission wavelength of their 10^{-4}-M rhodamine 101 solution was found to be 8 nm longer than the peak wavelength obtained in the limit of zero optical density. Reducing the sample size does not ameliorate these problems because the surface of the sample cell promotes nonradiative relaxation; an analogous issue arises in the anti-Stokes fluorescence cooling of gases. Embedding the dye molecules in a polymer host may eliminate this problem if the quantum efficiency remains high and if photodegradation of the dye does not occur. This suggestion emphasizes the greater practicality of cooling solids rather than fluids.

C. SEMICONDUCTORS

In the early 1960s, gallium arsenide was identified as a candidate for fluorescence refrigeration. Keyes and Quist (1962) fabricated diodes by diffusing Zn into single-crystal n-type $GaAs$ to create a p-type overlayer. The forward current-voltage characteristics were measured at 298 and 77 K, and the current i was found to vary exponentially with $eV/2k_BT$ just below saturation. The luminescence spectra of the diode were measured and consisted of two peaks at both temperatures, with the higher energy peak—centered at 1.33 eV at 77 K and 1.44 eV at 298 K—identified with emission across the bandgap. On the other hand, the relative intensity of the broad, low-energy peak was sample- and temperature-dependent and hence was presumably not intrinsic. An absolute calibration of the emission intensity indicated that approximately 40% of the injected electrons gave rise to externally emitted bandgap photons at 298 K, an amount that fell to about 5% at 77 K because internal reflection trapped a large fraction of the radiation. After correcting for this effect, an internal quantum efficiency for bandgap emission of between 0.48 and 0.85 was deduced. The term *internal* quantum efficiency is here defined as the ratio of the number of photons emitted anywhere inside the material to the number of injected electrons. In contrast, the *external* quantum efficiency is the corresponding quantity in terms of photons that actually escape from the medium. Keyes and Quist found that the emission intensity varied linearly with the density of the injection current, saturating at about 2.5 kA/cm^2. At 77 K, the luminescence spectrum exhibited a weak tail extending well into the visible range. These researchers thus concluded that electroluminescence refrigeration with a cooling power of $i(V_F - V)$, where V_F is the average emitted photon energy in volts, would be feasible if the external quantum efficiency could be made large enough.

Further measurements related to this suggestion were made at temperatures down to 27 K by Dousmanis *et al.* (1963, 1964), as has been reviewed by Pankove (1975). Beginning with a thermodynamic analysis, these workers showed that the cooling efficiency η cannot exceed T/T_{B_h}, where T is the sample temperature and T_{B_h} is the fluorescence brightness temperature. The latter quantity is related to the emitted photon occupation number n_ν^h according to $n_\nu^h = 1/[\exp(h\nu/k_B T_{B_h}) - 1]$, with $h\nu$ the photon energy. (Note that the total rate of photon emission is $\int n_\nu^h \, d\nu = \eta_Q i/e$.) At low carrier currents i, $T_{B_h} \approx T$ and the cooling efficiency η is expected to be large, although this operating condition is not particularly useful as the cooling rate would be negligibly low. In the opposite limit of high currents, the lasing threshold of the *GaAs* diode is reached, corresponding to $T_{B_h} \to \infty$, and η falls to zero. At intermediate currents, however, the frequency of the spectral peak of the incoherent emission favorably rises with the drive current; it likewise rises by increasing the dopant concentration. On the other hand, Joule heating limits the cooling rate obtainable at large carrier currents. It is also clear that $\eta \to 0$ as $T \to 0$. These facts imply that compromises are required in selecting the operating current, dopant concentration, and operating temperature.

Dousmanis *et al.*'s best diode was prepared by epitaxial solution growth of an *n*-type *GaAs* layer on a *p*-type substrate. For an applied voltage V of 1.335 V (corresponding to a vacuum wavelength of $hc/eV = 927$ nm), an injection current of 5 mA, a diode temperature of 78 K, a *Zn* acceptor concentration of 3×10^{19} cm^{-3}, and a *Te* donor concentration of about 10^{19} cm^{-3}, the luminescence band peaked at 897 nm and about 94% of the emitted photons had energies exceeding eV. Taking this peak wavelength to correspond to the average emission energy implies a cooling efficiency of 3%, assuming unit quantum efficiency and zero Joule heating. To put it another way, 0.97 is the threshold value of the external quantum efficiency for cooling's to occur. However, because of the diode's ~0.5-Ω resistance and the commensurate i^2R loss, the threshold value of the external quantum efficiency rises to 0.99 for an operating current of 30 mA. By reducing the current to 10 mA, 250 mW of net cooling is expected at 78 K if $\eta_Q = 0.99$. Dousmanis et al., however, never reported a direct attempt at measuring the cooling effect.

The previous discussion suggests that one can do better through optical rather than electrical pumping of the sample, thereby avoiding Joule heating and eliminating the need for a *p-n* junction. Even so, a system with a near-unit external quantum efficiency is required. For this reason, no further experimental progress on laser cooling of semiconductors was achieved until recently. In 1993, a team of researchers at Bellcore (Schnitzer et al., 1993) demonstrated an internal fluorescence quantum efficiency of 99.7% ± 0.2% for a 500-nm thick, CVD-grown *AlGaAs/GaAs/AlGaAs* double heterostructure etched off of an *AlAs* release layer. The sample was Van der Waals bonded to an SiO_2-coated gold reflector and optically pumped at 780 nm using a cw *AlGaAs* laser. The active *GaAs* layer was *p*-doped and had a refractive index of 3.54, implying an emission escape cone due to total internal reflection encompassing only 2% of 4π steradians. Furthermore, even within the escape cone, the transmittance is only ~30%. Hence fluorescence was trapped between the sample surfaces, and consequently absorbed and reemitted ("reincarnated") many times before final escape. Such trapping multiplies the effect of parasitic losses such as those due to absorption by dark impurities, nonradiative electron-hole recombination, and absorption by the gold mirror. As a result, the measured external quantum efficiency was only 72%. From this measurement, however, the previously quoted value of 99.7% was deduced for the internal quantum efficiency. The Bellcore team further found that 23% of the lost quantum efficiency could be attributed to absorption by the *Au* reflector whereas only 5% derived from bulk parasitic effects. For purposes of solid-state photoluminescence cooling, the latter figure is encouragingly small.

These results motivated Gauck et al. (1995, 1997) to attempt bulk laser cooling of a direct-bandgap, III-V semiconductor heterostructure. The samples were grown by vapor-phase epitaxy at 1000 K and consisted of a nominally undoped *GaAs* active layer of $0.5-2-\mu$m thickness embedded between two passivating layers of 2- to 3-μm-thick $GaInP_2$. The basic idea is to pump the material in the Urbach absorption tail, thus creating cold-free carriers at the bottom of the conduction band and a complementary population of holes at the top of the valence band. Subsequently, intraband thermalization of the free carriers followed by radiative recombination would extract heat from the semiconductor lattice. By virtue of the picosecond time scale for the thermalization steps and the nanosecond time scale for radiative relaxation, a high density of cooling power could in principle be sustained with this approach.

This simple picture, however, is complicated by the existence of two nonradiative relaxation mechanisms intrinsic to the sample. On the one hand, surface recombination, whose rate increases linearly with the carrier density, dominates the nonradiative loss at low densities. On the other hand, Auger recombination, which increases with the cube of the density, dominates the loss at high densities. Because of the quadratic dependence of the radiative recombination rate on carrier density, a density that optimizes the fluorescence quantum efficiency must therefore lie at some intermediate value.

The best experimental results were obtained for a 0.5-μm-thick *GaAs* layer within which the optimum carrier density was calculated to be 8×10^{17} cm^{-3}. In order to enhance the escape probability of the emitted photons, the 0.5-mm diameter heterostructure was mounted in optical contact with the flat surface of a *ZnSe* hemisphere 4 mm in radius. This material was chosen for its very low absorption over the *GaAs* emission band, despite its nonideal index match to the sample— *ZnSe* has an index of refraction of only 2.50. The hemisphere was anti-reflection-coated on its curved surface and the sample was pumped from its back side with a titanium-sapphire laser. Monte Carlo simulations were used to estimate that this output coupler reduced the number of photon reincarnations from 40 to 6, thus greatly increasing the external quantum efficiency η_Q^{ext}. The sample and output-coupler combination were suspended *in vacuo* and a thermistor used to measure their temperature. With the pump laser tuned to the mean emission wavelength of $\lambda_F = 866$ nm, the observed heating rate gave a direct measure of η_Q^{ext}, namely 0.96. This value corresponded to an internal quantum efficiency of $(0.96)^{1/6} = 0.993$. Although impressive, the longest wavelength at which the sample could be pumped with the intensity necessary to excite the optimum carrier density was 888 nm—a wavelength shifted from λ_F by only 1.4 $k_B T$—and therefore net cooling was not observed. On the other hand, the heating rate was found to decrease with the expected linear dependence on increasing wavelength all the way to the 888-nm limit, indicating that free-carrier absorption must be negligible. Under the assumption that this continues to hold true out to a pump detuning from λ_F of 2.2 $k_B T$, thermal breakeven would be obtained with the existing external quantum efficiency. Conversely, breakeven would occur at the 888-nm pump wavelength by merely increasing the external quantum efficiency to 0.972. Net laser cooling of *GaAs* thus appears achievable in the near future.

D. RUBY

An anti-Stokes fluorescence cooler is essentially a laser run in reverse: a coherent, directional beam of light illuminates an active medium yielding higher-frequency, broadband, more or less isotropic radiation as output. This viewpoint has motivated many of the choices of working materials for laser cooling in this section. In particular, within a year of the first successful demonstration (Nelson and Boyle, 1962) of cw lasing in ruby $(Al_2O_3 : Cr^{3+})$, Tsujikawa and Murao (1963) proposed that optical cooling might be achieved in this material at temperatures near a few tens of kelvin. The cooling process would work by optical pumping from the Cr^{3+}

electronic ground state to the lower member of the ion's twofold ^2E excited-state manifold, followed by thermalization across the 29 cm^{-1} energy gap between these two member states. Subsequent emission from both levels (i.e., on the well-known R_1 and R_2 lines) would return the pumped ions to the ground state; the result of the cycle is thus extraction of heat from the host lattice. These researchers even suggested that cooling might be achievable in the millikelvin range by taking advantage of the 0.38 cm^{-1} splitting of the Cr^{3+} $^4A_2$ ground-state manifold.

Interestingly, Tucker (1961) at General Electric had previously observed the amplification of 9.3-GHz phonons resonant with this Zeeman-split pair of ground states when a population inversion was prepared by microwave pumping. In other words, stimulated phonon emission occurred as a result of a very strong spin-lattice coupling at this frequency. This concept has been recently extended to the amplification of 29-cm^{-1} phonons by selective excitation of the upper of the ^2E levels with a pulsed excimer-pumped dye laser (Fokker *et al.*, 1997a, 1997b). The ultimate goal of such experiments (Kittel, 1961) is to demonstrate "sound amplification by stimulated emission of phonons," to paraphrase the usual acronym. By contrast, an anti-Stokes ruby cooler would consist of optically pumping the lower ^2E level, rather than the upper one, and taking advantage of the strong spin-lattice coupling to absorb phonons rather than emit them.

Tsujikawa and Murao performed a detailed 4-level rate-equation analysis using the known R_1 lifetime and relative absorption cross sections for light polarized parallel or perpendicular to the trigonal axis of the host lattice. The dopant concentration was taken to be 500 ppm Cr_2O_3 by weight to give a peak absorption coefficient of 2 cm^{-1}. Further, Tsujikawa and Murao suggested silver plating the back side of the sample to enhance the pumping. The cooling rate was then computed in terms of the predicted cooling power as $\dot{T} = -\dot{Q}_c/C$, where C is the heat capacity of the sample, taken to be of the Debye T^3 form with $\theta_D = 800$ K. The paper ends with some general remarks about limiting factors such as direct multiphonon or tunnel-assisted relaxation, deexcitation via Cr^{3+} aggregates, ion-lattice equilibration times, and self absorption of the fluorescence. However, by and large, not enough data were available at the time to characterize these processes in detail.

In fact, two years later Nelson and Sturge (1965) published a detailed analysis of the spectroscopy of ruby that clearly rules out the possibility of laser cooling of this system. The emission spectrum (see their Fig. 5) shows the existence of phonon sidebands on the Stokes side; although weak, they are sufficiently broad that their integrated area is comparable to that of the R lines! These sidebands peak at about 400 cm^{-1} below the 14,400-cm^{-1} $^2E \rightarrow {}^4A_2$ transitions and extend down to almost 13,400 cm^{-1}. Hence fluorescence at the sideband frequencies with the combined emission of a photon and an average phonon will completely overwhelm the 29 cm^{-1} worth of cooling derived from the zero-phonon R_2 emission. More specifically, Nelson and Sturge defined the radiative efficiency of the R lines as the ratio of their integrated fluorescence intensity to the total integrated fluorescence intensity. This quantity depends on temperature, polarization relative to the c-axis, and chromium ion concentration. For optically thin samples, the efficiency rises smoothly with decreasing temperature, leveling off at about 0.49 for polarization parallel to the c axis and at 0.83 for the perpendicular polarization. Given the magnitude of the Stokes shift in the sidebands, these results definitively rule out ruby as a cooling candidate.

E. RARE-EARTH IONS IN SOLIDS

As in the case of ruby, the death knell of many otherwise promising doped solid-state coolers is Stokes' deactivation of the excited active ions via the phonon sidebands. An important advantage of the rare-earth ions is that their optically active $4f$ levels are well-shielded from the lattice by the filled, outer $5s$ and $5p$ shells. This greatly reduces the strength of the vibronic sidebands, sharpens the homogeneous lines (leading to larger absorption coefficients and hence more efficient pumping), and suppresses multiphonon nonradiative relaxation.

Kastler (1950) was the first to propose fluorescence cooling of solids doped with rare-earth ions. For an appropriately doped salt crystal, he suggested pumping the weak anti-Stokes vibronic sidebands of rare-earth electronic transitions, which is necessarily accompanied by the absorption of lattice energy. Cooling would then result, assuming near-unit fluorescence quantum efficiency for the zero-phonon transitions back to the ground state. Kastler suggested that the cooling effect, though small, could be enhanced by the use of thermal shields that pass the fluorescence but block the blackbody radiation from the environment. In principle, this proposal is sound. However, the absorption strength of rare-earth vibronic sidebands has generally proved too small to encourage actual experiments.

Yatsiv (1961), at the Second International Conference on Quantum Electronics, was the first to clearly spell out the prototypical cooling cycle depicted in Fig. 1. He considered two groups of energy levels, of which one or both had intragroup level spacings on the order of $k_B T$ (adjustable at low temperatures by application of an external magnetic field), whereas the groups themselves were separated from one another by a large intergroup energy gap. Specifically, a gap of at least 10,000 cm^{-1} was recommended, both for ease of optical pumping and to minimize the probability of nonradiative decay between the groups. A narrow-bandwidth source would be used to excite individual levels from the top of the ground-state group to the bottom of the excited group, so that no Stokes' emission would occur. Yatsiv suggested three pump sources for such an experiment: a high-intensity arc lamp from which a suitable frequency range is selected using a monochromator, a flash-lamp-pumped crystal identical to the cooling sample whose fluorescence spectrum is long-pass filtered, or, presciently, appropriate optical masers.

Under the assumption that the excited-state thermalization time is much shorter than the intergroup radiative relaxation times, Yatsiv performed a rate-equation analysis to derive the steady-state cooling rate. He specifically considered Gd^{3+}, whose large, ~33,000 cm^{-1} energy gap between the sharp $^8S_{7/2}$ ground-state and the $\{^6P_{7/2}, ^6P_{5/2}, ^6P_{3/2}\}$ excited states would guarantee negligible nonradiative relaxation. A Gd^{3+} concentration of 1 at% was recommended in order to balance the competing requirements of good pump absorption and minimal self absorption of the fluorescence. The radiative lifetime of Gd^{3+} out of the $^6P_{7/2}$ multiplet in stoichiometric, hydrated gadolinium chloride was known to be 7.8 ms, from which Yatsiv deduced values for the integrated absorption and emission cross sections. For a pump intensity of 1 mW/cm^2, he thus estimated a cooling-power density of 3 μW/cm^3 at a sample working temperature of 10–40 K.

The first solid-state laser-cooling experiment of any kind was performed in 1968 by Kushida and Geusic (1968) at Bell Laboratories. They chose to work with YAG + 1 at% Nd^{3+} because this material could be used both as the refrigeration sample and to generate the pump laser radiation; in their experiment, the lasing and cooling crystals were placed into the same optical cavity. The 0.1-inch-diameter by 2-inch-long cooling sample was supported by 3 needles inside an evacuable cell. The inner surfaces of this cell were gold coated, presumably to redirect scattered pump light back onto the sample, although this technique suffers the serious disadvantage of increasing the fluorescence self-absorption. When pumped at 1.064 μm, temperature changes of the sample were measured with an attached thermocouple and compared to those observed for an otherwise identical, undoped YAG rod.

For their experiment, Kushida and Geusic expected a volume cooling rate of $n_2 E_{cool}/\tau$, where n_2 is the excited-state population density, τ is the fluorescence decay time, and E_{cool} is the average absorbed phonon energy. Assuming thermal equilibrium among the emitting states and neglecting the small, intrinsic, nonradiative decay rates, an estimated value for E_{cool}/hc of 90 cm^{-1} was obtained by an appropriate sum of the net energy differences over all possible upper and lower levels. In vacuum, a sample temperature drop of 8.4 K below room temperature was thus predicted for a 100-W pump, whereas in air the predicted drop was 2.1 K. Under the latter conditions, the experiment yielded the expected time dependence of the sample's temperature but net cooling was not observed. Instead, only reduced heating was obtained: the final, steady-state temperature was found to be 0.6 K less than that of the undoped reference sample at a cavity power of 100 W. (Surprisingly, no cooling data is reported for the sample *in vacuo*.)

In search of an explanation for their results, Kushida and Geusic interpreted the ~2-K temperature increase they observed for the reference sample as being due to direct absorption of the laser light by nonfluorescent impurities. For example, 50 ppb of Dy^{3+} would account for the observed heating if this ion relaxed entirely by nonradiative processes. It was initially assumed that the doped sample would exhibit this same level of background heating. However, the fact that a temperature difference of only 0.6 K was measured between the sample and the reference rather than the calculated 2.1 K implied that other parasitic heating effects must also have afflicted the doped sample. In particular, a reduction in the fluorescence quantum efficiency to 0.995 was enough to account for the discrepancy. In turn, this quantum efficiency could have resulted from a 30 s^{-1} multiphonon decay rate across the 4700-cm^{-1} $^4F_{3/2} \rightarrow {}^4I_{15/2}$ energy gap of Nd^{3+}, a figure in good agreement with known nonradiative rates in other rare-earth ions. Kushida and Geusic mention that a similar nonradiative heating effect was seen when they attempted to laser cool the $^5I_7 \rightarrow {}^5I_8$ transition of YAG:Ho^{3+}. Unfortunately, no further details of this experiment are cited.

Chukova (1974), on the other hand, has argued that Kushida and Geusic's failure to obtain net cooling was due to an overly high intensity and an insufficiently narrow bandwidth of the pump source. She based these comments on purely thermodynamic calculations of the entropy fluxes of the laser and fluorescence radiation, obtaining a cooling efficiency of ~30% for a bandwidth of 10 MHz in the limit of zero pump intensity. This figure rises to over 60% as the laser line becomes infinitely sharp, but falls to zero for an intensity of about 10 W/mm². In fact, she believed that a temperature drop of 37 K should have been obtained with a pump intensity of 1 W/cm². However, this analysis appears to place too much weight upon idealized theoretical concepts. For example, it is clearly pointless to reduce the excitation bandwidth much below the absorption line's homogeneous width, which is certainly larger than 10 MHz at room temperature! All the more then, taking into consideration the strict, practical factors mentioned in the preceding paragraph, it is surprising and encouraging that Kushida and Geusic managed to get as close to absolute cooling as they did.

Nevertheless, no further experimental attempts to laser cool rare-earth-doped solids were performed until 1995, when net cooling of a Yb^{3+}-doped glass was successfully demonstrated at Los Alamos National Laboratory. A detailed discussion of these results follows.

IV. Laser Cooling of Ytterbium-Doped ZBLANP Glass

The trivalent ytterbium ion is an ideal dopant for anti-Stokes fluorescence cooling of a solid host: within the energy gap of typical insulators, the ion's energy-level structure consists of only two Stark multiplets, namely, the $^2F_{7/2}$ ground-state multiplet and the $^2F_{5/2}$ excited-state multiplet located about 1.3 eV above the ground state. In all but the highest-symmetry hosts, the ground multiplet is split into a quartet of doubly degenerate states—the so-called Kramers' doublets—whereas the excited-state multiplet is split into a triplet of such states (see Fig. 4 for the detailed level structure in ZBLANP). Because the energy gap between the two multiplets is so large, multiphonon relaxation across the gap occurs at a rate orders

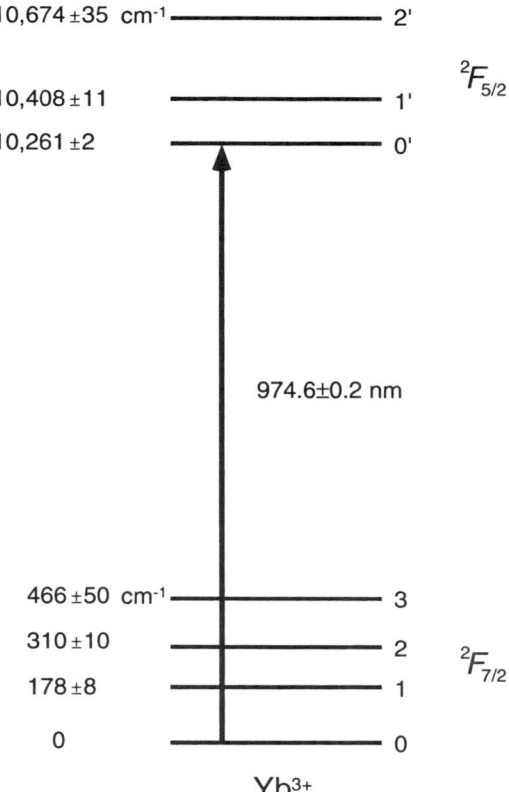

FIG. 4. Energy-level diagram for Yb^{3+} in a ZrF_4-BaF_2-LaF_3-AlF_3-NaF-PbF_2 (ZBLANP) glass host. The assigned energies are determined from our emission and absorption spectra measured at a sample temperature of 2 K.

of magnitude smaller than the radiative relaxation rate, γ_{rad}, of 10^2–10^3 s^{-1} (DeLoach et al., 1993; Hasz et al., 1993). Because of the spare energy-level structure, pairs of proximate Yb^{3+} ions do not participate in energy-transfer reactions that open nonradiative relaxation pathways out of the $^2F_{5/2}$ multiplet, an effect that is otherwise common for excited states in rare-earth-doped solids. These factors combine to produce an emission quantum efficiency for ytterbium that approaches unity, even for high dopant concentrations. The overlap of the Yb^{3+} $^2F_{7/2} \rightarrow {}^2F_{5/2}$ absorption band with the broad tuning range of the Ti:sapphire laser is of great advantage in performing experiments with this electronic species.

Figure 5 shows the absorption and emission spectra of Yb^{3+}-doped ZrF_4-

BaF_2-LaF_3-AlF_3-NaF-PbF_2 (ZBLANP) glass. The choice of this host material, a variation of the more common ZBLAN glass composition, is motivated by three considerations. The most important is that much effort has been devoted to producing high-purity ZBLAN-based glasses because of their potential applications in ultra-low-loss optical communications (Aggarwal and Lu, 1991). However, compared with silica—indeed compared with many other glass hosts (Zou and Toratini, 1995)—Yb^{3+} in ZBLAN glasses exhibits a greater anti-Stokes frequency shift for a given value of the absorption coefficient. Hence this material holds the promise of greater cooling power than other glass hosts. Finally, as a heavy-metal glass, the material's maximum phonon frequency of only 580 cm^{-1} (Deol et al., 1993) helps inhibit nonradiative multiphonon relaxation in both the Yb^{3+} ion itself and, perhaps more importantly, in other impurities that might quench the Yb^{3+} emission.

Another important property of this impurity system is that the emission spectrum at room temperature is independent of the pump wavelength and pump intensity. This implies that the 7-level Yb^{3+} system can be treated as homogeneously

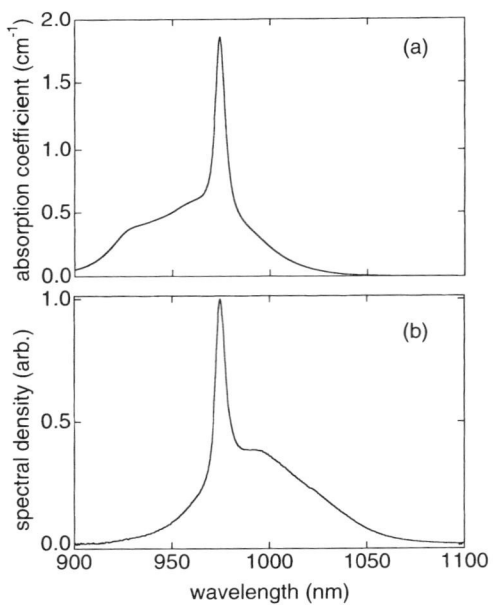

FIG. 5. (a) Absorption coefficient at 300 K for a ZBLANP host doped with 1 wt% Yb^{3+} or 2.42×10^{20} ions/cm^3. (b) Fluorescence spectrum of ZBLANP: Yb^{3+} at 300 K.

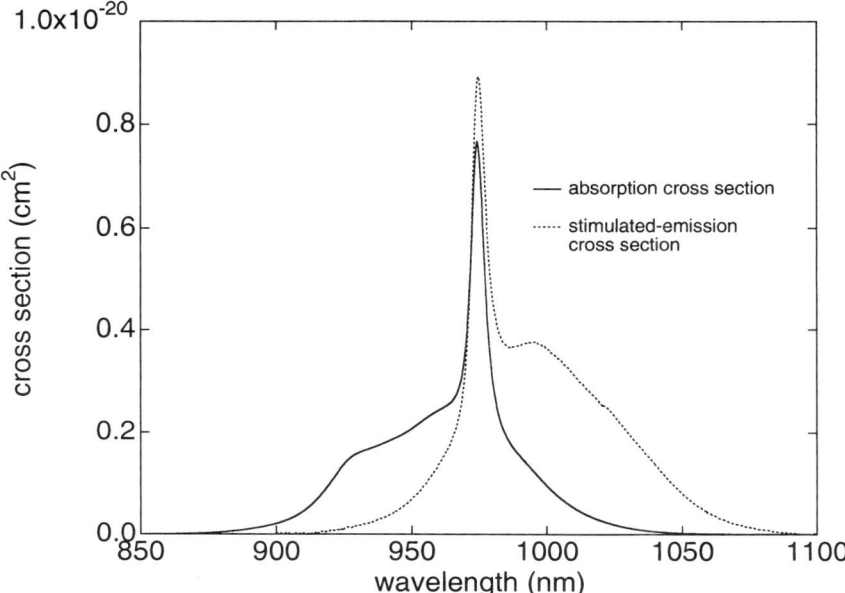

FIG. 6. Absorption and stimulated-emission cross sections of Yb^{3+} in a ZBLANP host. The stimulated-emission cross section is calculated from the fluorescence spectrum of Fig. 5(b) using the Füchtbauer-Ladenburg equation as quoted by DeLoach et al. (1993).

broadened and that it is therefore reducible to an effective two-level system. Within this two-level model, a single population density is associated with each of the ground- and excited-state manifolds. In turn, the absorption cross sections for individual transitions between various members of the ground- and excited-state multiplets are collapsed into a single wavelength-dependent equivalent that is applicable across the entire $^2F_{7/2} \to {}^2F_{5/2}$ band. An analogous stimulated-emission cross section is similarly defined. Figure 6 shows our measurements of these quantities at room temperature.

A. PHOTOTHERMAL DEFLECTION SPECTROSCOPY

An important technique for assessing the potential of condensed-matter systems to exhibit laser cooling and for diagnosing parasitic heating processes is photothermal deflection spectroscopy (Jackson et al., 1981). Aside from the ease with which such measurements can be performed, photothermal deflection spectros-

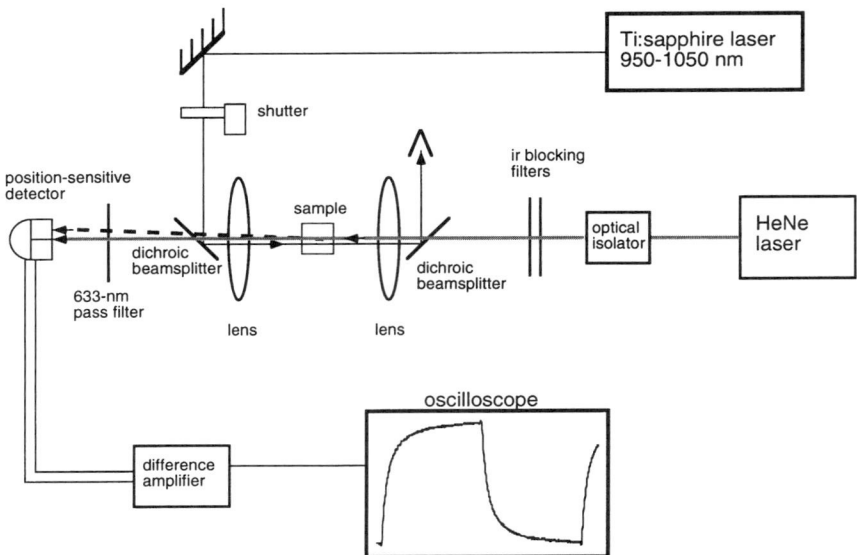

FIG. 7. Schematic of the photothermal deflection spectrometer. The pump beam (thin dark line) induces a refractive-index gradient in the sample that in turn deflects the *HeNe* probe beam (thick gray line). Detection of the deflected probe beam is accomplished with a dual split photodiode.

copy specifically addresses only microscopic aspects of the cooling mechanism as only the interior of the sample is probed.

A schematic of our spectrometer is shown in Fig. 7; the experiment works in the following way. A cw *Ti*:sapphire laser is optically chopped and focused through the sample, which has been well polished on the entrance and exit faces. A counterpropagating *HeNe* laser beam is focused through the same region of the sample as the pump beam but is slightly displaced from it in order to probe the region where the pump-induced thermal gradient is largest. The thermal gradient, in turn, gives rise to a refractive-index gradient that can be viewed as comprising a weak prism in the sample. Heating or cooling of the host lattice arising from nonradiative interactions between the host and the impurity ions will therefore deflect the *HeNe* probe beam when the pump beam is turned on. This deflection is detected with a position-sensitive detector optically isolated from the infrared pump beam; for the small magnitudes of deflection in our experiments, a dual split photodiode gives a linear response directly proportional to the magnitude of the index gradient generated in the sample. The deflection signal is then averaged with a digital oscilloscope or demodulated with a vector lock-in amplifier referenced to the optical chopper. A chopping frequency of between 0.5 and 1 Hz is used to ensure that thermal steady state is nearly attained within the sample.

With this configuration, a definitive signature of anti-Stokes fluorescence cooling is the 180° signal-phase shift expected when the pump wavelength is tuned from the Stokes to the anti-Stokes region of the absorption spectrum. Figure 8 shows such a transition when the pump wavelength is tuned from 980 to 1010 nm in a 300-K ZBLANP sample doped with 1 wt% Yb^{3+}.

More information about the cooling process can be obtained by measuring the amplitude of the deflection signal as a function of pump wavelength. Such a spectrum, normalized at each wavelength by the pump power, is shown in Fig. 9 by the solid circles. An interpretation of this spectrum is possible by considering the flow of energy into and out of the pumped volume. Excluding thermal conduction, the rate of energy accumulation per unit volume in the sample for a pump wavelength of λ is

$$\dot{\rho}(\lambda) = n_{7/2}\sigma_{abs}(\lambda)I - n_{5/2}\sigma_{se}(\lambda)I - \frac{hc}{\lambda_F}\gamma_{rad}n_{5/2} + \alpha_b I + \kappa n_{5/2} \quad (1)$$

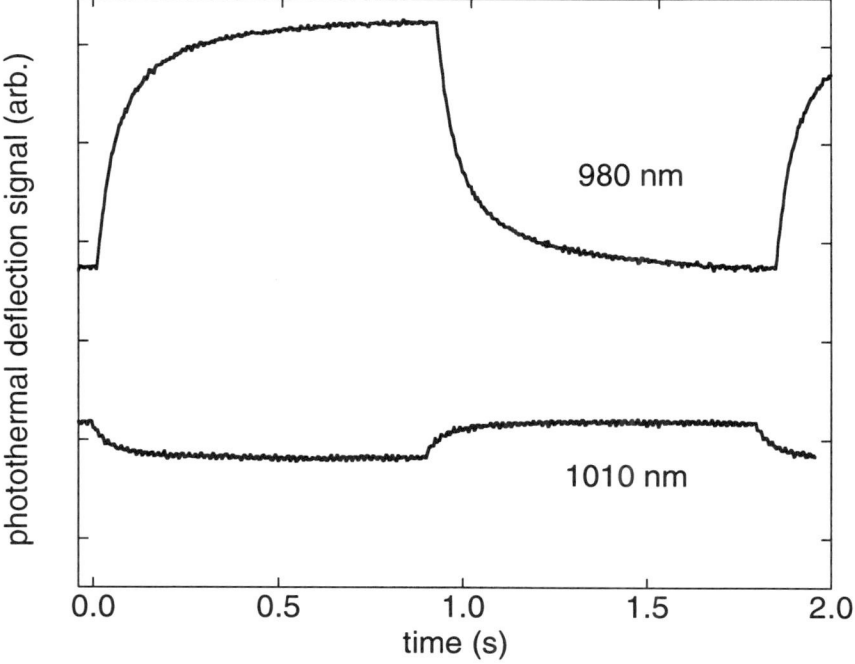

FIG. 8. Averaged signals obtained with the photothermal deflection apparatus of Fig. 7 for pump wavelengths of 980 and 1010 nm. The sample is ZBLANP + 1 wt% Yb^{3+}. The 180° phase difference between the signals indicates that a transition from laser heating to cooling has occurred.

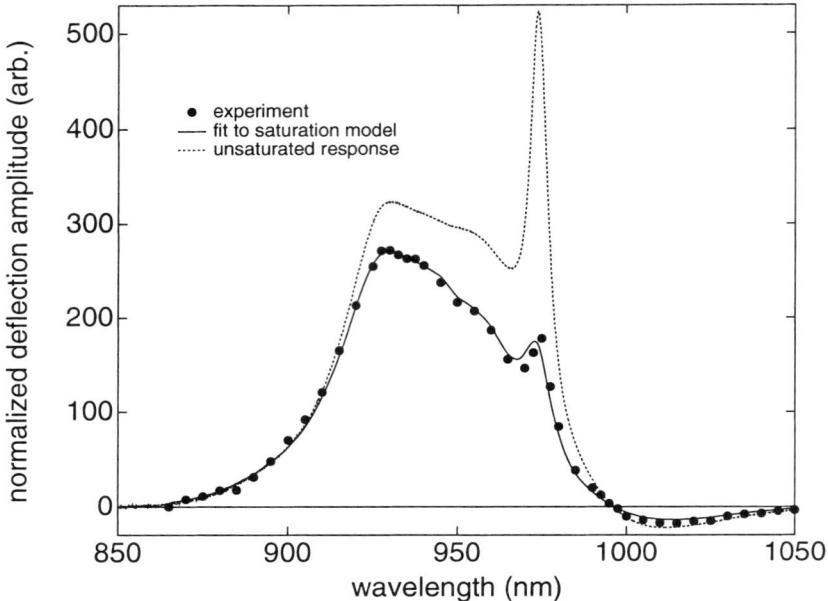

FIG. 9. Photothermal deflection spectrum of ZBLANP + 1 wt% Yb^{3+} at 300 K (solid circles), where the raw deflection amplitudes have been normalized by the pump power at each wavelength. The solid curve shows a fit of Eq. (5) to the data and indicates that optical saturation of Yb^{3+} occurs during the measurement. The dashed line shows the expected response in the unsaturated limit, as determined by Eq. (6), with the values of the needed parameters set to those found in the fit of Eq. (5).

where I is the pump intensity, $\sigma_{abs}(\lambda)$ and $\sigma_{se}(\lambda)$ are the absorption and stimulated-emission cross sections at the pump wavelength, respectively (see Fig. 6), α_b is a broadband, nonsaturable-absorption coefficient resulting from parasitic impurities such as transition metals, and κ is a heating rate resulting from nonradiative energy transfer from excited-state Yb^{3+} ions to quenching centers that subsequently relax exothermically. The quantity λ_F is of special significance in this expression; it is the wavelength of an emitted photon of average energy and is given by[2]

$$\frac{hc}{\lambda_F} = \langle h\nu \rangle = \frac{\int h\nu\, \Phi_\nu\, d\nu}{\int \Phi_\nu\, d\nu} = \frac{\int F_\nu\, d\nu}{\int \frac{F_\nu}{h\nu}\, d\nu} = hc\, \frac{\int F_\lambda\, d\lambda}{\int \lambda F_\lambda\, d\lambda} \qquad (2)$$

where Φ_ν is the emitted photon flux density (units of number per unit time per unit frequency interval), $F_\nu = h\nu\,\Phi_\nu$ is the Yb^{3+} emission spectral density (units of power per unit frequency interval), and $F_\lambda = \nu^2 F_\nu/c$ is the corresponding quantity in wavelength units. It is F_λ that is plotted in Fig. 5(b), as measured with a grating spectrometer. Finally, the factors $n_{7/2}$ and $n_{5/2}$ are the steady-state ground and excited-state population densities, respectively; they are pump-intensity dependent and must satisfy the equations

$$0 = \dot{n}_{5/2} = \frac{I\lambda}{hc}[n_{7/2}\sigma_{abs}(\lambda) - n_{5/2}\sigma_{se}(\lambda)] - \gamma_{rad}n_{5/2} \qquad (3)$$

and

$$n_{7/2} + n_{5/2} = N \qquad (4)$$

where N is the total number density of ions.

Combining Eqs. (1), (3), and (4) finally yields for the rate of laser-induced heat accumulation per unit pump intensity

$$\frac{\dot{\rho}}{I} = \frac{N\sigma_{abs}(I_s/I)(1 - \lambda/\lambda_F^*)}{1 + \sigma_{se}/\sigma_{abs} + I_s/I} + \alpha_b \qquad (5)$$

where I_s is a characteristic wavelength-dependent saturation intensity given by $I_s = hc\gamma_{rad}/(\lambda\sigma_{abs})$ and $\lambda_F^* = [1/\lambda_F - \kappa/(hc\gamma_{rad})]^{-1}$ is the effective mean emission wavelength. If the pump intensity is small ($I \ll I_s$), Eq. (5) simplifies to

$$\frac{\dot{\rho}_{I \ll I_s}}{I} = N\sigma_{abs}(1 - \lambda/\lambda_F^*) + \alpha_b \qquad (6)$$

Because the steady-state refractive-index gradient created by the pump beam is proportional to the rate of heat deposition (note that thermal conduction out of the pumped volume establishes an equilibrium index distribution), so is the deflection angle and hence the previous expressions can be applied as theoretical fits to the data of Fig. 9 with λ_F^*, α_b, and an overall scaling factor taken as adjustable parameters. Also required is an adjustable parameter a, the area of the pump spot within the probed volume; $P_{laser} = aI$ is therefore the measured pump power.

The solid line shown in Fig. 9 is a fit of Eq. (5) to the data. The excellent agreement between the model and the data validates our assertion that the Yb^{3+} system is effectively homogeneously broadened at room temperature. Also shown in the figure as the dotted line is the photothermal response that would have been obtained had no optical saturation occurred during the measurement. Although the fit determines values of all of the free parameters, a better determination of the most interesting parameters, λ_F^* and α_b, is obtained by restricting attention to the spectral region between 990 and 1050 nm where the zero-crossing occurs and where

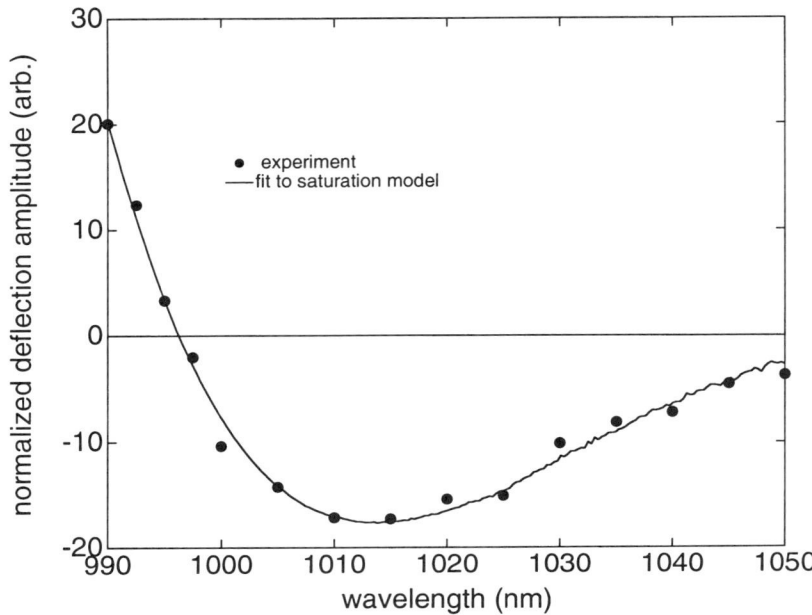

FIG. 10. Photothermal deflection spectrum of ZBLANP + 1 wt% Yb^{3+} at 300 K (solid circles) in the 990–1050 nm range. The data are normalized by the pump power. The solid curve shows a fit of Eq. (5) to the data with $\lambda_F^* = 995.3 \pm 0.3$ nm and $\alpha_b = (1.5 \pm 4.0) \times 10^{-5}$ cm^{-1}.

the data are most sensitive to the value of α_b. Over this limited wavelength range, the data and their accompanying fit are shown in Fig. 10. The fitting parameters are $\lambda_F^* = 995.3 \pm 0.3$ nm and $\alpha_b = (1.5 \pm 4.0) \times 10^{-5}$ cm^{-1}. Because the former quantity is in excellent agreement with the independently measured value of $\lambda_F = 995.5 \pm 2$ nm, we conclude that the heating rate due to energy transfer to quenching centers, κ, is indistinguishable from zero for a doping level of 1 wt%. Because the error bar on the otherwise small value of α_b is several times larger than the value itself, we also conclude that no evidence is found for nonradiative transitions of background parasitic impurities. In short, the measurements show that for a ZBLANP sample doped with 1 wt% Yb^{3+}, the photothermal deflection amplitudes are indistinguishable from their ideal values and therefore the emission quantum efficiency as measured *internal to the sample* must be very nearly unity.

An interesting question to ask with respect to these room-temperature measurements is whether comparable results can be obtained at lower temperatures. Figure 11 shows the absorption cross section of ZBLANP: Yb^{3+} at 300, 150, 100, and 50 K; emission spectra obtained at these same temperatures are shown in Fig. 12. The minimal shift in λ_F with temperature that is evident by inspection of the emission spectra, in combination with the persistent, albeit weakening, Yb^{3+} absorption beyond 1000 nm, indicate that anti-Stokes fluorescence cooling of ZBLANP: Yb^{3+} should be observable at temperatures at least as low as 100 K.

Figure 13 verifies this expectation by showing photothermal deflection measurements obtained for three different sample temperatures. In this case, the data are acquired in the unsaturated mode [Eq. (6)] and further are normalized with respect to both pump intensity and linear absorption coefficient; the expected signals are thus proportional to

$$\frac{\dot{\rho}_{I \ll I_s}}{N\sigma_{abs}I} = (1 - \lambda/\lambda_F^*) + \alpha_b/N\sigma_{abs} \tag{7}$$

This expression predicts an essentially linear dependence of the normalized deflection signal on the pump wavelength when the background absorption is small; the straight solid line drawn in the figure thus shows the expected photothermal deflection signal in the limit $\alpha_b \to 0$, with λ_F^* set for convenience to its 300-K value of 995.3 nm. The data demonstrate the potential for laser cooling of the ZBLANP: Yb^{3+} system at cryogenic temperatures.

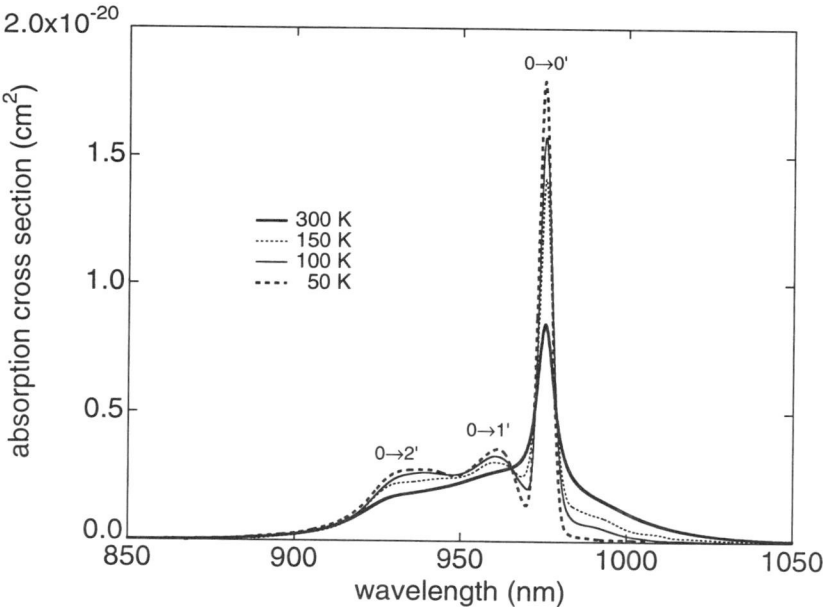

FIG. 11. Absorption cross sections for Yb^{3+} in a ZBLANP host at the indicated temperatures. Individual transitions between specific levels are indicated using the labeling scheme of Fig. 4.

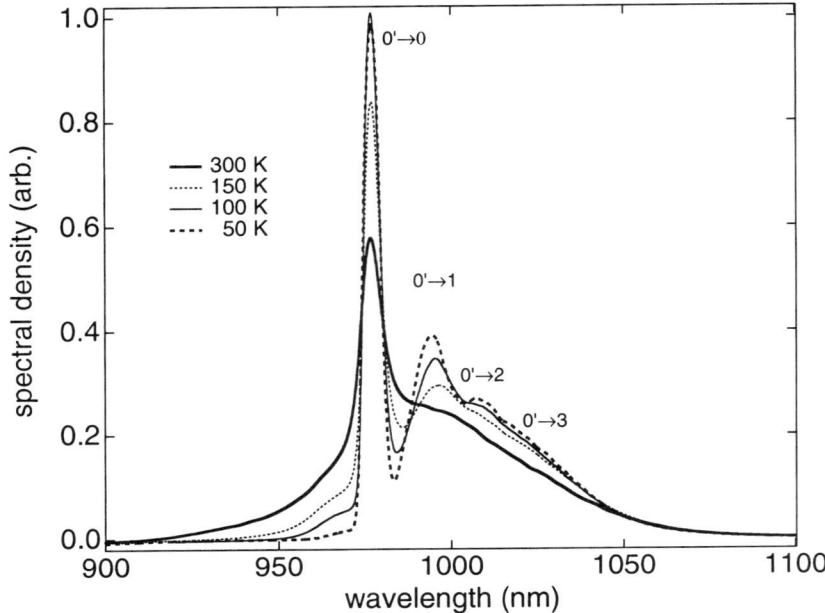

FIG. 12. ZBLANP: Yb^{3+} fluorescence spectra at the same temperatures as shown in Fig. 11. Individual transitions between specific levels are indicated using the labeling scheme of Fig. 4.

B. BULK COOLING EXPERIMENTS

The photothermal deflection results do not strictly allow one to claim that laser cooling of a bulk solid is guaranteed, as the technique does not reveal heating effects arising from interactions of either the pump beam or the emitted fluorescence with the surface of the sample. On the contrary, a net temperature reduction must be measured for the sample as a whole. For this purpose, a mutually compatible sample geometry and thermometric technique must be chosen to minimize extraneous heating. We developed two approaches to the measurement of net cooling of a solid.

In our initial investigations (Epstein *et al.*, 1995a), a $2.5 \times 2.5 \times 6.9$-mm^3 sample was supported in a vacuum chamber on a pair of thin glass slides. This configuration reduced the sample heating due to thermal conduction from the chamber walls to the point where the total load was dominated by the net absorption of room-temperature blackbody radiation. A thermometer in this experiment was constructed by attaching a 1-mm^2 piece of gold foil to the sample and painting the exposed surface with black paint. Measurement of the thermal emission from

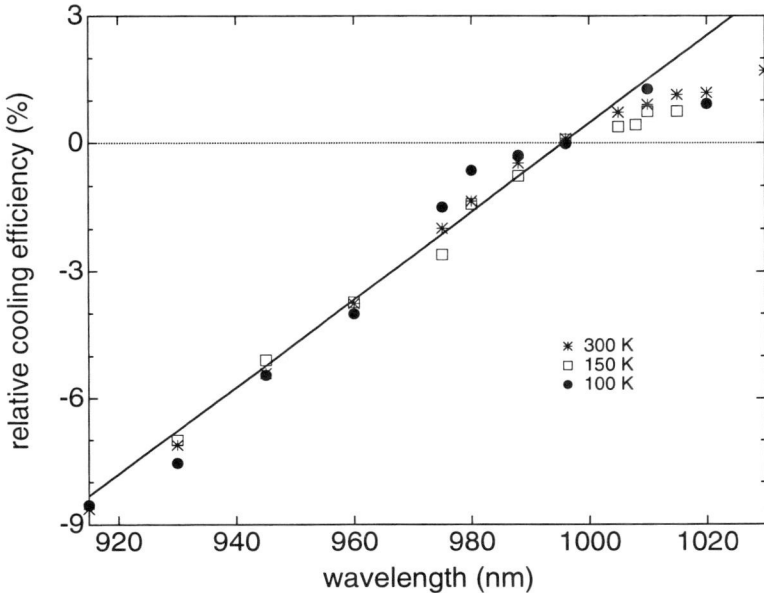

FIG. 13. Photothermal deflection measurements for ZBLANP + 1 wt% Yb^{3+} at 300, 150, and 100 K. The data are normalized with respect to the laser power and the absorption coefficient at the pump wavelength. Arbitrary vertical scale factors have been introduced for each data set so that the results for different temperatures can be directly compared. The solid line shows a plot of Eq. (7) for $\lambda_F^* = 995.5$ nm in the limit $\alpha_b \to 0$.

the foil with a liquid-nitrogen-cooled *InSb* camera gave the temperature change when a reference image captured at room temperature was compared with a second image collected after the sample had been exposed to the pump laser. To monitor and compensate for temperature drifts of the chamber during the experiment, a second, similarly prepared sample not exposed to the pump beam served as a standard. Calibration of the camera signal was accomplished by simultaneous measurement of a heated sample with both the camera and a thermocouple. We obtained a temperature drop of 0.3 K with this apparatus for a pump power of ~1 W at a laser wavelength of 1015 nm, thus demonstrating for the first time net cooling of a condensed material.

A significant disadvantage of the latter experiment, however, was that the maximum temperature change was limited by the small net value of the fluorescence cooling power compared with the radiative heat load from the environment. More quantitatively, the equilibrium temperature reached by the sample, T_S, satisfies the equation

$$P_{cool} = A \int \epsilon_\nu \pi [B_\nu(T_R) - B_\nu(T_S)] \, d\nu \tag{8}$$

where A is the surface area of the sample, $\pi B_\nu(T)$ is the radiative hemispherical energy emission rate per unit area at frequency ν from a blackbody at temperature T, ϵ_ν is the frequency-dependent emissivity of the sample (which for simplicity is taken to be temperature independent), and T_R is the temperature of the sample's environment. If the temperature difference $\Delta T = T_S - T_R$ between the sample and the environment is small, the above expression is approximated by

$$P_{cool} = -A \int \epsilon_\nu \pi \frac{\partial B_\nu(T)}{\partial T}\bigg|_{T_R} \Delta T \, d\nu = -4A\epsilon_{\mathit{eff}}\sigma_B T_R^3 \Delta T \tag{9}$$

where σ_B is the Stefan-Boltzmann constant and ϵ_{eff} is an effective emissivity defined such that $\epsilon_{\mathit{eff}} \to 1$ in the limit $\epsilon_\nu \to 1$.

Equation (9) shows that a reduction in the sample's surface area will increase the observed temperature drop for a given value of the cooling power. Motivated by this observation, a new experiment was designed using Yb^{3+}-doped ZBLANP optical fibers. The sample consisted of a 175-μm-diameter doped inner core surrounded by a lower-index undoped cladding that brought the total diameter to 250 μm; the length of the sample was ~ 1 cm. The numerical aperture of this multimode waveguide was 0.2. In order to avoid extraneous absorption of the emitted fluorescence, no protective polymer coating was applied to the fiber. Pump radiation from the Ti : sapphire laser was focused into the fiber core by a microscope objective mounted inside the vacuum space. Because the earlier thermometric technique is impractical for samples of such small size, we developed a noncontact approach that exploited the temperature dependence of the Yb^{3+} emission spectrum. The method improves upon the analogous technique employed by Clark and Rumbles (1996) in that it is the *shape* of the entire emission spectrum that is measured rather than its absolute magnitude at a select wavelength; in this way, use of a separate fluorescence excitation source is avoided and the sensitivity of the alignment is reduced. To measure the spectrum, a second, perpendicular microscope objective was mounted within the vacuum space. Outside the chamber, the collected fluorescence was imaged onto a silica-fiber bundle that fed a CCD optical multichannel analyzer. With this detection scheme, broadband emission spectra could be captured in a time span of between 30 and 300 s. Independent calibration of the temperature dependence of the emission spectrum was accomplished by back-filling the vacuum chamber with helium gas and measuring the spectrum at four separate calibration points in the vicinity of room temperature.

Figure 14 shows the best result we have obtained to date: in panel (a) are plotted two calibration spectra measured at temperatures of 303 K and 290 K; for emphasis, the inset shows the arithmetic difference between these two spectra when the amplitudes of the $0' \to 0$ peaks at 975 nm are normalized to unity. For a pump

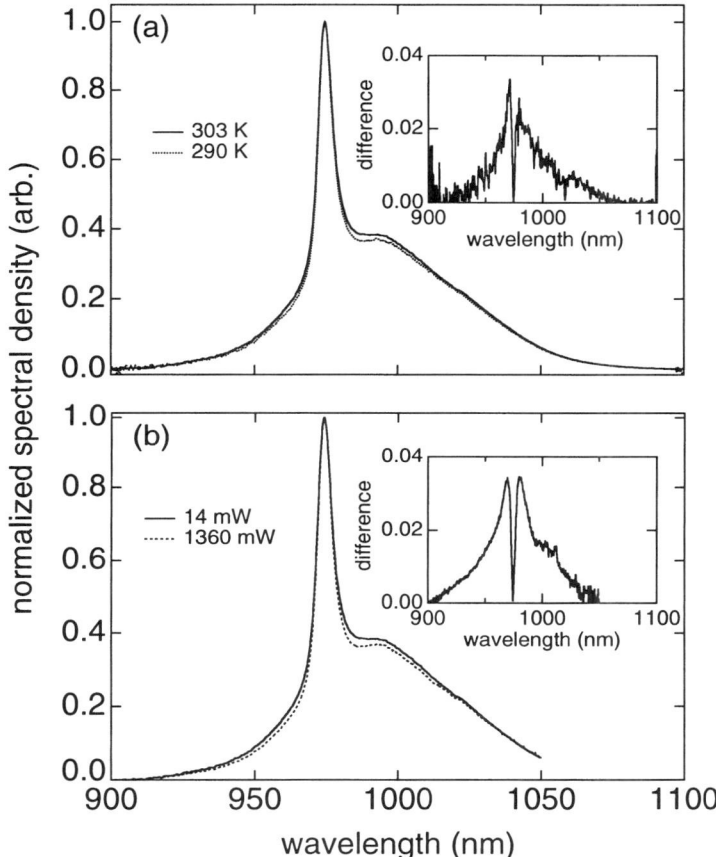

FIG. 14. Emission spectra for a ZBLANP + 1 wt% Yb^{3+} fiber of 250-μm total diameter demonstrating cooling of the fiber. (a) Temperature-calibration spectra, with their arithmetic difference plotted in the inset. The temperature difference between the two spectra is 13 K. (b) Spectra obtained for two values of the pump power at 1015 nm, with their difference plotted in the inset. Based on the calibration data in (a), the laser-induced temperature drop from room temperature is 21 K.

wavelength of 1015 nm, panel (b) shows the spectra obtained for two different values of the pump power incident on the fiber. The magnitude of the difference spectrum in the latter measurements, when compared with the reference, indicates a temperature drop of 21 K. This is an improvement on earlier work employing this fiber geometry (Mungan *et al.,* 1997b), for which a temperature drop of 16 K was obtained, and is currently the largest temperature decrease ever reported for a condensed-phase laser-cooling experiment.

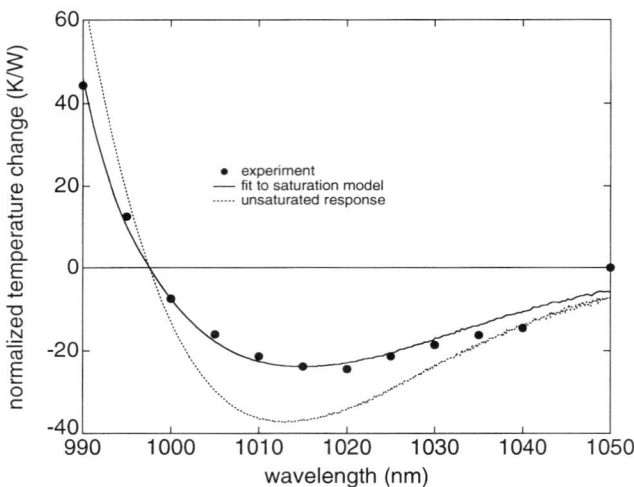

FIG. 15. Temperature changes, normalized by the pump power, measured as a function of pump wavelength (solid circles). The sample is a 250-μm diameter ZBLANP fiber doped with 1 wt% Yb^{3+}. The solid curve shows a fit of Eq. (10) to the data with fitting parameters of $\lambda_F^* = 997.6$ nm and $a_{eff} = 7.92 \times 10^{-5}$ cm^2.

As with photothermal deflection spectroscopy, measurement of the temperature change as a function of the pump wavelength exposes interesting details of the cooling process. The solid circles plotted in Fig. 15 show such a spectrum, where the temperature changes have been normalized by the pump power. In order to understand the spectrum, we apply an optical-pumping model derived from Eq. (5). Combining this equation with Eq. (9), the expected temperature drop of the fiber per unit pump power is given by

$$\Delta T/P_{laser} = \frac{\dot{\rho}/I}{4\pi D\epsilon_{eff}\sigma_B T_R^3} = \frac{\frac{N\sigma_{abs}(a_{eff}I_s/P_{laser})(1 - \lambda/\lambda_F^*)}{1 + \sigma_{se}/\sigma_{abs} + a_{eff}I_s/P_{laser}} + \alpha_b}{4\pi D\epsilon_{eff}\sigma_B T_R^3} \quad (10)$$

where P_{laser} is the pump power, D is the diameter of the fiber, and a_{eff} is an effective pump-spot area within the fiber core. Four unknowns appear in this expression: the effective mean emission-photon wavelength λ_F^*, the background absorption coefficient α_b, the effective pump area a_{eff}, and the effective emissivity ϵ_{eff}.

For the purposes of fitting the data of Fig. 15, we take $\alpha_b \equiv 0$, a choice well justified by the results of the photothermal deflection measurements. Rather than adopting ϵ_{eff} as a fitting parameter, however, an independent measurement of this quantity is possible by measuring the *time dependence* of the fluorescence spectrum after the fiber is suddenly exposed to the pump beam. For convenience, however, the fluorescence signal at a chosen wavelength is monitored instead, with an exponential relaxation toward equilibrium expected for small departures from the ambient temperature. In terms of the exponential time constant τ_c, the ZBLAN mass density $\rho_m = 4.31$ g cm^{-3}, and the specific heat $c_m = 0.596$ J g^{-1} K^{-1} (Hasz et al., 1993), the effective emissivity is given by [3]

$$\epsilon_{eff} = \frac{c_m \rho_m D}{16 \tau_c \sigma_B T_R^3} \quad (11)$$

Following multiple runs, the average value of the relaxation time constant for a 250-μm fiber was found to be 29.3 ± 0.6 s. This corresponds through Eq. (11) to an effective emissivity [4] of 0.90 ± 0.02.

With α_b and ϵ_{eff} now fixed, the solid line in Fig. 15 shows a fit of Eq. (10) to the temperature changes, with only λ_F^* and a_{eff} taken as adjustable parameters. The fitted values are 997.6 nm and 7.92 × 10^{-5} cm^2, respectively, indicating that the *external* fluorescence quantum efficiency is 99.8% and that the pump-spot diameter is only ~100 μm, somewhat smaller than the actual core diameter of 175 μm. The latter result implies that only lower-order modes of the fiber are occupied by the pump radiation and this was actually observed directly. The dashed curve in Fig. 15 shows the expected cooling spectrum in the unsaturated regime ($P_{laser} \to 0$); optical saturation due to the small spot size in the fiber has evidently limited the maximum observable temperature change in this experiment.

As a final check on the issue of optical saturation, Fig. 16 shows the power-normalized temperature changes measured in a 250-μm fiber as a function of

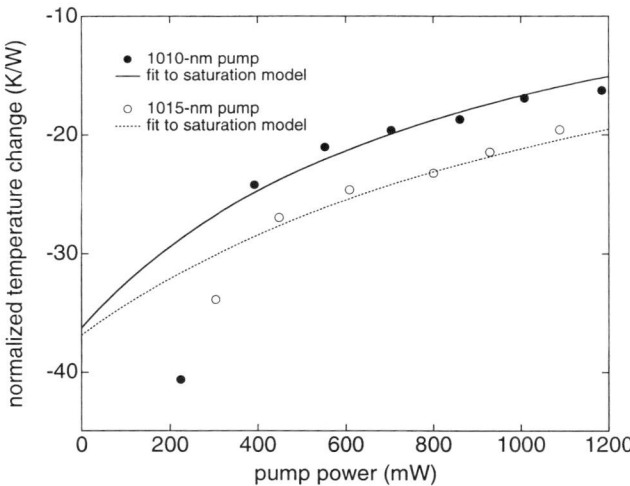

FIG. 16. Power-normalized temperature changes as a function of pump power at laser wavelengths of 1010 nm (solid circles) and 1015 nm (open circles). The solid and dashed curves show fits of Eq. (10) to the data. The measurements further support the optical saturation model developed in the text.

pump power for two different pump wavelengths. Also shown are fits of Eq. (10) to the two data sets, with λ_F^* taken to be 997.6 nm, so that the only adjustable parameter is the effective area a_{eff}. The larger power-normalized temperature changes observed for the lower pump powers further confirm the effect of optical saturation in this experiment; mode scrambling of the input radiation might therefore improve the pump's filling of the fiber core and thereby yield larger temperature drops for a given laser power.

V. Fundamental Limits

In this section we examine from three different perspectives the issues that affect the fundamental limits of laser-cooled condensed matter. In the first, we focus attention on *microscopic* aspects of the problem by developing a thermodynamic picture of the cooling process. This analysis is partly motivated by its academic interest but it will also offer a few insights into how microscopic parameters of the impurity-host system might be manipulated to optimize thermodynamic efficiencies. In the second perspective, we identify physical mechanisms that limit the minimum temperature that might be obtained in a condensed-matter laser-cooling experiment. Quantitative estimates of the minimum temperature will be derived. Finally, the last perspective focuses on more macroscopic aspects of the cooling problem and addresses the possibility of constructing a useful optically pumped cryogenic refrigerator based on the ZBLANP:Yb^{3+} system.

A. Thermodynamics

In general, a collection of impurity ions distributed within a solid host can be viewed as an energy-conversion device capable of performing some useful function at the necessary expense of generating waste heat. As such, condensed-phase cooling experiments are open to conventional thermodynamic analysis. Energy and entropy are accepted by the device as inputs, then transformed in accord with the device's working details, thereby yielding work or refrigeration, and finally rejected as outputs in quantities constrained by the first law of thermodynamics and in forms constrained by the second law. In the following development of a thermodynamic description of laser-cooled solids, we first offer a picture of energy and entropy flow in an energy-conversion device that is sufficiently general to include input and output radiation fields. We then argue that a simpler and more familiar description is possible, with one caveat, that will display two advantages. The first of these is that the analysis will not require the explicit introduction of terms for the energy and entropy of propagating light fields; the second advantage is that the simpler description will help focus attention on those aspects of laser cooling that are most relevant to optimizing the physics of the cooling process.

1. General Refrigerator

We begin with Fig. 17, which shows a refrigeration process for which a flow of pump energy at the rate $\dot{E}_p$, with a corresponding rate of entropy flow $\dot{S}_p$, into a converter at temperature T leads to the extraction of heat at the rate $\dot{Q}_c$ from a low-temperature reservoir at temperature T_c. The refrigerator is also presumed to deposit exhaust energy and entropy at rates of $\dot{E}_h$ and $\dot{S}_h$, respectively, into an output reservoir. In analogy with the presentation of Landsberg and Tonge (1980), we have here adopted a broader picture of refrigeration than that found in elementary texts by allowing for the possibility that the pump source could be more general than a supplier of pure work and the output reservoir might be more general than a receiver of pure heat. Hence, insofar as the pump and output reservoirs are concerned, entropy flows are taken as fundamental quantities instead of the thermodynamic temperatures. It is this generalization that specifically enables a thermodynamic analysis that includes the possibility of a pump source and an exhaust in the form of electromagnetic radiation. Returning to Fig. 17, note that we have allowed for the possibility that energy and entropy might accumulate in the converter at the rates $\dot{E}$ and $\dot{S}$ and that unspecified irreversible processes internal to the converter generate entropy at the rate of $\dot{S}_g$.

With these definitions, the energy and entropy flows into the converter must therefore satisfy

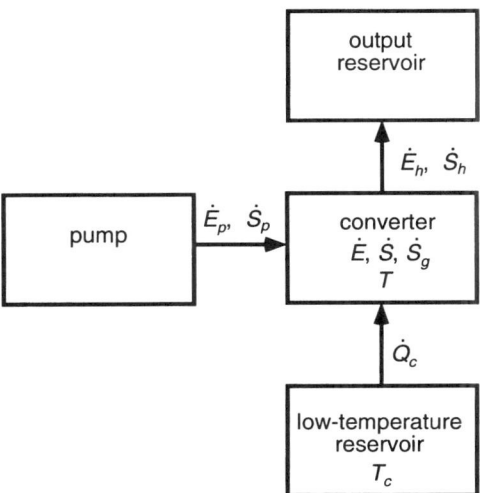

FIG. 17. A generalized refrigerator. Both energy and entropy are accepted by the device as inputs and ejected as outputs. The refrigerator operates at temperature T, and accumulates energy and entropy at the rates and $\dot{E}$ and $\dot{S}$, respectively. Due to irreversible processes, excess entropy is produced at the rate $\dot{S}_g$.

$$\dot{E} = \dot{E}_p + \dot{Q}_c - \dot{E}_h \tag{12}$$

and

$$\dot{S} = \dot{S}_p + \dot{Q}_c/T - \dot{S}_h + \dot{S}_g \tag{13}$$

In the latter equation, $\dot{S}_g$ simply expresses the rate of entropy generation that occurs irreversibly anywhere in the system. In the optical context, possible phenomena contributing to $\dot{S}_g$ are nonradiative relaxation within the converter (in other words, irreversible absorption of the pump laser light), and line-broadened or multidirectional radiative emission to the environment. Therefore, in the case of a completely reversible refrigerator, $\dot{S}_g = 0$. Assuming the converter to be in steady-state operation ($\dot{E} = \dot{S} = 0$), Eqs. (12) and (13) can be manipulated to yield the first-law efficiency, or coefficient of performance (COP), of the refrigerator:

$$\eta_1 \equiv \frac{\dot{Q}_c}{\dot{E}_p} = \frac{T(1/T_{F_h} - 1/T_{F_p}) - T\dot{S}_g/\dot{E}_p}{1 - T/T_{F_h}} \tag{14}$$

where $T_{F_p} = \dot{E}_p/\dot{S}_p$ and $T_{F_h} = \dot{E}_h/\dot{S}_h$ are purely algebraic substitutions known as *flux temperatures* whose introduction is motivated by a desire to derive Carnot-like expressions for thermodynamic efficiencies in this more general context (Landsberg and Tonge, 1980). Maximum first-law efficiency is obtained for a reversible refrigerator and is given by

$$\eta_1^{rev} = \frac{T(1 - T_{F_h}/T_{F_p})}{T_{F_h} - T} \tag{15}$$

2. A Three-Level Optical Refrigerator

In order to proceed further, a more specific model is now needed for the converter, pump, and output reservoir. Figure 18 shows a generic optically pumped refrigerator comprised of a three-level system with states labeled 0, 1, and 2 plus couplings to the pump source, low-temperature reservoir, and output reservoir. The pump can be taken to be any source of narrow-band electromagnetic radiation resonant with the $0 \rightarrow 1$ transition (of transition energy E) whereas the output reservoir, which serves only as a receiver of fluorescence from the $2 \rightarrow 0$ transition (of energy $\epsilon + E$), can be taken as free space. The low-temperature reservoir is simply the host lattice within which the three-level impurity atoms are embedded; it couples levels 1 and 2, which are spaced by energy ϵ, but is otherwise idealized to be decoupled from level 0. By virtue of the 1–2 coupling, and because the populations in levels 1 and 2 are taken to be rapidly equilibrated with respect to each other, heat is subsumed by the converter at temperature T_c, and thus $T \equiv T_c$.

All that remains in order to apply Eqs. (14) or (15) to the computation of the coefficient of performance for this optical refrigerator is to find explicit expressions for the rates of pump and output energy and entropy flow when the applicable

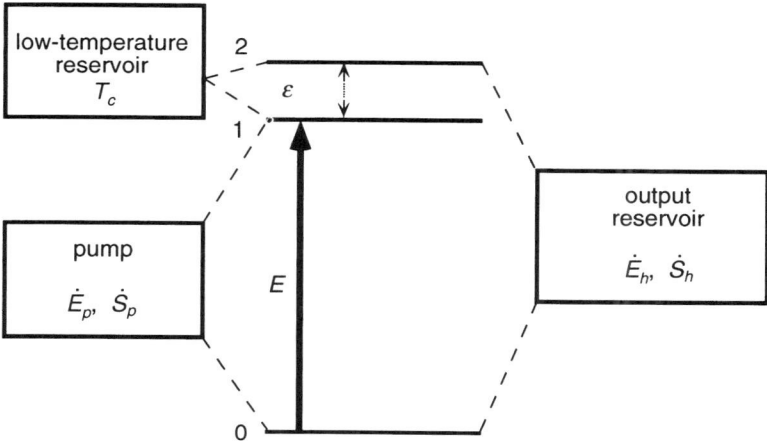

FIG. 18. Generic laser-pumped refrigeration scheme based on a three-level model. Pairs of levels are identified with the components of the generalized refrigerator of Fig. 17; the pump and output reservoir may exchange radiation resonant with the two-level systems to which they are coupled.

radiation fields are not necessarily in equilibrium. This statement means that for the purposes of deriving thermodynamic quantities for the refrigerator, individual photon-mode occupation numbers for either the pump radiation or the output radiation are not required to be Bose-Einstein distributed with a unique thermodynamic temperature; instead, individual radiation modes can be taken as independent, decoupled oscillators with any one of which is affiliated well-defined energy and entropy fluxes. In this way, a sum of the energy and entropy fluxes over all of the occupied modes yields well-defined totals for these quantities, in much the same way that the total energy and entropy are well defined for nuclear spins and a solid lattice even when the spin and lattice temperatures are unequal. With this understanding, the fluxes (for unpolarized fields) are formally given by

$$\dot{E}_i = \int I_\nu^i(\hat{\mathbf{n}}, \mathbf{r})\cos\theta \, d\nu \, d\Omega \, da;$$

$$I_\nu^i(\hat{\mathbf{n}}, \mathbf{r}) = \frac{c}{4\pi} u_\nu^i; \qquad u_\nu^i = \frac{8\pi h \nu^3}{c^3} n_\nu^i$$

(16)

and

$$\dot{S}_i = \int \Xi_\nu^i(\hat{\mathbf{n}}, \mathbf{r})\cos\theta \, d\nu \, d\Omega \, da; \qquad \Xi_\nu^i(\hat{\mathbf{n}}, \mathbf{r}) = \frac{c}{4\pi} s_\nu^i;$$

$$s_\nu^i = \frac{8\pi k_B \nu^2}{c^3}[(1 + n_\nu^i)\ln(1 + n_\nu^i) - n_\nu^i \ln(n_\nu^i)]$$

(17)

where $i = p$ or h for the pump and output radiation fields, respectively, and $I^i_\nu(\hat{\mathbf{n}}, \mathbf{r})$ is a quantity known as the specific intensity (Mihalas, 1978); at the frequency ν, it is the rate of radiation energy flow at the position $\mathbf{r}$ in the direction $\hat{\mathbf{n}}$ through a unit area within unit frequency interval into unit solid angle. The factor u^i_ν is the radiation energy density at frequency ν that would be found in a blackbody cavity exhibiting the specific intensity $I^i_\nu(\hat{\mathbf{n}}, \mathbf{r})$, an identification that allows the assignment of a brightness temperature for the radiation at all values of ν, $\hat{\mathbf{n}}$, and $\mathbf{r}$. The factor n^i_ν is the photon occupation number corresponding to this energy density; it is related to the brightness temperature, $T_{B_i}(\nu, \hat{\mathbf{n}}, \mathbf{r})$ according to $n^i_\nu = 1/[\exp(h\nu/k_B T_{B_i}) - 1]$. Finally, the entropy terms $\Xi^i_\nu(\hat{\mathbf{n}}, \mathbf{r})$ and s^i_ν are defined in analogous fashion to the energy terms $I^i_\nu(\hat{\mathbf{n}}, \mathbf{r})$ and u^i_ν. Hence Eq. (16) provides a means for finding n^i_ν as a function of ν, $\hat{\mathbf{n}}$, and $\mathbf{r}$, and with this knowledge Eq. (17) provides a means for computing $\dot{S}_i$. The integrals are taken over any surface that bounds the refrigerator, with da being the element of surface area, $d\Omega$ the element of solid angle, and θ the angle between $\hat{\mathbf{n}}$ and the local surface normal.

3. A Simpler Approach to the Three-Level Optical Refrigerator

Although Eqs. (16) and (17) provide a complete prescription for computing the needed thermodynamic quantities, their use requires detailed spectroscopic knowledge of the pump and especially the output radiation fields. Although a laser (with $\dot{S}_p \to 0$ in the limit of infinite intensity, zero linewidth, or perfect beam quality) can be taken as an idealized pump source, the degree of line broadening assumed for the $2 \to 0$ output transition will have a significant effect on the fluorescence flux temperature. Moreover, because for condensed phases line broadening and luminescence emission into multiple directions in space necessarily involve irreversible processes, calculation of the COP of the refrigerator will be complicated by the requisite use of Eq. (14).

We therefore adopt for the remainder of this discussion—at a small expense to be described later—a simpler approach with the advantages that direct consideration of radiation fields is unnecessary and that intrinsically irreversible processes affiliated with these fields will be removed. By insisting that the refrigerator only interact reversibly with the pump and output reservoirs, emphasis remains placed on the microscopic aspects of the refrigerator's internal operation. Equation (15) will therefore prove adequate to the task of calculating a coefficient of performance with which to compare the COP derived from a specific physical model of the refrigerator. Such a comparison will reveal the best strategies for thermodynamic optimization. This new picture will also prove more familiar as heat alone will be exchanged between the refrigerator and the pump, the low-temperature reservoir, and the high-temperature reservoir.

The new picture, the essence of which was apparently first discussed by Scovil and coworkers (Scovil and Schulz-DuBois, 1959; Geusic et al., 1967) and by

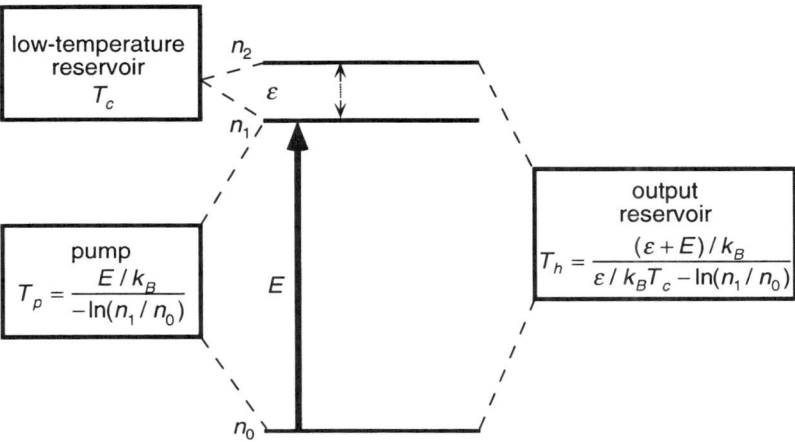

FIG. 19. Simplification of the three-level laser-pumped refrigerator, as depicted in FIG. 18, where the pump and output reservoirs have been replaced with heat baths that interact only with the indicated two-level systems.

Weinstein (1960), is shown in Fig. 19: the optical pump source is replaced by a heat bath that exchanges energy with the refrigerator only through the 0–1 transitions; it may be idealized as an ensemble of harmonic oscillators with resonant frequencies closely bunched around the $0 \rightarrow 1$ transition frequency. Because only levels 0 and 1 are involved in this interaction, the populations in these levels, n_0 and n_1, determine a well-defined thermodynamic temperature,

$$T_{01} = \frac{E/k_B}{-\ln(n_1/n_0)} \qquad (18)$$

If the temperature of the pump bath, T_p, is set equal to T_{01}, a reversible exchange of energy is assured between the pump and the two-level system comprised of the 0 and 1 states. The output reservoir is constructed in a similar way with a second well-defined thermodynamic temperature given by

$$T_{02} = \frac{(\epsilon + E)/k_B}{\epsilon/k_B T_c - \ln(n_1/n_0)} \qquad (19)$$

As with the pump reservoir, reversible exchange is assured for $T_h = T_{02}$.

With these definitions for the nature and temperature of the pump and output heat baths under reversible operation of the refrigerator, Eq. (15) for the COP becomes

$$\eta_1^{rev} = \frac{T_c(1 - T_h/T_p)}{T_h - T_c} \qquad (20)$$

Note that in this expression, absolute thermodynamic temperatures have replaced the flux temperatures called for in Eq. (15) precisely because these quantities are identical when the pump and output reservoirs are comprised of conventional heat baths. Substitution for T_p and T_h with the help of Eqs. (18) and (19) gives

$$\eta_1^{rev} = \frac{\epsilon}{E} \qquad (21)$$

a result we could have foreseen by noting that for reversible operation heat ϵ is removed from the low-temperature reservoir only through a net investment of heat E from the pump.

The simple result of Eq. (21) has been obtained with one sacrifice in generality: by eliminating the consideration of entropy and possible heat removal produced by line broadening, we have explicitly eliminated the possibility that broadening due to phonon collisions or to spectral diffusion within an inhomogeneously broadened line can contribute to the cooling effect. Most simply, an example of this exception is a two-level impurity system with a homogeneously broadened upper state. Optical pumping with a narrow-band source *below* the center frequency of the transition will cool the host lattice as a consequence of phonon collision broadening. Although this represents in principle a viable mechanism for optical refrigeration, one that might best be analyzed in terms of Eq. (14) and the required entropy calculations of one or both of the pump and output radiation fields, we continue to assume that line broadening makes little contribution to refrigeration in the three-level system of interest here.

4. A Model Three-Level Optical Refrigerator

Figure 20(a) now shows an explicit physical model for pumping and relaxation processes in a three-level system used as a refrigerator. We assume that the pump source is a laser of photon energy $E = h\nu$; while radiative relaxation out of the two excited states occurs at rates of γ_1 and γ_2, respectively, nonradiative relaxation (as a consequence of impurity quenching or pure multiphonon relaxation to the ground state) occurs at the rate Γ for both excited states, and g_0, g_1, and g_2 are the level degeneracies. The double-headed arrow between levels 1 and 2 denotes the nonradiative thermalization of these two states, which we assume to occur at a rate much larger than all other rates in the model. Hence levels 1 and 2 are always thermally equilibrated with respect to each other. With these assumptions and definitions, the equations that the population densities in the three levels must satisfy are

$$\dot{n}_1 + \dot{n}_2 = \frac{I\sigma_{abs}}{E}\left(n_0 - \frac{g_0}{g_1}n_1\right) - (\Gamma + \gamma_1)n_1 - (\Gamma + \gamma_2)n_2 \qquad (22a)$$

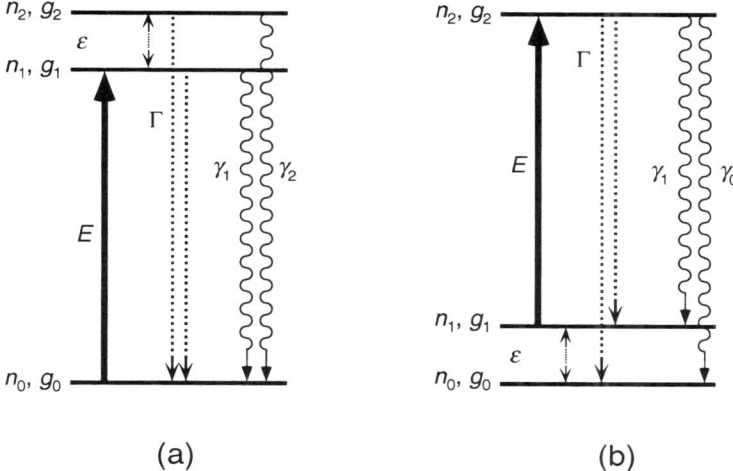

FIG. 20. Explicit models for the pump and relaxation processes in the optically pumped three-level refrigerator for two different arrangements of the energy levels. Purely radiative transitions are shown as solid lines whereas nonradiative transitions are shown as dashed lines.

$$n_2 = \frac{g_2}{g_1} n_1 e^{-\epsilon/k_B T_c} \quad (22b)$$

and

$$n_0 + n_1 + n_2 = N \quad (22c)$$

where $\dot{n}_1$ and $\dot{n}_2$ denote the time rates of change of the level 1 and level 2 population densities, respectively, I is the intensity of the pump laser, and σ_{abs} is the absorption cross section for the $0 \to 1$ transition. Note that the second equation expresses our assumption of fast thermal equilibration of the two excited states (with a host lattice of temperature T_c), while the third expresses the conservation of the total number density of atoms, N.

Expressions for the rate of heat extraction per unit volume by this model refrigerator, $\dot{Q}'$, and the corresponding required rate at which work must be supplied, $\dot{W}'$, in terms of the level population densities are

$$\dot{Q}' = \epsilon \gamma_2 n_2 - \Gamma E n_1 - \Gamma(E + \epsilon) n_2 \quad (23)$$

and

$$\dot{W}' = E \gamma_1 (I/I_s) \left(n_0 - \frac{g_0}{g_1} n_1 \right) \quad (24)$$

where $I_s = \gamma_1 E/\sigma_{abs}$ is the saturation intensity of the $0 \to 1$ transition. In the latter equation the rate of supplied work is taken to be that portion of the laser power that is actually *absorbed* by the three-level impurities; in this way, emphasis remains placed on the purely microscopic aspects of the refrigerator's operation.

Assuming cw laser excitation of the system, steady-state population densities are achieved and Eqs. (22) can be easily solved for the quantities n_0, n_1, and n_2. Substitution of these steady-state solutions into Eqs. (23) and (24) yields for the model coefficient of performance of the refrigerator

$$\eta_1' = \frac{\dot{Q}'}{\dot{W}'} = \frac{\epsilon}{E} \frac{(\gamma_2 - \Gamma)(g_2/g_1)e^{-\epsilon/k_B T_c} - \Gamma(E/\epsilon)[1 + (g_2/g_1)e^{-\epsilon/k_B T_c}]}{\gamma_1 + \gamma_2(g_2/g_1)e^{-\epsilon/k_B T_c} + \Gamma[1 + (g_2/g_1)e^{-\epsilon/k_B T_c}]} \quad (25)$$

and therefore for the second-law efficiency

$$\eta_2 = \frac{\eta_1'}{\eta_1^{rev}} = \frac{(\gamma_2 - \Gamma)(g_2/g_1)e^{-\epsilon/k_B T_c} - \Gamma(E/\epsilon)[1 + (g_2/g_1)e^{-\epsilon/k_B T_c}]}{\gamma_1 + \gamma_2(g_2/g_1)e^{-\epsilon/k_B T_c} + \Gamma[1 + (g_2/g_1)e^{-\epsilon/k_B T_c}]} \quad (26)$$

Note that the latter result recalls Eq. (21) for the coefficient of performance under reversible operation. If nonradiative transitions to the ground state are negligible, $\Gamma \to 0$, then the above equations become

$$\eta_1' = \frac{\epsilon}{E} \frac{(\gamma_2 g_2/\gamma_1 g_1)e^{-\epsilon/k_B T_c}}{1 + (\gamma_2 g_2/\gamma_1 g_1)e^{-\epsilon/k_B T_c}} \quad (27)$$

and

$$\eta_2 = \frac{(\gamma_2 g_2/\gamma_1 g_1)e^{-\epsilon/k_B T_c}}{1 + (\gamma_2 g_2/\gamma_1 g_1)e^{-\epsilon/k_B T_c}} \quad (28)$$

Inspection of Eqs. (25) through (28) lead to the following conclusions on the optimization of the refrigerator:

1. In the presence of nonradiative transitions to the ground state, both increased first- and second-law efficiencies are obtained by decreasing E, the ground-state to excited-state energy gap. This amounts to decreasing the temperature of the output reservoir. In the absence of nonradiative transitions to the ground state, only the first-law efficiency is improved by decreasing E.
2. Increased first- and second-law efficiencies are obtained by increasing the ratio $\gamma_2 g_2/\gamma_1 g_1$. This is illustrated in Fig. 21, which plots both efficiencies as functions of the level splitting ϵ for various values of this ratio. Note in the figure that for $\gamma_2 g_2/\gamma_1 g_1 = 10$, second-law efficiencies greater than 50% are obtainable in the range of splittings that optimizes the first-law efficiency.
3. In the presence of nonradiative transitions to the ground state, the mini-

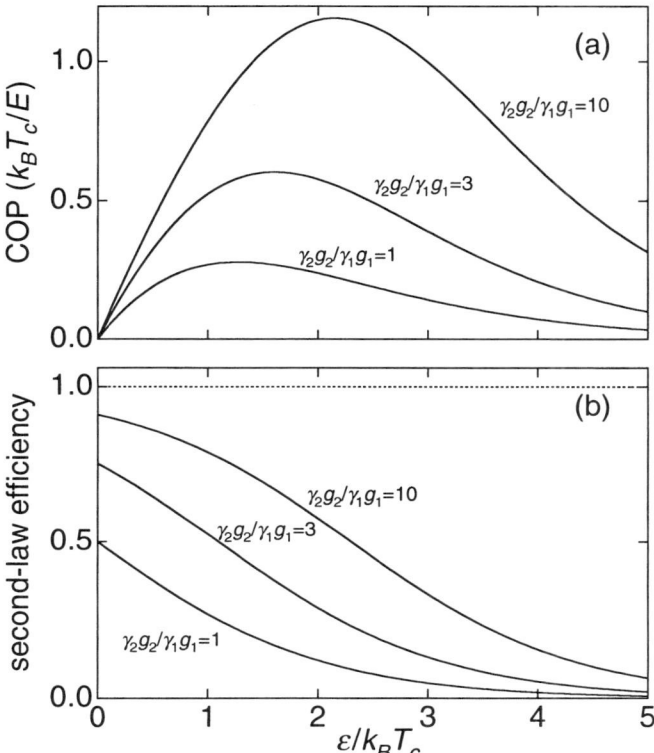

FIG. 21. (a) Coefficients of performance (or first-law efficiencies), in units of $k_B T_c/E$, for the three-level refrigerator of Fig. 20(a). The curves are plotted as a function of the upper-manifold level splitting ϵ at fixed operating temperature T_c and fixed intermanifold energy gap E. (b) Second-law efficiencies for the same model. Curves are derived from Eqs. (27) and (28) and given for three different ratios of the product of the level degeneracy and radiative relaxation rate in the two excited states.

mum obtainable temperature depends on the value of Γ, as determined by Eq. (25).

It is interesting to compare these results with a different three-level model for the refrigerator, as shown in Fig. 20(b). In analogy with Eqs. (22), the population densities must satisfy

$$\dot{n}_0 + \dot{n}_1 = -\frac{I\sigma_{abs}}{E}\left(n_1 - \frac{g_1}{g_2}n_2\right) + (2\Gamma + \gamma_0 + \gamma_1)n_2 \quad (29a)$$

$$n_1 = \frac{g_1}{g_2}n_0 e^{-\epsilon/k_B T_c} \quad (29b)$$

and as before,

$$n_0 + n_1 + n_2 = N \tag{29c}$$

The corresponding rates of heat extraction and supplied work are

$$\dot{Q}' = \epsilon\gamma_0 n_2 - \Gamma(2E + \epsilon)n_2 \tag{30}$$

and

$$\dot{W}' = E\gamma_1(I/I_s)\left(n_1 - \frac{g_1}{g_2}n_2\right) \tag{31}$$

and in steady state the coefficient of performance reduces to

$$\eta_1' = \frac{\epsilon}{E}\frac{\gamma_0 - \Gamma(2E/\epsilon + 1)}{\gamma_0 + \gamma_1 + 2\Gamma} \tag{32}$$

Because $\eta_1^{rev} = \epsilon/E$ is still the coefficient of performance under reversible operation for this new model, the second-law efficiency becomes

$$\eta_2 = \frac{\gamma_0 - \Gamma(2E/\epsilon + 1)}{\gamma_0 + \gamma_1 + 2\Gamma} \tag{33}$$

These results differ strikingly from those obtained with the model of Fig. 20(a). Although the effective reduction in temperature of the output reservoir that is commensurate with a decrease in E still yields the expected improvements in first- and second-law efficiencies, these efficiencies are no longer temperature dependent and moreover no longer exhibit a fundamental limiting temperature when Γ is nonzero, at least as long as $\dot{Q}'$ is initially positive. Gone, however, is the advantage of large relative degeneracies, g_1/g_0, in the narrowly spaced levels—second-law efficiencies near unity require larger values of the ratio γ_0/γ_1 than is necessary for the corresponding ratio in the first model. All of these differences are a consequence of the temperature-dependent population that resides in level 1 of the second model: Because the absorption of pump light decreases as the temperature is lowered, no power is absorbed that cannot be exploited in extracting heat from the host lattice. In short, the refrigerator becomes transparent as it cools, thus offering no opportunity for nonradiative transitions to heat the lattice.

The thermodynamic analysis of the two simple models just described illustrates essential features of laser-cooled solids when interlevel thermalization is responsible for heat removal from the host lattice. Although beyond the scope of this review, a combination of the two models, with perhaps the addition of more realistic incorporation of parasitic nonradiative processes, should prove useful in the analysis of Yb^{3+}-based systems as well as of general impurity-doped insulators and even semiconductors.

B. Minimum Temperature

An initial estimate of the minimum temperature to which a solid may be laser cooled can be obtained by equating the rate of heat generation by nonradiative mechanisms to the temperature-dependent cooling power predicted by a chosen model of the cooling process. For this purpose, we will suppose that the three-level model previously presented can be applied at all temperatures and further that no nonradiative transitions contribute to heating of the host lattice ($\Gamma = 0$). Instead, unavoidable inelastic optical processes intrinsic to the undoped solid host will determine the heat load that the refrigerator must support at low temperatures.

The primary mechanisms (Lines, 1991) that we will consider are (1) multiphonon absorption, which dominates optical loss in the host lattice on the low-frequency side of the transmission window; (2) indirect electronic interband absorption, which dominates the loss on the high-frequency side and is known as the Urbach edge; and (3) lattice Raman scattering, which dominates all other inelastic light-scattering phenomena (such as Brillouin scattering). Exciton transitions, impurity-induced effects on the intrinsic electronic or multiphonon absorption, and free-carrier absorption will be presumed to be too weak or too limited in frequency range to make a significant contribution to the heat load.

Figure 22 shows in graphical form estimates of the heat load contributed by the three mechanisms just listed for three types of solids: silica glass, heavy-metal fluoride glass, and crystalline KCl. The meaning of the units nW/cm W on the ordinate axis is that for 1 W of incident pump power driving the cooling process, absorption or inelastic scattering will contribute the indicated heat load per centimeter of path length traveled by the driving beam. For multiphonon absorption, the heat load is given by P_{load}[nW/cm W] $\approx 10^9 \, \alpha_{MP}(\nu, T)$, where

$$\alpha_{MP}(\nu, T) = A_0 \frac{[n(\nu_0) + 1]^N}{n(\nu) + 1} \exp(-b\nu/\nu_0) \tag{34}$$

is the N-phonon absorption coefficient (in cm^{-1}) with $N \equiv \nu/\nu_0$, ν_0 is a typical phonon frequency, and $n(\nu) = 1/[\exp(h\nu/k_B T) - 1]$ is the Bose-Einstein occupation number for a phonon of frequency ν at temperature T (Bendow, 1991). The parameters A_0 and b are purely material-dependent constants. Note that this expression was originally derived for multiphonon absorption in crystals, although analogous frequency and temperature dependences are commonly observed in disordered materials as well (Bendow, 1991). The Urbach edge in crystals likewise shows an exponential frequency dependence

$$\alpha_{Ur}(\nu, T) = A_0' \exp[b'(\nu - \nu_g)/k_B T] \tag{35}$$

where the threshold for direct absorption occurs at frequency ν_g. In this expression, the indicated temperature dependence is only approximately obeyed (Lines, 1991). For glasses, the frequency dependence of indirect interband absorption is more complicated but still tends to exhibit overall an approximately exponential

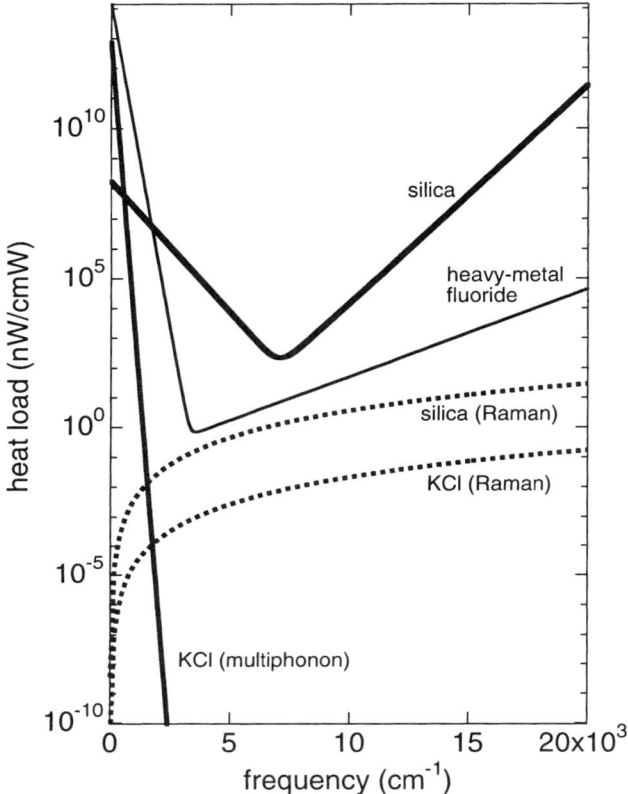

FIG. 22. Intrinsic optical loss expressed in terms of the heat generated per unit path length per unit pump power as a function of the laser frequency for three typical optical materials. Solid V-shaped curves are the sum of the multiphonon optical absorption and the Urbach-edge absorption except for the *KCl* curve, where the Urbach edge is too weak to be seen. Dashed lines show the heat load due to Raman scattering. All curves are for materials at 300 K. References: silica-glass multiphonon and Urbach edge absorption (Lines, 1991); heavy-metal-fluoride-glass multiphonon and Urbach-edge absorption (Bendow, 1991); *KCl* multiphonon absorption (Boyer et al., 1975); silica Raman scattering (Heiman et al., 1979); *KCl* Raman scattering (Gallo et al., 1991).

form. In contrast to multiphonon absorption, only a weak temperature dependence is observed for the electronic edge in glasses (Lines, 1991).

Referring again to Fig. 22, the V-shaped curves show the sum of the multiphonon and Urbach absorption spectra for both silica and heavy-metal fluoride glasses at room temperature, whereas for crystalline *KCl*, the electronic absorption is so weak in the region below 20,000 cm^{-1} that only the multiphonon absorption is shown.

The curves representing the heat load due to Raman scattering from the host phonons are given by

$$P_{load}[\text{nW/cm W}] \approx 10^9 \cdot 4\pi \left.\frac{d\sigma}{d\Omega}\right|_{\nu_R} \left(\frac{\nu}{\nu_R}\right)^4 \left(\frac{\nu_{\text{Stokes}}}{\nu}\right) \tag{36}$$

where $(d\sigma/d\Omega)|_{\nu_R}$ is the Stokes-Raman cross section, measured at the reference pump frequency ν_R, (with units of cm^{-1} sr^{-1}) integrated over polarization and Stokes frequency (neglecting the much weaker anti-Stokes scattering from the phonons); the average Raman-Stokes shift is ν_{Stokes}. The scattering direction is taken to be perpendicular to the pump laser beam (of frequency ν) with the factor of 4π introduced under the simplifying approximation that the scattering is isotropic.

Figure 22 immediately reveals that greater heat loads are expected for disordered host materials; at room temperature they are limited by multiphonon and Urbach-edge absorption, depending on the pump frequency. In the zero-temperature limit, extrapolation by eye of the electronic edge absorption (which is only weakly temperature dependent) to low frequencies shows that this mechanism dominates the heat load at most frequencies of interest. For crystals, Raman scattering limits the heat load at all but the lowest frequencies.

An estimate of the minimum attainable temperature is given by

$$T_{min} \approx P_{load} h\nu/k_B \alpha_{ions} \tag{37}$$

where ν is the frequency of the pump laser and α_{ions} is the absorption coefficient of the dopant three-level systems with an energy splitting $\epsilon = k_B T_{min}$ optimized for operation at the minimum temperature. Thus for $\nu = 10{,}000$ cm^{-1}, $\alpha_{ions} = 1$ cm^{-1}, and $P_{load} = 0.02$ nW/cm W (the Raman limit for KCl), $T_{min} \approx 1$ mK. In strongly absorbing samples, T_{min} may fall into the microkelvin range. These figures are lower bounds on the temperature to which a solid material might be optically cooled, at least for mechanisms providing $\sim k_B T_{min}$ of heat extraction per cycled pump photon. Note that for the three-level model, however, the energy splitting ϵ corresponding to 1 mK is only a few tens of MHz; therefore, absorption and emission broadening effects and the interlevel thermalization rate must be compatible with cooling across such a small energy gap in order for such low temperatures to be obtainable in practice.

C. Maximum Cooling Power and Optical Refrigerators

A natural application of laser-cooled solids is in the construction of optically pumped cryogenic refrigerators. In this context, the most important performance characteristics are the absolute efficiency of the cooling process (defined as the ratio of the cooling power to the *incident* laser power) and the volume density of cooling power that might be obtained in the refrigerator's working substance. We will specifically discuss the performance of an optical refrigerator based on the ZBLANP:Yb^{3+} system.

Equation (5) for the rate of energy accumulation per unit volume in the sample can be recast in the following way:

$$\dot{\rho}_{cool}(I, \nu, T) = \frac{N\gamma_{rad}(T)h\nu[\nu_F(T)/\nu - 1]}{1 + \sigma_{se}(\nu, T)/\sigma_{abs}(\nu, T) + I_s(\nu, T)/I} \quad (38)$$

where a sign change has been introduced so that positive values of $\dot{\rho}_{cool}$ indicate cooling of the material, the loss terms κ and α_b have been set to zero, ν is the pump frequency, and $h\nu_F$ is the average energy of an emitted photon. All functional dependences of the various factors are shown explicitly; in particular, the stimulated-emission and absorption cross sections are strongly temperature dependent, owing to the exponential dependence of the populations of the individual levels composing the Yb^{3+} upper and lower Stark-split multiplets. Temperature-dependent line broadening will also affect these factors.

It is useful at this point to take advantage of a relationship first derived by McCumber (1964) for the ratio of the stimulated-emission and absorption cross sections in broadband systems:

$$\frac{\sigma_{se}(\nu, T)}{\sigma_{abs}(\nu, T)} = \frac{Z_{7/2}(T)}{Z_{5/2}(T)} \exp[h(\nu_{00'} - \nu)/k_B T] \quad (39)$$

where $Z_i(T)$ is the partition function for the multiplet ^2F_i, with the zero of energy defined at the lowest-lying level of the multiplet, and $\nu_{00'} = 10{,}261$ cm^{-1} is the frequency corresponding to transitions between these two zero levels.

Substitution of this relation into Eq. (38) yields

$$\dot{\rho}_{cool} = \frac{N\gamma_{rad}h\nu[\nu_F/\nu - 1]}{1 + \dfrac{Z_{7/2}(T)}{Z_{5/2}(T)} \exp[h(\nu_{00'} - \nu)/k_B T] + I_s/I} \quad (40)$$

This last result allows us to calculate the expected cooling-power density for any pump intensity and any temperature as a function of the pump frequency. In the limit $I \to \infty$, and choosing the pump frequency to maximize the cooling power, the maximum cooling-power density can be obtained as a function of temperature. Figure 23 shows this quantity for a concentration of 1 wt% Yb^{3+} or equivalently $N = 2.42 \times 10^{20}$ ions/cm^3. The maximum power density of $\sim$50 W/cm^3 at 300 K falls to $\sim$1 W/cm^3 at 100 K and to $\sim$0.1 W/cm^3 at 77 K.

Although these values of the cooling density are revealing, they overestimate what is obtainable in practice because they derive from infinite pump intensity. Equation (40) can be used to plot the expected cooling-power density as a function

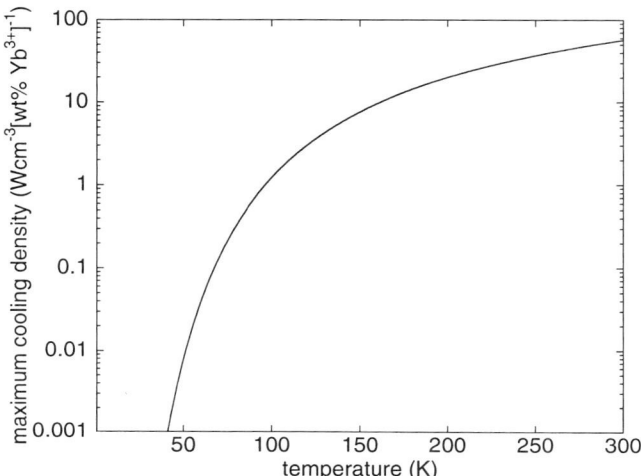

FIG. 23. Theoretical maximum cooling-power density expected for a ZBLANP host doped with 1 wt% Yb^{3+}. The curve is derived from Eq. (40) in the limit $I \to \infty$.

of pump intensity for a given temperature of the cooling element. Figure 24 shows this result, where for each temperature and laser intensity the pump wavelength has been chosen so as to maximize the cooling power. (The data of Fig. 11 were used in these calculations.)

One can obtain a better feeling for the meaning of these cooling-density curves by considering a cubic cooling element of 1-cm^3 volume. If this element is illuminated with a 100-W laser whose beam is in the shape of a square 1 cm on a side, then for a 1-wt%-doped sample the expected cooling power will be ~200 mW at 300 K, ~10 mW at 100 K, and ~2 mW at 77 K. If we now take advantage of the fact that ZBLANP can support a maximum Yb^{3+} concentration of 3 wt%, and also design the cooling element into an optical cavity with a Q of 10 for the pump radiation (this increases the circulating intensity in the cooling element from 100 W to ~1000 W), the cooling powers increase to ~6 W at 300 K, ~240 mW at 100 K, and to ~60 mW at 77 K. The latter values correspond to efficiencies of 6.0%, 0.24%, and 0.060%, respectively. These figures may be compared with those of a typical Stirling-cycle refrigerator, which produces ~1 W of cooling power at 77 K with an efficiency of ~1%.

Another perspective on the cooling efficiency is given in Fig. 25, which plots the quantity $\dot{\rho}_{cool}/I$ as a function of the pump intensity I. The curves clearly indicate the values of the intensity at which the efficiency begins to fall off as a consequence of optical saturation. Finally, Fig. 26 plots the cooling efficiency $\dot{\rho}_{cool}/I$

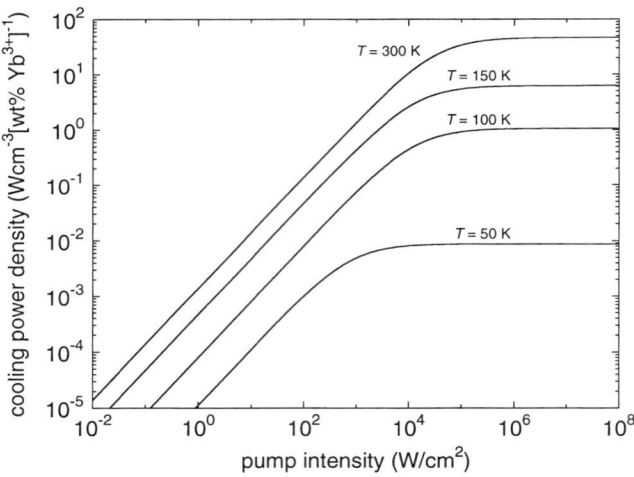

FIG. 24. Theoretical cooling power density for ZBLANP + 1 wt% Yb^{3+} as a function of pump intensity for four different temperatures. The curves are derived from Eq. (40) and measurements of the Yb^{3+} absorption cross section (see Fig. 11).

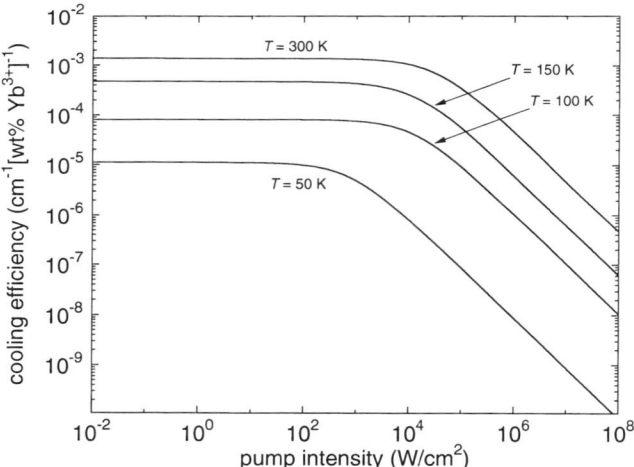

FIG. 25. Theoretical cooling efficiencies for ZBLANP + 1 wt% Yb^{3+} as a function of pump intensity for four different temperatures. The curves are derived from Eq. (40) and measurements of the Yb^{3+} absorption cross section (see Fig. 11); they show the decline in efficiency that results from optical saturation.

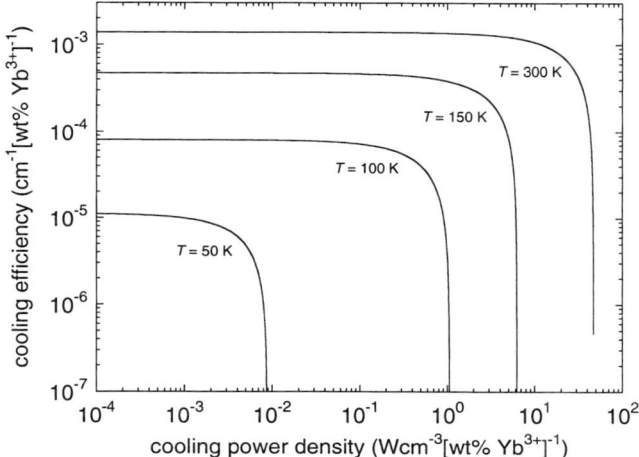

FIG. 26. Theoretical cooling efficiencies for ZBLANP + 1 wt% Yb^{3+} as a function of cooling-power density for four different temperatures. The curves are derived from Eq. (40) and measurements of the Yb^{3+} absorption cross section (see Fig. 11); they show the range of cooling-power densities over which the cooling efficiency is uncompromised by optical saturation.

as a function of the cooling-power density obtained at the pump intensity I. This figure reveals the range of cooling-power densities available before optical saturation begins to compromise the efficiency.

The preceding, general analysis indicates the potential of optical cryogenic refrigeration, but it also makes clear that great challenges remain in developing a practical and competitive device. Efficient refrigerators most likely will require the preparation of new host materials that support higher concentrations of Yb^{3+}. They also may require a cooling cycle based on other active dopants and the corresponding development of new, high-power, compact, and rugged laser sources.

VI. Conclusions and Prospects

Despite its 70-year history, the concept of anti-Stokes photoluminescence cooling of matter has only recently become practical, as a consequence of the laser pump sources and ultra-high-purity materials developed in the past 10 years. Further advances will continue to depend heavily on these factors. As far as the Yb^{3+} species is concerned, of particular interest will be the study of host materials that allow higher concentrations of the active ion; in this way, greater cooling densities can be achieved and hence lower temperatures for a given heat load from the surroundings. Crystalline hosts are the best prospects in this regard; they offer not only the possibility of continuously adjustable concentrations all the way up to the stoichiometric limit but also sharper absorption lines than in glasses and therefore increased absorption of the pump radiation.

Indeed, with an eye toward developing optically pumped cryogenic refrigerators, simple scaling of the ZBLANP: Yb^{3+} results up to 100% concentration yields at a working temperature of 100 K an upper limit on the cooling power of about 1 W in a 1-cm^3 sample pumped with 100 W. This estimate ignores the effects of radiation trapping by reabsorption, which becomes serious at high concentrations and requires that special sample geometries be used. The practical problem of managing 100 W of fluorescence radiating into all directions around a small cooling element should also not be underestimated. Nevertheless, newly developed high-power semiconductor diode lasers (Razeghi, 1995) operating at wavelengths near 1015 nm with reasonably high electrical-to-optical conversion efficiencies increase the prospects of developing a practical device. Furthermore, recent spectroscopic studies of Yb^{3+}-doped crystals (DeLoach et al., 1993) help identify several host materials (e.g. $LiYF_4$) favorable to anti-Stokes fluorescence cooling.

The greatest challenge, however, to the advance of Yb^{3+}-based laser cooling is the reduction of parasitic quenching centers that provide nonradiative deexcitation pathways for the excited state. It is informative to consider more closely how such heat-generating processes work (Kaplyanskii and Macfarlane, 1987). An Yb^{3+} ion in the excited $^2F_{5/2}$ state can interact through electric-multipole coupling with neighboring ions. If a near resonance exists between the Yb^{3+} $^2F_{5/2} \to ^2F_{7/2}$ transition and a comparable transition in the neighbor (be it a second Yb^{3+} ion or a quenching impurity), radiationless exchange of energy from one ion to the other is possible. In the event of dipole-dipole coupling, this exchange occurs at the rate $W \propto R^{-6}$, where R is the distance between the donor and acceptor ions. Even for a very low concentration of quenching centers, the Yb^{3+} excitation can hop from site to site until a quenching center is encountered. As the Yb^{3+} concentration N goes up, the hopping rate will increase as N^2, causing the quenching to increase as N, assuming that the concentration of quenching centers remains fixed. The problem is aggravated if the multipole coupling is of higher order than dipole-dipole. Indeed, evidence for Yb^{3+} quadrupole-quadrupole coupling—with a hopping rate $W \propto R^{-10}$ and hence an accompanying $N^{2.3}$ increase in the quenching rate with concentration—has already been reported for a silicate-glass host (Brundage and Yen, 1986). For ZBLANP, the multipole order of the energy-transfer coupling is unknown. These considerations underscore the difficulties ahead: Both new standards or methods of purification—an especially challenging task for rare-earth species owing to their similar chemistry—and new diagnostics capable of detecting impurity concentrations below the parts-per-million level will be needed as the concentration of the active ion is increased. Materials scientists must make a lie of the old saw, "To a chemist, nothing is clean."

Beyond laser cooling of Yb^{3+} ions, the level structure of Tm^{3+} (Sanghera and Aggarwal, 1993) shows promise of yielding net cooling by optical pumping of the

$^3H_6 \rightarrow {}^3F_4$ transition near 1.9 μm. The excited state is expected to exhibit high fluorescence quantum efficiency in low-phonon hosts, despite the smaller energy gap than that of Yb^{3+}, whereas the ion's higher-lying states appear to have energies for which upconversion effects between pairs of Tm^{3+} ($^3F_4$) ions are inhibited, especially at low temperatures. Indeed, the ion may be more suited to operation at cryogenic temperatures than Yb^{3+}. The absence, however, of high-power, tunable, cw lasers in the 2-μm spectral region presents an obstacle to experiments with this species, although the $Co:MgF_2$ laser at least makes possible pulsed photothermal deflection measurements.

Outside the realm of impurity-doped insulators, recent experiments with semiconductors make net anti-Stokes fluorescence cooling of these materials appear achievable (Gauck et al., 1997). Due to the nanosecond time scale for radiative recombination in direct-gap semiconductors, cooling densities exceeding those of rare-earth-doped insulators are likely, provided that the problem of fluorescence escape from these high-refractive-index materials can be solved. Other new ideas for laser cooling of solids include the cooling of a magnetic spin system (Kalachev et al., 1996; Kalachev and Samartsev, 1997), of a phonon mode (Andrianov and Samartsev, 1996b), or of Frenkel excitons (Andrianov and Samartsev, 1996a). Andrianov and Samartsev's (1997b) suggestions for exploiting superradiant phenomena offer further food for thought.

Finally, another long-term idea for improving laser cooling of solids takes advantage of photonic band gap materials (Joannopoulos et al., 1995). The idea is to fabricate a refractive-index structure within the cooling element that would establish a three-dimensional photonic gap immediately below the pump frequency. In this way, emission at Stokes frequencies falling within the gap would be inhibited, thereby enhancing the cooling power derived from the anti-Stokes emission. Continuing in this speculative vein, Garret et al. (1997) have recently demonstrated the creation of squeezed phonon states in crystalline $KTaO_3$. Inspired by these results, Burin et al. (1997) have identified a new approach to laser cooling of vibrating atoms arrayed within an optical lattice. Although the degree of squeezing observed in the $KTaO_3$ experiment was too small to be practical as a cooling technique, future advances may make these ideas applicable to conventional solids. A condensed-phase realization of the ideas of Lloyd (1997a, 1997b) on coherent pumping and on a quantum-mechanical Maxwell's demon (see Section II.C) are at the limit of at least our experimental sobriety. Nevertheless, these imaginative ideas offer plenty of inspiration for continued research.

VII. Acknowledgments

One of us (TRG) thanks with great pleasure T. W. Darling, A. Migliori, G. W. Swift, and W. H. Zurek for helpful discussions on thermodynamics, and especially X. Luo and M. D. Eisaman for their outstanding experimental contributions to this work. CEM acknowledges enlightening interactions with M. I. Buchwald, B. C. Edwards, and R. I. Epstein and the support of the College of Science and Technology at UWF. This work was performed in part under the auspices of the United States Department of Energy.

VIII. Notes

1. Erickson (1972) incorrectly claimed that Vavilov's Law states that the *quantum efficiency* is independent of pump wavelength; Vavilov clearly never believed this (see Section II.A).
2. This corrects the erroneous definition used in our previous work (Mungan *et al.*, 1997c), although only a 0.5-nm increase in λ_F is obtained in applying the correct expression.
3. Due to the small amount of PbF_2 (2–3%) in the ZBLANP composition, values of the mass density and specific heat for ZBLAN are used as approximations to the corresponding values for ZBLANP.
4. A value of 1.0 was assumed in our previous work (Mungan *et al.*, 1997b).

IX. References

Aggarwal, I. D. and Lu, G. (1991). *Fluoride glass fiber optics.* Academic Press (San Diego).
Anderson, M. H., Ensher, J. R., Matthews, M. R., Weiman, C. E., and Cornell, E. A. (1995). Observation of Bose-Einstein condensation in a dilute atomic vapor. *Science* 269, 198.
Andrianov, S. N. and Samartsev, V. V. (1996a). Exciton mechanism of laser cooling in solid-state systems. *Laser Physics* 6(5), 949.
Andrianov, S. N. and Samartsev, V. V. (1996b). Laser cooling of the phonon mode in a molecular crystal. *Laser Physics* 6(4), 759.
Andrianov, S. N. and Samartsev, V. V. (1997a). Laser cooling of matter in condensed phase. *Laser Physics* 7(4), 1.
Andrianov, S. N. and Samartsev, V. V. (1997b). Optical superradiation and laser cooling. *Laser Physics* 7(1), 1.
Bartana, A., Kosloff, R., and Tannor, D. J. (1997). Laser cooling of internal degrees of freedom II. *J. Chem. Phys.* 106, 1435.
Bendow, B. (1991). Transparency of Bulk Halide Glasses. In I. D. Aggarwal and G. Lu, Eds. *Fluoride glass fiber optics* (p. 85). Academic Press (San Diego).
Berdahl, P. (1985). Radiant refrigeration by semiconductor diodes. *J. Appl. Phys.* 58, 1369.
Boyer, L. L., Harrington, J. A., Hass, M., and Rosenstock, H. B. (1975). Optical absorption by alkali halides: Possible structure in the multiphonon region. In S. S. Mitra and B. Bendow, Eds. Optical properties of highly transparent solids. *Optical physics and engineering* (p. 59). Plenum Press (New York).
Bradley, W. E. (1959). Electronic cooling device and method for the fabrication thereof. U.S. Patent. No. 2,898,743 (Philco Corporation).
Brundage, R. T. and Yen, W. M. (1986). Energy transfer among Yb^{3+} ions in silicate glass. *Phys. Rev. B* 34, 8810.
Burin, A. L., Birman, J. L., Bulatov, A., and Rabitz, H. (1997). Squeezing atomic vibrations in an optical lattice. *Phys. Rev. Lett.* (submitted).
Chang, M. S., Elliott, S. S., Gustafson, T. K., Hu, C., and Jain, R. K. (1972). Observation of anti-Stokes fluorescence in organic dye solutions. *IEEE J. Quantum Electronics* 8, 527.

Chu, S. (1991). Laser manipulation of atoms and particles. *Science* 253, 861.
Chukova, Yu. P. (1969a). Maximum light yield of luminescent light sources. *Opt. Spectrosc.* 26, 251.
Chukova, Yu. P. (1969b). Thermodynamic limit of the luminescence efficiency. *JETP Lett.* 10, 294.
Chukova, Yu. P. (1971). Thermodynamic limit to the efficiency of broadband photoluminescence. *Bull. Acad. Sci. USSR Phys. Ser.* 35, 1349.
Chukova, Yu. P. (1974). Influence of excitation-line characteristics on efficiency of spectral conversion of energy by ions of trivalent neodymium in yttrium-aluminum garnet. *Bull. Acad. Sci. USSR Phys. Ser.* 38, 57.
Chukova, Yu. P. (1976). The region of thermodynamic admissibility of light efficiencies larger than unity. *Sov. Phys. JETP* 41, 613.
Clark, J. L. and Rumbles, G. (1996). Laser cooling in the condensed phase by frequency up-conversion. *Phys. Rev. Lett.* 76, 2037.
Cohen-Tannoudji, C. N. and Phillips, W. D. (1990). New mechanisms for laser cooling. *Phys. Today* 43(10), 33.
DeLoach, L. D., Payne, S. A., Chase, L. L., Smith, L. K., Kway, W. L., and Krupke, W. F. (1993). Evaluation of absorption and emission properties of Yb^{3+} doped crystals for laser applications. *IEEE J. Quantum Electron.* 29, 1179.
Deol, R. S., Hewak, D. W., Jordery, S., Jha, A., Poulain, M., Baró, M. D., and Payne, D. N. (1993). Improved fluoride glasses for 1.3 μm optical amplifiers. *J. Non-Cryst. Solids* 161, 257.
Djeu, N. (1978). Laser cooling of gases by reradiation at higher frequency transitions. *Opt. Commun.* 26, 354.
Djeu, N. and Whitney, W. T. (1981). Laser cooling by spontaneous anti-Stokes scattering. *Phys. Rev. Lett.* 46, 236.
Dousmanis, G. C., Mueller, C. W., and Nelson, H. (1963). Effect of doping on frequency of stimulated and incoherent emission in GaAs diodes. *Appl. Phys. Lett.* 3, 133.
Dousmanis, G. C., Mueller, C. W., Nelson, H., and Petzinger, K. G. (1964). Evidence of refrigerating action by means of photon emission in semiconductor diodes. *Phys. Rev.* 133, A316.
Epstein, R. I., Buchwald, M. I., Edwards, B. C., Gosnell, T. R., and Mungan, C. E. (1995a). Observation of laser-induced fluorescent cooling of a solid. *Nature (London)* 377, 500.
Epstein, R. I., Edwards, B. C., Buchwald, M. I., and Gosnell, T. R. (1995b). Fluorescent refrigeration. U.S. Patent. No. 5,447,032 (University of California).
Erickson, L. E. (1972). On anti-Stokes luminescence from Rhodamine 6G in ethanol solutions. *J. Lumin.* 5, 1.
Fokker, P. A., Dijkhuis, J. I., and de Wijn, H. W. (1997a). Stimulated emission of phonons in an acoustical cavity. *Phys. Rev. B* 55, 2925.
Fokker, P. A., Koster, W. D., Dijkhuis, J. I., de Wijn, H. W., Lu, L., Meltzer, R. S., and Yen, W. M. (1997b). Stimulated emission of phonons in a ruby fiber. *Phys. Rev. B* 56, 2306.
Gallo, P., Massacurai, V., Ruocco, G., and Signorelli, G. (1991). Absolute two-phonon Raman cross section in potassium chloride. *Phys. Rev. B* 43, 14268.
Garrett, G. A., Rojo, A. G., Sood, A. K., Whittaker, J. D., and Merlin, R. (1997). Vacuum squeezing of solids: Macroscopic quantum states driven by light pulses. *Science* 275, 1638.
Gauck, H., Gfroerer, T. H., Renn, M. J., Cornell, E. A., and Bertness, K. A. (1997). External radiative quantum efficiency of 96% from a *GaAs/GaInP* heterostructure. *Appl. Phys. A* 64, 143.
Gauck, H. G., Renn, M. J., Cornell, E. A., and Bertness, K. A. (1995). Laser refrigeration in the solid state. Postdeadline Papers of the Quantum Electronics and Laser Science Conference, Baltimore, MD, #QPD16.
Gerthsen, P. and Kauer, E. (1965). The luminescence diode acting as a heat pump. *Phys. Lett.* 17, 255.
Geusic, J. E., Schulz-DuBois, E. O., and Scovil, H. E. D. (1967). Quantum equivalent of the Carnot cycle. *Phys. Rev.* 156, 343.
Geva, E. and Kosloff, R. (1996). The quantum heat engine and heat pump: An irreversible thermodynamic analysis of the three-level amplifier. *J. Chem. Phys.* 104, 7681.

Hänsch, T.W. and Schawlow, A. L. (1975). Cooling of gases by laser radiation. *Opt. Commun.* 13, 68.
Hasz, W. C., Whang, J. H., and Moynihan, C. T. (1993). Comparison of physical properties of ZrF_4- and HfF_4-based melts and glasses. *J. Non-Cryst. Solids* 161, 127.
Haynes, J. R. and Briggs, H. B. (1952). Radiation produced in germanium and silicon by electron-hole recombination. *Phys. Rev.* 86, 647.
Heiman, D., Hellwarth, R. W., and Hamilton, D. S. (1979). Raman scattering and nonlinear refractive index measurements of optical glasses. *J. Noncryst. Solids* 34, 63.
Jablonski, A. (1933). Efficiency of anti-Stokes fluorescence in dyes. *Nature (London)* 131, 839.
Jackson, W. B., Amer, N. M., Boccara, A. C., and Fournier, D. (1981). Photothermal deflection spectroscopy and detection. *Appl. Opt.* 20, 1333.
Joannopoulos, J. D., Meade, R. D., and Winn, J. N. (1995). *Photonic Crystals.* Princeton University Press (Princeton).
Kafri, O. and Levine, R. D. (1974). Thermodynamics of adiabatic laser processes: Optical heaters and refrigerators. *Opt. Commun.* 12, 118.
Kalachev, A. A., Karamyshev, S. B., and Samartsev, V. V. (1996). Laser cooling of the spin system in Van Vleck paramagnetics. *Laser Physics* 6(1), 27.
Kalachev, A. A. and Samartsev, V. V. (1997). Specific features of local data erasure and laser cooling in Van Vleck paramagnetics. *Laser Physics* 7(2), 1.
Kaplyanskii, A. A. and Macfarlane, R. M. (1987). *Spectroscopy of Solids Containing Rare Earth Ions.* North Holland (Amsterdam).
Kastler, A. (1950). Quelques suggestions concernant la production optique et la détection optique d'une inégalité de population des niveaux de quantification spatiale des atomes: Application à l'expérience de Stern et Gerlach et à la résonance magnétique. *J. Phys. Radium* 11, 255.
Ketskeméty, I. and Farkas, É. (1970). Neuere Überlegungen bezüglich der oberen Schranke der Fluoreszenzausbeute. *Acta Phys. Chem. Szeged (Hungary)* 16, 77.
Keyes, R. J. and Quist, T. M. (1962). Recombination radiation emitted by gallium arsenide. *Proc. I.R.E.* 50, 1822.
Kittel, C. (1961). Phonon masers and the phonon bottleneck. *Phys. Rev. Lett.* 6, 449.
Kushida, T. and Geusic, J. E. (1968). Optical refrigeration in Nd-doped yttrium aluminum garnet. *Phys. Rev. Lett.* 21, 1172.
Landau, L. (1946). On the thermodynamics of photoluminescence. *J. Phys. (Moscow)* 10, 503.
Landsberg, P. T. and Evans, D. A. (1968). Thermodynamic limits for some light-producing devices. *Phys. Rev.* 166, 242.
Landsberg, P. T. and Tonge, G. (1980). Thermodynamic energy conversion efficiencies. *J. Appl. Phys.* 51, R1.
Lehovec, K., Accardo, C. A., and Jamgochian, E. (1951). Injected light emission of silicon carbide crystals. *Phys. Rev.* 83, 603.
Lehovec, K., Accardo, C. A., and Jamgochian, E. (1953). Light emission produced by current injected into a green silicon-carbide crystal. *Phys. Rev.* 89, 20.
Lines, M. E. (1991). Interaction of light with matter: Theoretical overview. In P. Klocek, Ed. *Handbook of infrared optical materials* (*Optical engineering*, p. 71). Marcel Dekker (New York).
Lloyd, S. (1997a). Quantum optical refrigeration. *Phys. Rev. A* (submitted).
Lloyd, S. (1997b). A quantum-mechanical Maxwell's demon. *Phys. Rev. A* 56, 3374.
Mazurenko, Yu. T. (1965a). Some properties of lasers from the point of view of thermodynamics. *Opt. Spectrosc.* 19, 85.
Mazurenko, Yu. T. (1965b). A thermodynamic treatment of the process of photoluminescence. *Opt. Spectrosc.* 18, 24.
McCumber, D. E. (1964). Einstein relations connecting broadband emission and absorption spectra. *Phys. Rev.* 136, A954.
Mihalas, D. (1978). *Stellar Atmospheres* (Chap. 1). W. H. Freeman (San Francisco).

Miniscalco, W. J. (1993). Optical and electronic properties of rare earth ions in glasses. In M. J. F. Digonnet, Ed. *Rare earth doped fiber lasers and amplifiers* (p. 19). Marcel Dekker (New York).

Mungan, C. E., Buchwald, M. I., Edwards, B. C., Epstein, R. I., and Gosnell, T. R. (1997a). Internal laser cooling of Yb^{3+}-doped glass measured between 100 and 300 K. *Appl. Phys. Lett.* 71, 1458.

Mungan, C. E., Buchwald, M. I., Edwards, B. C., Epstein, R. I., and Gosnell, T. R. (1997b). Laser cooling of a solid by 16 K starting from room temperature. *Phys. Rev. Lett.* 78, 1030.

Mungan, C. E., Buchwald, M. I., Edwards, B. C., Epstein, R. I., and Gosnell, T. R. (1997c). Spectroscopic determination of the expected optical cooling of ytterbium-doped glass. *Mat. Sci. Forum* 239–241, 501.

Mungan, C. E. and Gosnell, T. R. (1996). Comment on "Laser cooling in the condensed phase by frequency up-conversion." *Phys. Rev. Lett.* 77, 2840.

Nelson, D. F. and Boyle, W. S. (1962). A continuous operating ruby optical maser. *Appl. Opt.* 1, 181.

Nelson, D. F. and Sturge, M. D. (1965). Relation between absorption and emission in the region of the R lines of ruby. *Phys. Rev.* 137, A1117.

Newman, R. (1953). Optical studies of injected carriers. II: Recombination radiation in germanium. *Phys. Rev.* 91, 1313.

Oraevsky, A. N. (1996). Cooling of semiconductors by laser radiation. *J. Russ. Laser Research* 17, 471.

Pankove, J. I. (1975). *Optical processes in semiconductors* (p. 193). Dover (New York).

Prigogine, I. (1954). *Introduction to thermodynamics of irreversible processes*. (3rd ed., p. 17). Wiley (New York).

Pringsheim, P. (1929). Zwei Bemerkungen über den Unterschied von Lumineszenz- und Temperaturstrahlung. *Z. Physik* 57, 739.

Pringsheim, P. (1946). Some remarks concerning the difference between luminescence and temperature radiation: Anti-Stokes fluorescence. *J. Phys. (Moscow)* 10, 495.

Raman, C. V. (1928). A change of wavelength in light scattering. *Nature (London)* 121, 619.

Razeghi, M. (1995). InGaAsP-based high power laser diodes. *Optics & Photonics News* 6(8), 16.

Rivlin, L. A. and Zadernovsky, A. A. (1997). Laser cooling of semiconductors. *Opt. Commun.* 139, 219.

Rumbles, G. and Clark, J. L. (1996). Rumbles and Clark reply. *Phys. Rev. Lett.* 77, 2841.

Sanghera, J. S. and Aggarwal, I. D. (1993). Rare earth doped heavy-metal fluoride glass fibers. In M. J. F. Digonnet, Ed. *Rare earth doped fiber lasers and amplifiers* (p. 423). Marcel Dekker (New York).

Schnitzer, I., Yablonovitch, E., Caneau, C., and Gmitter, T. J. (1993). Ultrahigh spontaneous emission quantum efficiency, 99.7% internally and 72% externally, from AlGaAs/GaAs/AlGaAs double heterostructures. *Appl. Phys. Lett.* 62, 131.

Scovil, H. E. D. and Schulz-DuBois, E. O. (1959). Three-level masers as heat engines. *Phys. Rev. Lett.* 2, 262.

Stepanov, B. I. and Gribkovskii, V. P. (1968). *Theory of luminescence* (p. 322). Iliffe Books (London).

Stokes, G. G. (1852). On the change of refrangibility of light. *Phil. Trans.* 143, 463.

Tauc, J. (1957). The share of thermal energy taken from the surroundings in the electroluminescent energy radiated from a $p-n$ junction. *Czech. J. Phys.* 7, 275.

Treanor, C. E., Rich, J. W., and Rehm, R. G. (1967). Vibrational relaxation of anharmonic oscillators with exchange-dominated collisions. *J. Chem. Phys.* 48, 1798.

Tsujikawa, I. and Murao, T. (1963). Possibility of optical cooling of ruby. *J. Phys. Soc. Jpn.* 18, 503.

Tucker, E. B. (1961). Amplification of 9.3-kMc/sec ultrasonic pulses by maser action in ruby. *Phys. Rev. Lett.* 6, 547.

Vavilov, S. (1945). Some remarks on the Stokes law. *J. Phys. (Moscow)* 9, 68.

Vavilov, S. (1946). Photoluminescence and thermodynamics. *J. Phys. (Moscow)* 10, 499.

Weinstein, M. A. (1960). Thermodynamic limitation on the conversion of heat into light. *J. Opt. Soc. Am.* 50, 597.

Wineland, D. J. and Itano, W. M. (1987). Laser cooling. *Phys. Today* 40(6), 34.

Wood, R. W. (1928). Anti-Stokes radiation of fluorescent liquids. *Phil. Mag.* 6, 310.
Yardley, J. T. (1971). Population inversion and energy transfer in CO lasers. *Appl. Opt.* 10, 1760.
Yatsiv, S. (1961). Anti-Stokes fluorescence as a cooling process. In J. R. Singer, Ed. *Advances in quantum electronics* (p. 200). Columbia University Press (New York).
Zadernovskii, A. A. and Rivlin, L. A. (1996). Laser cooling of a semiconductor (optical heat engine). *Quantum Electronics* 26, 1100.
Zander, C. (1991). Abkühlung einer Farbstofflösung durch Anti-Stokes-Fluoreszenz. Ph.D. thesis, Universität Siegen.
Zander, C. and Drexhage, K. H. (1995). Cooling of a dye solution by anti-Stokes fluorescence. In D. C. Neckers, D. H. Volman, and G. von Bünau, Eds. *Advances in photochemistry* (Vol. 20, p. 59). Wiley (New York).
Zou, X. and Toratini, H. (1995). Evaluation of the spectroscopic properties of Yb^{3+}-doped glasses. *Phys. Rev. B* 52, 15889.

OPTICAL PATTERN FORMATION

L. A. LUGIATO

Istituto Nazionale di Fisica per la Materia,
Università di Milano, Milano, Italy

M. BRAMBILLA

Istituto Nazionale di Fisica per la Materia, Dipartimento
Interateneo di Fisica del Politecnico di Bari, Bari, Italy

A. GATTI

Istituto Nazionale di Fisica per la Materia,
Università di Milano, Milano, Italy

I. Introduction	229
II. General Features About OPF	233
A. Dynamical Equations with Diffraction	233
B. Systems with Translational Symmetry. The Mechanisms for Pattern Formation	239
C. Systems with Single Feedback Mirror	249
D. Analogy with Hydrodynamics. Vortices and Other Defects	252
E. Cavities with Spherical Mirrors	259
F. Applicative Aspects	265
G. Further Issues in OPF	268
III. OPF and Solitary Structures in Cavities	269
A. Localized Structures in Optics	270
B. Exciting LS and Spatial Solitons (SS)	271
1. Spatial Soliton Control	274
C. Spatial Solitons in Semiconductor Models	276
1. OPF in Semiconductor Models	277
IV. Quantum Fluctuations and Optical Pattern Formation	278
A. Quantum Models for χ^2 Media Including Diffraction	279
B. A Langevin Approach to Quantum Fluctuation Dynamics	282
C. Quantum Images in the Near Field	285
1. Spatial Structure of Squeezed States	289
D. Quantum Images in the Far Field	291
E. Final Considerations on Quantum Aspects	294
V. Acknowledgments	295
VI. References	295

I. Introduction

The field of optical pattern formation (OPF) studies the spatial and spatio-temporal phenomena that arise in the structure of the electromagnetic field in the planes orthogonal with respect to the direction of propagation. Most theoretical treatments of the interaction between matter and radiation introduce the plane wave approximation, that is, they assume that the electric field is uniform in each transverse plane. In this way, the time evolution equations depend only on one spatial variable, the longitudinal variable z, which corresponds to the direction of propagation. By dropping the plane wave approximation, one opens the door to the fascinating world of pattern formation. In the paraxial approximation this corresponds to adding, in the time evolution equation of the electric field E, a term proportional to $i\nabla_\perp^2 E = i(\partial^2/\partial x^2 + \partial^2/\partial y^2) E$, which describes diffraction of radiation (i is the imaginary unit); this term couples the different points of the transverse plane (x, y), as it is necessary for pattern formation.

The interaction of light with linear inhomogeneous media can give rise to structures of interesting and remarkable complexity (Berry, 1981; Berry, Nye, and Wright, 1979; Baranova *et al.*, 1981). However, the field of OPF studies mainly the interaction with nonlinear media, where the phenomena emerge spontaneously as a consequence of an instability; another name that is commonly used to designate OPF is "transverse nonlinear optics." Historically, the broad interest in OPF emerged as a natural evolution of the previous development of the field of optical instabilities and chaos, when the main attention shifted gradually from purely temporal effects to spatio-temporal phenomena. The evolution was made possible also by the spectacular increase of simulational capabilities on available computers. For both the fields of optical instabilities and OPF, continuous inspiration arose from the formulation of general disciplines as Haken's synergetics (Haken, 1977) or Prigogine's theory of dissipative structures (Nicolis and Prigogine, 1977; Nicolis, 1995).

The existence of transverse effects is well known since the earliest days of laser physics. In order to obtain the simple Gaussian transverse structure that is desired for most applications, one introduces apertures in the laser cavity. Otherwise, the system spontaneously generates more or less complex configurations. These phenomena were, however, mostly considered as undesirable or difficult to control. A basic understanding, though, was achieved with the analysis of diffractive effects upon propagation of Gaussian beams with phenomena such as self-focusing, self-defocusing, and self-trapping (Akhmanov *et al.*, 1972).

Despite the existence of an important early literature on this subject (see Refs. 8–22 in Lugiato, 1994), systematic investigations on OPF have been activated only in the 1980s. Great impulse was given by the attention devoted to the case of nonlinear materials contained in optical cavities; in the mean field limit, such systems are described by sets of partial differential equations with two spatial variables plus time, exactly as in the case of two-dimensional patterns in hydrodynamics and nonlinear chemical reactions (Lugiato and Lefever, 1987; Lugiano and Oldano, 1988; Luciano, Oldano, and Narducci, 1988). On the other hand, the study of systems with a single feedback mirror (Giusfredi *et al.*, 1988; Firth, 1990;

D'Alessandro and Firth, 1992) produced the best compromise between simplicity of theoretical treatment and accessibility to experimental realization.

We can enlist at least six different configurations that have been considered in studying OPF:

1. Mirrorless configuration with unidirectional propagation. This case is usually studied only at steady state, that is, without the derivative with respect to time, and is often connected with the analysis of bright or dark solitons that arise in the nonlinear Schrödinger equations (or generalizations thereof) from the balance between nonlinearity and diffraction (Haelterman and Sheppard, 1994; Horak *et al.*, 1995; Kivshar and Xiaoping Yang, 1994; Law and Schwartzlander, 1994; Rosanov *et al.*, 1994).
2. Mirrorless configuration with counterpropagating waves. The analysis of this case has led to the first theoretical prediction and experimental observation of hexagonal patterns in an optical system (Firth and Parè, 1988; Grynberg *et al.*, 1988).
3. Single feedback mirror configuration.
4. Nonlinear medium in an optical resonator.
5. Nonlinear optical systems having two-dimensional field rotation in the optical feedback loop (Akhmanov *et al.*, 1992; Vorontsov *et al.*, 1994; Zheleznikh *et al.*, 1994).
6. Laser arrays (Wang and Winful, 1988; Otsuka, 1990; Jacobsen *et al.*, 1991).

The simplest models for studying OPF assume translational symmetry in the transverse plane. This implies that, if there are mirrors, they must be plane mirrors, and if there is a driving field, it must have a plane wave configuration. This is the most fundamental setting, because it allows for studying the spontaneous onset of a pattern from a homogeneous state with breaking of the translational symmetry, as a consequence of an instability arising from the interplay of nonlinearity and diffraction. In this case one can perform analytically both the calculation of the homogeneous stationary solution and its linear stability analysis. When there is a cavity, the modes of the empty cavity correspond to the plane waves tilted with respect to the propagation direction, with a continuous frequency spectrum.

The assumption of translational symmetry implies that the system is infinitely extended in the transverse directions, and this is conveniently formalized by using periodic boundary conditions. Clearly, this kind of model is strongly idealized, because in practice one always has a beam confined to a certain region of the transverse plane. A confinement can be introduced in different ways, for example, by including into the model the transverse shape of the medium or, if an input beam is present, by considering a Gaussian or a flat-top transverse profile. In this case, everything must be calculated numerically. When the transverse dimensions of the confinement region are much larger than the length that characterizes the spatial modulation of the pattern, one can recover qualitatively the results of the idealized model.

In the cavity case, one can obtain the transverse confinement by considering spherical instead of planar mirrors. In this case the system has only rotational symmetry; the modes of the empty cavity correspond to Gauss-Laguerre functions, with a discrete frequency spectrum. An important advantage of this configuration is that the number of modes in play can be controlled by the geometric parameters of the cavity. For example, by reducing the Fresnel number one can cut off Gauss-Laguerre modes of high order. By varying the radius of curvature of the spherical mirrors and their distance, one controls the frequency spacing between adjacent transverse modes. When only a small number of Gauss-Laguerre modes are relevant, one can conveniently describe the temporal evolution by means of a set of ordinary differential equations for the modal amplitudes. On the other hand, when the Fresnel number is large and the frequency spacing is small with respect to the width of the cavity resonances (as one has for quasi-planar, quasi-confocal, or quasi-concentric cavities), the dynamics is governed by a large number of Gauss-Laguerre modes. In this case the set of modal equations becomes unmanageable, and it is better to solve the partial differential equations for the field as a whole. Again, one recovers qualitatively the results of the idealized model (Harkness *et al.*, 1997) which, in this way, become boundary independent. The overwhelming variety of phenomena one meets in OPF displays several similarities with those, for example, in hydrodynamics and nonlinear chemical reactions, where diffusion and not diffraction couples to the nonlinearity to govern the arising patterns. A formal analogy between laser equations and hydrodynamics has been formalized (Akhmanov *et al.*, 1972; Brambilla, Lugiato, *et al.*, 1991). In definite and small domains of the parameter space, the exact dynamical equations can be approximated by Ginzburg-Landau (Coulet *et al.*, 1989; Oppo *et al.*, 1991; Staliunas, 1993; Mandel *et al.*, 1993; Tlidi *et al.*, 1993; Lega *et al.*, 1994), Kuramoto-Shivashinsky (Lefever *et al.*, 1989; Wang *et al.*, 1993; Huyet and Rica, 1996), Newell-Whitehead (Lega *et al.*, 1994), or Swift-Hohenberg (Tlidi *et al.*, 1994) equations; pattern formation in a wide variety of fields has been described by these types of equations. Visual analogies with hydrodynamics have also been found in several experiments (see for example Fig. 22 in Section II).

Undoubtedly, hydrodynamics or nonlinear chemical reactions have a much longer tradition than optics in the study of pattern formation (Chan and Lefever, 1995; Mertens *et al.*, 1997). However, radiation-matter interaction is fundamental in physics and chemistry, and this is already a strong motivation for studying pattern formation in optics. In addition, optics presents two special features that are interesting and stimulating. First, optical systems are very fast and have a large frequency bandwidth; hence they lend themselves naturally for applications such as telecommunications and information technology. The relevant example of useful application of optical structures is already provided by solitonic transmission in optical fibers. The investigations of transverse nonlinear optics offer, in principle, the possibility of an approach to parallel optical information processing, by encoding information in the transverse structure of the electric field.

The second special feature is that optical systems are macroscopic or mesoscopic, yet they are capable of displaying interesting quantum effects, even at room temperature. The recent years have witnessed the start of investigations on quantum aspects in transverse optical structures.

This article is divided into three parts. The first and largest part is general and devoted to the illustration of the key concepts necessary for understanding the nature of the investigations in the field of OPF and their developments. The second part concerns the topic of spatial solitons in cavities as a tool to define an interesting applicative avenue for this field. The third part deals with the quantum features of nonlinear optical patterns.

The field of OPF has developed remarkably in recent years, to such an extent that a complete review is by now an exceedingly difficult task, especially in a rather limited space. Therefore, this article does not aim to completeness, but only wishes to provide a guideline. Unavoidably, the emphasis on various aspects and results is influenced by the authors' personal taste; we apologize for any omission.

Prior reviews of this field are (Arecchi, 1991; Lugiato, 1992; Weiss, 1992; Arecchi, 1994; Firth, 1995; Lugiato, 1994) and contain reference lists updated to 1994, including general references on the field of optical instabilities. Recent volumes that discuss optical instabilities are (Newell and Maloney, 1992; Khanin, 1995; Mandel, 1997). In order to limit the number of references in this article, we include mainly those that are essential for the discussed issues. Some additional references to papers published after 1994 are also included in Section II.G; nevertheless, the list will presumably appear incomplete.

II. General Features About OPF

A. DYNAMICAL EQUATIONS WITH DIFFRACTION

In the field of nonlinear optics the Maxwell-Bloch equations (MBE) play a role analogous to that of the Navier-Stokes equations in hydrodynamics; they provide the paradigm for the resonant interaction between matter and radiation by focusing on the simplest picture of a collection of two-level atoms. We shall limit our attention to the case of homogeneously broadened atomic systems and to unidirectional propagation.

The MBE are derived in the dipole, semiclassical, slowly varying envelope/paraxial and rotating wave approximations. To the purpose of writing them we set

$$\mathcal{E}(x, y, z, t) = \frac{\hbar\sqrt{\gamma_\| \gamma_\perp}}{\mu} \frac{1}{2} \{F(x, y, z, t)e^{i(k_0 z - \omega_0 t)} + \text{c.c.}\} \quad (\text{II.1})$$

$$\mathcal{P}(x, y, z, t) = \frac{\sqrt{\gamma_\parallel}}{\gamma_\perp} \mu\rho \{P(x, y, z, t)\, e^{i(k_0 z - \omega_0 t)} + \text{c.c.}\} \qquad \text{(II.2)}$$

and assume that the envelope functions F and P vary in space and time much more slowly than the associated exponential factors. $\mathcal{E}$ denotes the linearly polarized electric field and $\mathcal{P}$ is the macroscopic atomic polarization density. The parameter μ denotes the modulus of the atomic dipole moment and ρ is the atomic density; $\gamma_\perp$ and $\gamma_\parallel$ are the atomic relaxation rates for the polarization and population difference, respectively. The constant $\gamma_\perp$ coincides with the atomic linewidth. The symbol ω_0 denotes the reference frequency, which can be selected with a certain degree of arbitrariness, and $k_0 = \omega_0/c$ is the corresponding wave number. For systems that are driven by a stationary external field, the natural choice for ω_0 is the frequency of the driving field; in the case of the laser we will take $\omega_0 = \omega_a$ where ω_a is the Bohr transition frequency of the two-level atoms.

The MBE have the form

$$\frac{1}{2ik_0} \nabla_\perp^2 F + \frac{\partial F}{\partial z} + \frac{1}{c}\frac{\partial F}{\partial t} = \alpha P \qquad \text{(II.3)}$$

$$\frac{\partial P}{\partial t} = \gamma_\perp [FD - (1 + i\Delta)P] \qquad \text{(II.4)}$$

$$\frac{\partial D}{\partial t} = -\gamma_\parallel \left[\frac{1}{2}(F^*P + FP^*) + D - \chi(x, y, z)\right] \qquad \text{(II.5)}$$

The gain parameter α of the electric field per unit length is given by

$$\alpha = \frac{2\pi\omega_0 \mu^2 \sigma \rho}{\hbar c \gamma_\perp} \qquad \text{(II.6)}$$

where σ is the unperturbed population inversion per atom induced by the incoherent pump mechanism; in the absence of pump, so that there is no population inversion, σ equals -1 and $-\alpha$ acquires the meaning of the absorption coefficient per unit length of the electric field. The atomic detuning parameter Δ is defined as

$$\Delta = \frac{\omega_a - \omega_0}{\gamma_\parallel} \qquad \text{(II.7)}$$

The variable D in Eq. (II.5) corresponds to the difference between the probability of finding a single atom in the upper and in the lower state, divided by σ. The function $\chi(x, y, z)$ is such that $0 \leq \chi \leq 1$ and vanishes outside the sample. More explicitly, it equals unity inside the sample for nonpumped systems, whereas it describes the spatial profile of the gain region for a system with population inversion.

Next, let us consider the case where the atomic sample is contained in a unidirectional ring cavity (Lugiato and Narducci, 1992). The feedback action of the cavity can be expressed in terms of appropriate boundary conditions (in z) for the field envelope $F(x, y, z)$. The problem constituted by the MBE plus boundary conditions is very hard to treat because it involves three spatial coordinates in addition to time. Some authors (Moloney and Gibbs, 1982; Berre et al., 1994) have introduced simplifying assumptions, such as the adiabatic elimination of the atomic polarization, and have cast the problem in the form of one or two partial differential equations associated with a map; in doing that they generalized the procedure introduced by Ikeda (1979) to include the transverse degrees of freedom.

We followed a different procedure in order to arrive at a problem that involves two spatial coordinates instead of three, so that it is much less demanding in terms of CPU time and more similar to the problems analyzed in hydrodynamics and nonlinear chemical reactions. Similarly, we take advantage of the mean field limit already extensively utilized in the plane wave models (Bonifacio and Lugiato, 1978).

Let us consider first the case of a cavity with plane mirrors. If we assume that the atomic linewidth $\gamma_\perp$ is much smaller than the free spectral range $c/\mathscr{L}$, where $\mathscr{L}$ is the length of the ring cavity, only one longitudinal mode of the cavity is relevant, and we indicate its frequency by ω_c. Correspondingly, the longitudinal configuration of the electric field is now fixed and we can write

$$\mathscr{E}(x, y, z, t) = \frac{\hbar\sqrt{\gamma_\parallel \gamma_\perp}}{2\mu} \{A(x, y, t)e^{ik_z z}e^{-i\omega_0 t} + c.c.\} \tag{II.8}$$

with $k_z = \omega_c/c$. The transverse modes of the cavity have the form $A(x, y) \propto \exp(i\vec{k}_\perp \cdot \vec{x})$ with $\vec{x} = (x, y)$ and $\vec{k}_\perp = (k_x, k_y)$. Their frequencies are given by $\omega_{k_\perp} = c(k_z^2 + k_\perp^2)^{1/2}$ with $k_\perp^2 = k_x^2 + k_y^2$, and in the paraxial approximation $k_\perp \ll k_z$ one has

$$\omega_{k_\perp} = \omega_c + \frac{ck_\perp^2}{2k_z} \tag{II.9}$$

with a continuous frequency distribution from ω_c to ∞. Consistent with the adopted single longitudinal mode approximation, one assumes that only the modes such that $ck_\perp^2/2k_z$ is much smaller than the free spectral range are relevant.

Specifically, the mean field limit assumes that the intensity transmissivity coefficient T of the input/output cavity mirror is small, that is, $T \ll 1$ and correspondingly, these other quantities are of order T:

$$\alpha L = \mathcal{O}(T) \qquad \frac{\omega_c - \omega_0}{c/\mathscr{L}} = \mathcal{O}(T) \tag{II.10}$$

$$\frac{\gamma_\perp}{c/\mathscr{L}} = \mathcal{O}(T) \qquad \frac{ck_\perp^2 \mathscr{L}}{2ck_z} = \mathcal{O}(T) \tag{II.11}$$

where L is the length of the atomic sample.

In the limit $T \ll 1$ the quantities A, P, D become practically uniform along the cavity and one obtains the following mean field model (Lugiato and Oldano, 1988; Lugiato, Oldano, and Narducci, 1988)

$$\frac{\partial A}{\partial t} = -\kappa\{(1 + i\theta)A - A_I(x, y) - 2C P - ia\nabla^2_\perp A\} \quad \text{(II.12)}$$

$$\frac{\partial P}{\partial t} = \gamma_\perp [AD - (1 + i\Delta)P] \quad \text{(II.13)}$$

$$\frac{\partial D}{\partial t} = -\gamma_\parallel \left[\frac{1}{2}(A^*P + AP^*) + D - \chi(x, y)\right] \quad \text{(II.14)}$$

where κ is the cavity damping constant (or cavity linewidth)

$$\kappa = \frac{cT}{2\mathscr{L}} \quad \text{(II.15)}$$

c and θ are dimensionless parameters given by

$$C = \frac{\alpha L}{T} \qquad \theta = \frac{\omega_c - \omega_0}{\kappa} \quad \text{(II.16)}$$

and

$$a = \frac{c}{2k_z\kappa} \quad \text{(II.17)}$$

We call C the pump parameter in the case with population inversion, whereas we call $|C|$ the bistability parameter in the case of unpumped medium; θ is the cavity detuning parameter. In writing (II.15) and (II.16) we assumed for definiteness that the cavity has only one input/output mirror. The symbol $A_I(x, y)$ denotes a stationary input field that may be injected into the cavity. The parameter a determines the spatial scale of the pattern; apart from trivial factors the characteristic length is given by $(\lambda L/T)^{1/2}$, where λ is the wavelength. The quantity $(\lambda \mathscr{L})^{1/2}$ is the diffraction length in the paraxial approximation; the factor T is the length enhancement due to the cavity. Out of the paraxial approximation, the diffraction length coincides with the wavelength, and the pattern develops over a finer scale.

Let us turn now to the case of a cavity with spherical mirrors. In this case, because of the cylindrical symmetry, we will use the polar coordinates $r = (x^2 + y^2)^{1/2}$ and φ. The modes of the empty cavity are given by (Yariv, 1989)

$$f_{pli}(r, \varphi, z) = \tilde{f}_{pl}(r, z) \exp\{(z/z_0)[r^2/w^2(z)] - \theta_{pl}(z)\} \cdot \begin{pmatrix} \cos(l\varphi), & i = 1 \\ \sin(l\varphi), & i = 2 \end{pmatrix} \quad \text{(II.18)}$$

$$\tilde{f}_{pl}(r, z) = \frac{2}{w(z)(2^{\delta_{l0}}\pi)^{1/2}} \sqrt{\frac{p!}{(p+l)!}} \left[\frac{2r^2}{w^2(z)}\right]^{l/2} \quad \text{(II.19)}$$

$$L_p^l\left(\frac{2r^2}{w^2(z)}\right) e^{r^2/w^2(z)} \quad \text{(II.20)}$$

where $p, l = 0, 1, 2, \ldots$,

$$w(z) = w_0 \sqrt{1 + \left(\frac{z}{z_0}\right)^2}, \quad z_0 = \frac{\pi w_0^2}{\lambda} \quad \text{(II.21)}$$

$$\theta_{pl} = (2p + l + 1)tg^{-1}\left(\frac{z}{z_0}\right) \quad \text{(II.22)}$$

w_0 and z_0 denote the beam waist and the Rayleigh range, respectively; L_p^l is the Laguerre polynomial of indicated indices. In order to obtain that the system operates with a single longitudinal mode but several transverse modes, we consider quasi-planar mirrors, for which the relevant cavity frequencies are given by

$$\omega_{pl} = \omega_c + (2p + l)\eta \quad \text{(II.23)}$$

where the intermode frequency spacing depends on the curvature of the mirrors and on their distance (Prati *et al.*, 1994); note that modes with the same value of $2p + l$ are frequency degenerate. The mean field limit is defined as before, only replacing the last of conditions (II.11) by

$$(2p + l)\eta \frac{\mathscr{L}}{c} = \mathbb{O}(T) \quad \text{(II.24)}$$

and the resulting set of equations reads (Lugiato *et al.*, 1990)

$$\frac{\partial A}{\partial t} = -\kappa\bigg\{(1 + i\theta)A - A_I(x, y) - 2C\,P$$

$$-i\frac{\eta}{\kappa}\left(1 - \frac{r^2}{w_0^2} + \frac{w_0^2}{4}\nabla_\perp^2\right)A\bigg\} \quad \text{(II.25)}$$

$$\frac{\partial P}{\partial t} = \gamma_\perp[AD - (1 + i\Delta)P] \quad \text{(II.26)}$$

$$\frac{\partial D}{\partial t} = -\gamma_\parallel\left[\frac{1}{2}(A^*P + AP^*) + D - \chi(x, y)\right] \quad \text{(II.27)}$$

where κ, C, θ are defined by (II.15) and (II.16), and in polar coordinates

$$\nabla_\perp^2 E = \frac{\partial^2}{\partial r^2} + \frac{1}{r}\frac{\partial}{\partial r} + \frac{1}{r^2}\frac{\partial^2}{\partial \varphi^2} \quad \text{(II.28)}$$

It can be noted that Eqs. (II.25–II.27) can be used also for a quasi-confocal or quasi-concentric cavity (Lugiato et al., 1990), provided that the sample length L is much smaller than the Rayleigh range z_0 (thin medium), and the sample is located at cavity center; $A(r, \varphi, t)$ describes the transverse variation of the field in the sample. In the quasi-concentric case η is negative. In the quasi-confocal case η can be positive or negative, but one must include the additional constraint that A must be even with respect to the parity transformation $\vec{x} \to -\vec{x}$.

Two special cases of the model (II.12–II.14) obtained in appropriate limits have been also extensively analyzed in the literature. Both include the adiabatic elimination of atomic variables for $\kappa \ll \gamma_\perp, \gamma_\parallel$, and refer to the case of no population inversion, that is, $\sigma = -1$, $C = -|C|$. The first corresponds to the purely absorptive case $\Delta = 0$ (Lugiato and Aldano, 1988; Firth and Scroggie, 1994)

$$\frac{\partial A}{\partial t} = -\kappa \left[(1 + i\theta)A - ia\nabla_\perp^2 A - A_I + \frac{2|C|A}{1 + |A|^2} \right] \quad \text{(II.29)}$$

whereas the other corresponds to the opposite purely dispersive limit (for a derivation see, for example, Lugiato et al., 1994); by appropriate normalization of A, it reads (Lugiato and Lefever, 1987)

$$\frac{\partial \tilde{A}}{\partial t} = -\kappa[\tilde{A} - \tilde{A}_I - i(\zeta|\tilde{A}|^2 - \theta)\tilde{A} - ia\nabla_\perp^2 \tilde{A}] \quad \text{(II.30)}$$

where $\zeta = +1$ or -1 corresponds to the self-focusing or defocusing case $\Delta < 0$ or $\Delta > 0$, respectively.

Equation (II.30) also describes the case of a pure Kerr medium, that is, of a refractive $\chi^{(3)}$ nonlinearity. On the other hand, the case of $\chi^{(2)}$ media has also been considered in research on OPF. Here we write a model that describes a degenerate optical parametric oscillator with plane mirrors (Oppo et al., 1994) and represents a generalization of the model of (Drummond et al., 1980) to include the transverse degrees of freedom. In this case the $\chi^{(2)}$ medium in the cavity induces the partial transformation of a pump field A_0 of frequency $2\omega_s$, which is injected into the cavity, into a signal field A_1 of frequency ω_s. In terms of appropriately normalized fields (Lugiato and Strini, 1980), the model reads

$$\frac{\partial}{\partial t} A_0 = -\kappa_0[1 + i\theta_0)A_0 - A_I(x, y) + A_1^2 - ia_0 \nabla_\perp^2 A_0] \quad \text{(II.31)}$$

$$\frac{\partial}{\partial t} A_1 = -\kappa_1[(1 + i\theta_1)A_1 - A_1^* A_0 - ia_1 \nabla_\perp^2 A_1] \quad \text{(II.32)}$$

where κ_0 and κ_1 are the cavity damping rates of the two fields, respectively; a_0, a_1 are defined as

$$a_0 = \frac{c^2}{4\gamma_0 \omega_s n^2}, \quad a_1 = \frac{c^2}{2\gamma_1 \omega_s n^2} \quad \text{(II.33)}$$

with n being the refractive index of the medium, which is the same for the two fields, as it is necessary for phase-matched degenerate emission (Boyd, 1992). It is assumed that the medium length equals the cavity length. The cavity detuning parameters are defined as

$$\theta_0 = \frac{\omega_{c0} - 2\omega_s}{\kappa_0}, \qquad \theta_1 = \frac{\omega_{c1} - \omega_s}{\kappa_1} \qquad \text{(II.34)}$$

where ω_{c1}, ω_{c0} are the frequencies of the two longitudinal cavity modes closest to ω_s, $2\omega_s$, respectively.

All the models (II.29), (II.30), and (II.31, II.32) can be straightforwardly generalized to the case of spherical mirrors.

Recent works that perform detailed stability analyses and numerical simulations beyond the mean field limit are (Berre *et al.*, 1994; Berre *et al.*, 1997).

B. Systems with Translational Symmetry. The Mechanisms for Pattern Formation

Let us now focus on the case when the system has translational and rotational symmetry; this requires in particular that the input field A_I, if existing, does not depend on (x, y). We review now the basic mechanisms that govern the formation of a spatial pattern from a homogeneous state.

Let us indicate by X a generic dynamical variable in the time evolution equations in play; in case a variable is complex, X can denote also the complex conjugate of the variable itself (e.g., X can be A, A^*, P, P^*, or D in the set of equations (II.12–II.14). Because of translational symmetry in the transverse plane, the dynamical equations admit a spatially homogeneous stationary solution (there might be more than one, but we will ignore this in our discussion); let us denote by X_s the stationary value of X. Whether this solution is stable or not is analyzed by linearizing the equations of motion around the stationary solution and checking the eigenvalues of these linearized equations. Whereas an eigenvalue with positive real part indicated instability, a stable solution only has eigenvalues with negative real part.

Specifically, one sets $X(x, y, t) = X_s + \delta X(x, y, t)$, where δX represents a small perturbation from the stationary state, and one linearizes the dynamical equations with respect to δX. Because the coefficients are independent from x, y, one focuses on solutions of the form

$$\delta X(x, y, t) = \delta X_k(t) \exp[i(\vec{k}_\perp \cdot \vec{x})] \qquad \text{(II.35)}$$

where $\vec{k}_\perp$ is the same for all variables X. This amounts to assuming that the perturbation is modulated with wavelength $2\pi/k_\perp$, which explains the term "modulational instabilities" used for those instabilities that arise for wavevector $k_\perp \neq 0$.

The ansatz (II.35) corresponds to performing a Fourier analysis in space; in the case of optics the Fourier transform of the intracavity field corresponds to the far field configuration. This is strictly true only for cavities with the output mirror different from the input mirror, because in this case the field immediately out of the cavity is proportional to the intracavity field. When the input and the output mirrors coincide, to obtain the field immediately out of the cavity one must add to the intracavity field the part of the input field reflected by the input/output mirror (see Eq. (IV.1) in Section IV). If the input field is uniform, this addition does not change the configuration of the pattern in the near field, whereas it affects only the central point ($k_\perp = 0$) in the far field.

With (II.35), the linearized equations become ordinary differential equations with constant coefficients, so that one can set

$$\delta X_k(t) = \overline{\delta X_k} \exp[\lambda t] \tag{II.36}$$

where λ is the same for all variables X. Ansatz (II.36) leads to a set of linear algebraic equations with the form of an eigenvalue equation with eigenvalue λ. This is an algebraic equation of order equal to the number n of variables X

$$\sum_{i=0}^{n} c_i(k_\perp^2, \beta)\lambda^i = 0 \tag{II.37}$$

where the coefficients c_i depend on the transverse wavevector and on one control parameter β (which may be, for example, the input field A_I or the parameter $|C|$) having in mind to keep the other parameters fixed. Because of the rotational symmetry, the coefficients depend only on the modulus of $\vec{k}_\perp$: as a matter of fact, $\nabla_\perp^2 \exp(i\vec{k}_\perp \cdot \vec{x}) = -k_\perp^2 \exp(i\vec{k}_\perp \cdot \vec{x})$.

In the discussion of the eigenvalue equation (II.37), it is best to focus on the boundary of the stability domain in the plane $(k_\perp, \beta)$, which is defined by the condition Re(λ) = 0. By setting $\lambda = i\nu$ and taking into account that the coefficients c_i are real, by dividing real and imaginary terms one obtains two algebraic equations of the form

$$\sum_{i=0}^{(n-1)/2} d_i(k_\perp^2, \beta)\nu^{2i} = 0, \quad \nu \sum_{i=0}^{(n-1)/2} e_i(k_\perp^2, \beta)\nu^{2i} = 0 \tag{II.38}$$

where we assumed for definiteness that n is odd. Note that $d_0(k_\perp^2, \beta) = c_0(k_\perp^2, \beta)$. There are, in general, two kinds of boundaries. The first is characterized by the condition $\nu = 0$ and, in the plane $(k_\perp^2, \beta)$, corresponds to the line $c_0(k_\perp^2, \beta) = 0$. The second boundary is obtained by eliminating ν between the two former equations; in this case ν is different from zero. Of course, it may happen that one boundary or the other (or both) does not exist. In Fig. 1 we draw qualitatively the two boundaries, assuming that for small enough β the stationary solution is stable. By increasing β one hits the minimum of the instability domain, which is called the critical point; its coordinates k_c, β_c correspond to the critical wavevector and

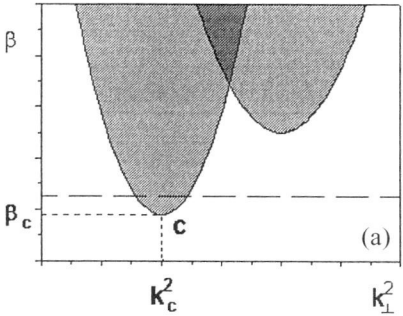 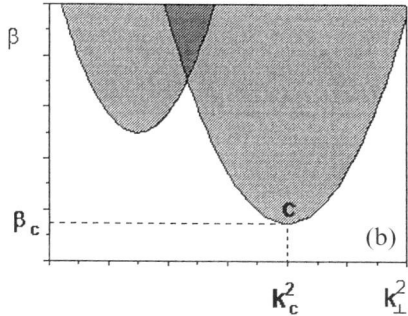

FIG. 1. Qualitative shape of the instability domain (shaded region) in the plane of the modulus square of transverse wave vector and of the control parameter β. The point C of coordinates (k_c^2, β_c) is the critical point.

to the instability threshold, respectively. In Fig. 1a the critical point belongs to the boundary $\nu = 0$, whereas in Fig. 1b it lies on the boundary $\nu \neq 0$. Some general qualitative features of the solution, which emerges immediately beyond threshold, can be specified when this solution develops with continuity from the stationary solution, which has become unstable. When this happens and $k_c \neq 0$, in the case of Fig. 1a one has the onset of a stationary pattern. Those instabilities that lead to the formation of a stationary spatial pattern are usually called Turing instabilities (Turing, 1952). On the other hand, in the case of Fig. 1b, with $k_c \neq 0$ one has a dynamical pattern characterized by the temporal frequency ν.

Let us analyze first the case of Fig. 1a. When β is larger than β_c, there is a range of unstable modes that tend to grow; if we look at the plane of transverse wavevector (k_x, k_y) there is an annulus of unstable modes (Fig. 2) for the value of β indicated by the dashed line in Fig. 1a. Immediately above threshold, the annulus reduces to a circle with radius equal to k_c; this is called the critical circle (Fig. 3a). If we consider the whole wavevector (k_x, k_y, k_z), we have a cone of unstable wavevectors (Fig. 3b).

It may happen that all the waves of the critical cone are emitted so that, in the far field, one has a circle that corresponds to the critical circle. In this case, the near field also has circular symmetry and it is called ring pattern (see Fig. 5c).

Most commonly, however, the nonlinearity of the system causes a spontaneous breaking of the rotational symmetry such that only a subset of the modes of the critical circle build up, suppressing the other ones. Some simple examples of possible far fields, obtained for various systems closely above the instability threshold, are shown in Fig. 4; for all these cases the far field consists of a number of spatially separated beams, corresponding to the spots over the critical circle. Of course the orientation of these patterns is arbitrary due to the rotational symmetry.

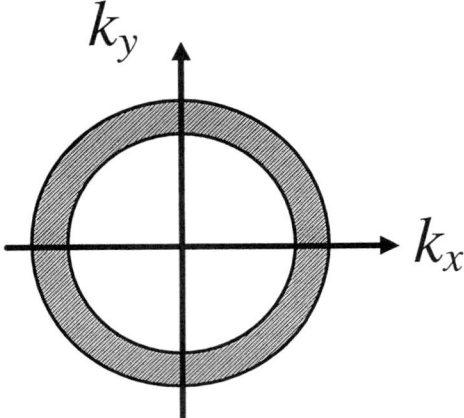

FIG. 2. Annulus of unstable wave vectors for the value of β corresponding to the dashed line in Fig. 1a.

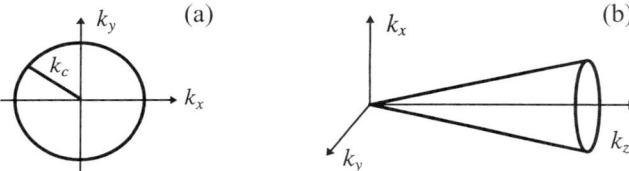

FIG. 3. (a) Critical circle in the plane (k_x, k_y). (b) Critical cone in the space (k_x, k_y, k_z).

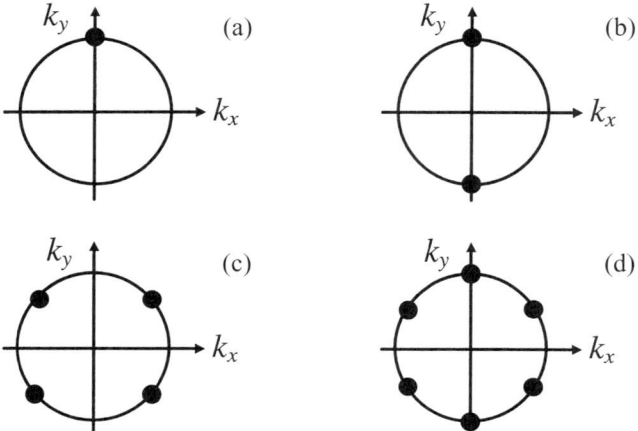

FIG. 4. The simplest examples of far field patterns. In case (a) only one dominant mode survives. In (b) there are two modes with opposite wave vectors, in case (c) four modes evolve, and in (d) one has the formation of hexagons.

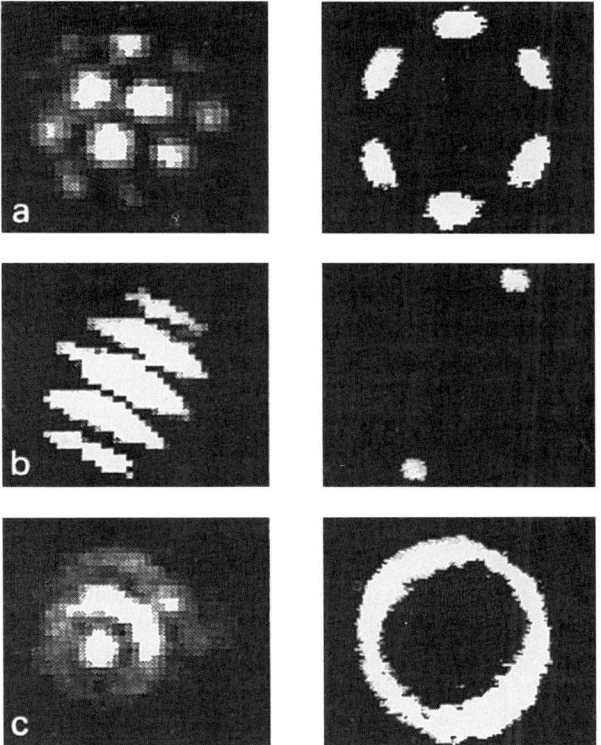

FIG. 5. Near field (left) and far field (right) patterns for (a) hexagons, (b) rolls, (c) rings. Reprinted from *Phys. Rep. 219.* In L. A. Lugiato, Spatio-temporal structures I. (p. 303) (1992), with kind permission from Elsevier Science - NL. Sara Burgerhartstraat 25, 1055 KV Amsterdam, The Netherlands.

The corresponding near field pattern is given by a linear superposition of plane waves

$$\delta X(x, y) = \sum_l c_l \exp[i \vec{k}_l \cdot \vec{x}] \qquad (II.39)$$

where the sum is extended to the spots in the far field pattern. The exact solution has in general contributions also from spatial harmonics and combinations of the wavevectors $\vec{k}_l$, but these corrections are negligible when close to threshold. Examples of patterns observed in experiments of Grynberg and collaborators in the near and far field are given in Fig. 5.

Clearly there is a competition between the different patterns that can evolve from the Turing instability, and the question is which one among the different possibilities is realized for a specific system. The conceptually simplest way to answer is to use a computer and solve the dynamical equations numerically. In two transverse dimensions the problem of numerically solving partial differential equations, in which $\nabla_\perp^2$ has a purely imaginary coefficient, is rather tricky. To our experience, the hopscotch integration method for such systems may lead to erroneous results, while the split-step technique has been found more reliable. In general, it is wise to check numerical results by comparing them with those obtained by analytical methods, or by other numerical techniques that directly calculate the stationary solution bypassing the integration in time.

A feeling for the origin of the numerical difficulty can be obtained by considering the Kerr medium model (II.30), which, due to its simplicity, is a paradigm for Turing instabilities in optics. The presence of an instability of this kind was demonstrated analytically and numerically (Lugiato and Lefever, 1987) in the case of one transverse dimension, which can be forced by imposing a planar waveguide configuration; the relationship with the pioneering work by Moloney, Gibbs, McLaughlin, Newell (Moloney and Gibbs, 1982; McLaughlin et al., 1983) was established in (Ouazzardini et al., 1988); further results can be found in (Haeltermann et al., 1991) and in references quoted therein. In two dimensions, however, the situation is much more delicate, because if one drops the terms $-\kappa[\tilde{A} - \tilde{A}_I + i\theta\tilde{A}]$, Eq. (II.30) reduces to the nonlinear Schrödinger equation (NLS). It is well known that the NLS predicts the formation of solitons that arise from the balance of nonlinearity and diffraction, but in two dimensions these solitons are unstable because they decay or collapse catastrophically. The additional dissipative terms in Eq. (II.30) can lead to the formation of hexagonal patterns (Firth et al., 1992) or of spatial solitons (Scroggie et al., 1994) but these solutions are stable only in narrow parameter domains (Tlidi et al., 1996; Firth and Lord, 1996), and often one finds time-dependent patterns even if the instability of the homogeneous stationary solution has $\nu = 0$.

A technique used to calculate analytically the patterns that emerge beyond the critical point is called nonlinear stability analysis or bifurcation analysis (Newell and Moloney, 1992; Manneville, 1990). Basically, this method treats the quantity $(\beta - \beta_c)^{1/2}$ as a perturbative smallness parameter. For each simple pattern (e.g., rolls, squares, hexagons) one considers the ansatz (II.39) and derives perturbatively from the exact model a set of time-evolution equations for the coefficients c_l. These equations allow one to calculate the possible stationary patterns, and to check their stability with respect to the onset of other patterns (for example, the stability of rolls against the onset of hexagons and vice versa).

In the case of nonlinear optics, however, the consideration of the nonlinear interaction process can provide in a very straightforward way an intuitive picture of the scenario that emerges. Let us illustrate this point by three examples.

(a) In the laser case, we must consider Eqs. (II.12–II.14) with $A_I = 0$, $\Delta = 0$. The critical point is just the laser threshold. Here there is a situation of maximum competition between transverse modes and it is not surprising that only one mode survives; the near field has exactly the form $\exp(i\vec{k}_c \cdot \vec{x})$. The corresponding far

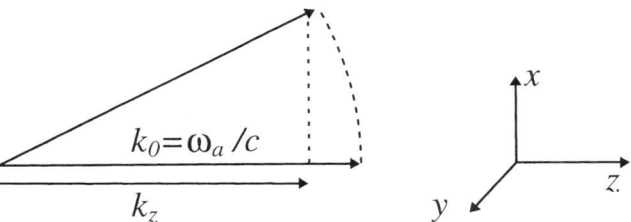

FIG. 6. A longitudinal wave with $k_0 = \omega_a/c > k_z = \omega_c/c$ is detuned from resonance. By tilting the wave one achieves exact resonance.

field consists of a single spot that may be emitted on or off axis. To understand what happens, let us substitute $A \propto \exp(i\vec{k}_\perp \cdot \vec{x})$ in Eq. (II.12); the effect of the term with $\nabla_\perp^2$ is to produce the total detuning of mode $k_\perp$

$$\theta_{k_\perp} = \theta + ak_\perp^2 = \frac{\omega_{k_\perp} - \omega_0}{\kappa} \tag{II.40}$$

where we used Eqs. (II.9), (II.16), and (II.17). Because in the laser case we have taken ω_0 equal to the atomic transition frequency ω_a, the mode with the largest gain is that with the smallest detuning $\theta_{k_\perp}$. When $\theta \geq 0$ the detuning is minimum for $k_\perp = 0$, hence the axial mode is emitted. If $\theta < 0$ instead, the modes with

$$k_\perp = k_c = \sqrt{-\theta/a} \tag{II.41}$$

are exactly on resonance, and therefore an off-axial wave onto the critical circle is emitted. An equivalent way to understand the origin of this tilted wave emission is illustrated in Fig. 6. This mechanism was pointed out in (Jakobsen et al., 1992), and has a general validity, provided that one takes into account the shift of the cavity frequency ω_c caused by the linear and nonlinear dispersions, if any. This mechanism basically coincides with that for which the spontaneous emission of an atom in a cavity occurs off-axis when the atom-cavity detuning has the appropriate sign (Dawling et al., 1991).

Note that the intensity distribution of the near field of the laser above threshold is uniform also for $k_c \neq 0$. In raising the pump parameter C beyond threshold, however, one meets further instabilities that give rise to intensity patterns of increasing complexity (Jakobsen et al., 1994; Lega et al., 1994).

(b) In the case of the degenerate optical parametric oscillator (OPO), the quantum mechanical interaction governing the dynamics is

$$H_{INT} = i\hbar \frac{g}{2} \iint dx\, dy\, [B_1^\dagger(\vec{x})^2 B_0(\vec{x}) - \text{h.c.}] \tag{II.42}$$

where B_0, B_1 are the operators associated with the pump and signal field, respectively; H_{INT} describes the annihilation of one pump photon accompanied by the

FIG. 7. Roll pattern in the near field.

creation of two signal photons and vice versa. With respect to the signal field, the OPO is a special kind of laser with the characteristics of emitting photons in pairs. Similarly to the laser, the OPO has a threshold. Below threshold, according to the semiclassical equations (II.31 and II.32), at steady state the signal field is zero. Again, the critical point is the OPO threshold, and the selection of the critical wave-vector k_c is governed by the tilted wave mechanism applied to the signal field, that is, by the resonance condition $\theta_1 + a_1 k_c^2 = 0$ (Oppo et al., 1994). The basic difference from the laser case is that, because photons are created in pairs, the far field for $\theta_1 < 0$ consists in two spots with opposite wave vectors (case (b) of Fig. 4). The two waves interfere with each other and, in the near field, one obtains

$$A_1(\vec{x}) = \sigma \exp[i(\varphi_+ + \vec{k}_c \cdot \vec{x})] + \sigma \exp[i(\varphi_- - \vec{k}_c \cdot \vec{x})]$$
$$= 2\sigma \cos\left(\vec{k}_c \cdot \vec{x} + \frac{\varphi_+ - \varphi_-}{2}\right) \exp\left[i\frac{\varphi_+ + \varphi_-}{2}\right] \quad \text{(II.43)}$$

so that the intensity distribution $|A_1|^2$ is a roll pattern, as indicated in Fig. 7. If the input field A_I is increased further beyond threshold, secondary instabilities lead to zig-zag and dynamical patterns (Oppo et al., 1994).

In the case of the nondegenerate OPO, the signal (idler) near field immediately above threshold has the form of a single travelling wave proportional to $\exp[i\vec{k}_c \cdot \vec{x}]$ ($\exp[-i\vec{k}_c \cdot \vec{x}]$), like in the laser, instead of the standing wave configuration (II.43) (Longhi, 1996).

(c) In the case of the Kerr medium, the interaction Hamiltonian is given by

$$H_{INT} = \bar{g} \iint d^2x \, [B^\dagger(\vec{x})]^2 \, [B(\vec{x})]^2 \qquad (\text{II}.44)$$

so that an expansion of the field operator $B(\vec{x})$ in Fourier space leads to a combination of four wavevectors $\vec{k}$ with conservation of the total transverse wavevector in the simultaneous annihilation of two photons and creation of two photons. That is, we have four-wave-mixing processes (Boyd, 1992). In particular there is a process of annihilation of two photons of the input field and the emission of two symmetrically tilted photons. This leads directly to a far-field configuration with three spots: two opposite spots as in the OPO and a central spot corresponding to the mode with $k = 0$ (input field). The near field is given by (II.43) with the addition of a constant contribution from the mode $k = 0$, and corresponds again to a roll pattern. The contribution of the homogeneous mode, however, destabilizes the roll pattern according to the following consideration. The first four-wave-mixing process generates, out of a pair of photons with $k = 0$ (Fig. 8), two other photons (1 and 4) with opposite transverse wavevector. However, then a second four-wave-mixing process can create two additional photons (2 and 6) out of 0 and 1, or the pair 3 and 5 out of 0 and 4, which gives a hexagonal structure. This argument was elucidated by Grynberg for the case of counterpropagating waves (Firth and Parè, 1988; Grynberg et al., 1988), but holds also in other situations

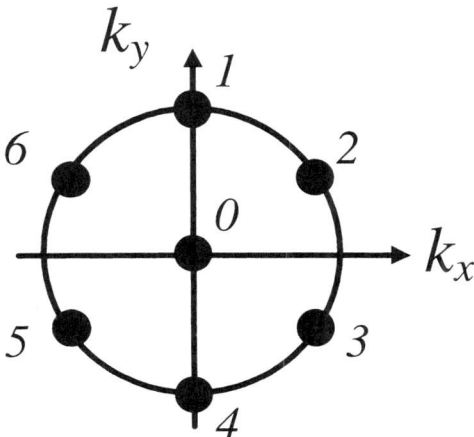

FIG. 8. Generation of a hexagonal far field (see text).

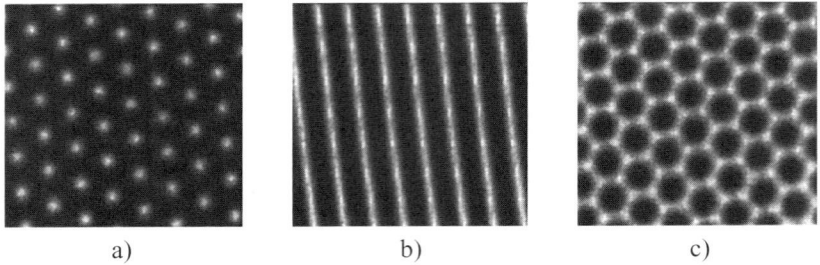

FIG. 9. Example of the change in the stable pattern observed for $\theta = -1$ as I is varied at a fixed value of $C (= 4.4)$. (a) H^+ for $I = 2.9$, (b) rolls for $|A|^2 = 3.3$, and (c) H^- for $|A|^2 = 4.5$. (From Firth and Scroggie, 1994)

such as, for example, the cavity model (II.30). Another more formal argument for the onset of hexagonal patterns was offered by Firth: If one considers a term of the form $A|A|^2$ and sets $A = A_s + \delta A$, then one obtains for δA quadratic terms that are characteristic for hexagon formation (Haken, 1977; Manneville, 1990).

As a final point in our illustration of steady-state patterns, we give an example of how one meets different patterns by continuously varying a control parameter. We refer to the case of the absorptive model (II.29) and the results have been obtained in (Firth and Scroggie, 1994). When the input field intensity $I + |A_I|^2$ is increased above the instability threshold, a hexagonal pattern H^+ is realized (Fig. 9a). Increasing I far from threshold, one finds a stable branch of rolls (Fig. 9b) and, further on, a honeycomb pattern H_- (i.e., a hexagonal pattern with π phase shift among the three sets of tilted standing waves).

Let us now come to the case of $\nu \neq 0$ illustrated in Fig. 1b, in which as we said one has the formation of a dynamical pattern when $k_c \neq 0$ and the solution immediately above the critical point develops with continuity from the unstable homogeneous state. The simplest dynamical intensity pattern is a drifting roll, obtained by adding (νt) to the argument of the cosine in Eq. (II.43). Solutions of this kind have been numerically predicted (Haelterman and Vitrant, 1992) and experimentally observed in $\chi^{(3)}$ media (Penna and Giusfredi, 1993; Petrossian et al., 1995), by using an input field tilted with respect to the mirror(s). When the tilt angle exceeds a certain minimum (Ackemann, 1997), one generates a drifting instability of this sort.

Beyond this simple case, one can find numerically a rich variety of dynamical patterns, but it is possible to appreciate them adequately only with the help of a movie, hence we will not give examples here (see for example Logvin et al., 1996).

We close this section with an important remark. The mechanisms for optical pattern formation that we illustrated are identical to those that govern these phenomena in all other fields. Also in hydrodynamics, for example, the pattern selection is determined by appropriate combinations of wavevectors on a critical circle, dictated by the nonlinearity. As shown by Eqs. (II.42) and (II.44), however, in nonlinear optics the nonlinearity is associated with the simultaneous absorption and emission of a number of photons. This circumstance creates correlation of a quantum nature, and is the very origin of the quantum aspects of optical patterns that will be discussed in Section IV.

C. Systems with Single Feedback Mirror

In the framework of models with translational symmetry, the systems with single feedback mirror have played a relevant role. As described by Fig. 10, a thin slice of Kerr medium is irradiated on one side by a stationary plane wave field, which freely propagates beyond the slice to a mirror, which generates a counterpropagating beam in the Kerr slice. If we indicate by φ the nonlinear phase shift the radiation field undergoes upon crossing the slice, the dynamics inside the slice are governed by the time evolution equation

$$\tau \frac{\partial \varphi}{\partial t} = l^2 \nabla_\perp^2 \varphi + \xi(I_0 + I_1) \tag{II.45}$$

where τ is the response time and l the diffusion length of the Kerr excitation,

$$F_0 = \sqrt{I_0} \, e^{i\Psi_0} \tag{II.46}$$

is the input field, and I_1 is the intensity of the counterpropagating field F_1; it is assumed that diffusion washes out the grating formed by the two counterpropagating waves. Another hypothesis is that the medium is sluggish, that is, it cannot respond on the roundtrip time scale. Therefore the free propagation of the field from the slice to the mirror and back is described by the equation

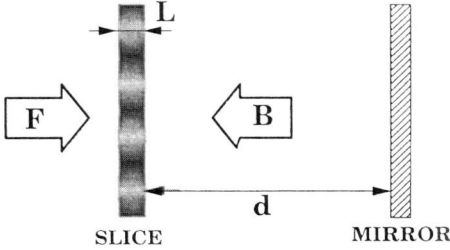

FIG. 10. Schematic diagram of the model. A thin slice of Kerr material, of thickness L, is illuminated from the left. The mirror, at distance d, reflects this field back through the slice, thus closing the feedback loop. (From D'Alessandro and Firth, 1992)

$$\frac{\partial F}{\partial z} = \pm \frac{1}{2ik_z} \nabla_\perp^2 F \tag{II.47}$$

without any time-dependent term (this is an adiabatic elimination of the field); the minus (plus) sign corresponds to propagation to the right (left). The quantity I_1 in Eq. (II.45) is obtained by calculating the field $F_1 = \sqrt{I_1}\, e^{i\Psi_1}$ after the roundtrip with the initial condition

$$F(z = 0) = F_0 e^{i(\varphi_0 + \varphi)} \tag{II.48}$$

where φ_0 is the linear phase shift induced by the medium. The value of $F_1(x, y)$ obtained in this way is a nonlinear functional of $\varphi(x, y)$, given by a Fresnel formula.

This model was introduced by Firth (1990) and extensively analyzed by D'Alessandro and Firth (1992). It was inspired on the one hand by the pioneering experiment of Giusfredi et al. (1988), and on the other hand by the previous experience on the problem of counterpropagating waves in Kerr media (Firth and Parè, 1988). In that case, the time evolution equations include both nonlinearity and diffraction, which leads to computational complexity. In the single mirror case, however, the two elements are physically and mathematically separated; because of the sample thinness, diffraction is neglected in the Kerr medium, Eq. (II.45), and is confined in the free propagation, Eq. (II.47).

The mechanism responsible for pattern formation is simply explained as follows (Firth, 1991). Assume that a fluctuation creates a small spatial phase modulation in the sample, that is,

$$F(x, y) = F_0\{1 + i\epsilon \cos kx\} \tag{II.49}$$

where the i (imaginary unit) encapsulates the $\pi/2$ phase shift between carrier and signal characteristic of phase modulation. Now, as one sees from Eq. (II.47), propagation generates a phase slippage relative to the carrier, at a rate $k^2/(2k_z)$ per unit distance. This introduces an element of amplitude modulation. Next, the amplitude modulation gives rise to phase modulation in the Kerr medium (see Eq. (II.45)), and in this way one has the feedback loop needed for pattern formation.

Patterns arise for both self-focusing and self-defocusing media, and may be static or oscillatory, with a period of two roundtrip times. Hexagonal patterns are commonplace here (see the buildup in time from noise shown in Fig. 11); the occurrence of periodic structures has also been understood by using fractional Talbot effect of Fourier optics (Tamburrini et al., 1994). Increasing the input field, the hexagons "melt" and give rise to a turbolent-like behavior.

In experiments, however, the input beam has a Gaussian shape, and this introduces interesting boundary effects, which have been studied in (Papoff et al., 1993). The main result is that the system can realize patterns that are not possible

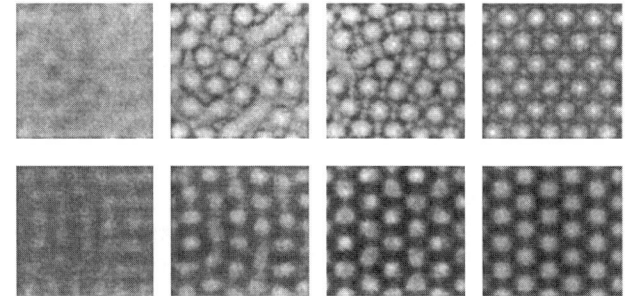

FIG. 11. Backward field intensity in a focusing (top) and defocusing (bottom) medium. Gray scale, from white (high intensity) to black (low intensity) is used. Time increases from left to right. (From D'Alessandro and Firth, 1992)

with a plane input, as, for example, pentagonal patterns; fivefold-symmetric patterns cannot form infinitely periodic lattices in a two-dimensional space.

This kind of system has allowed for several experimental realizations up to now. Extensive investigations with thin layers of liquid crystals (Macdonald and Eichler, 1992; Tamburrini *et al.*, 1993) and with sodium vapor in a buffer gas atmosphere (Ackemann *et al.*, 1995; Lange *et al.*, 1996; Logvin *et al.*, 1997) have been performed. Figure 12 shows some experimental observations of Lange and collaborators from (Ackemann *et al.*, 1995), obtained by gradually increasing the input intensity and compared with the numerical simulations from an appropriate microscopic model of the medium. Above the instability threshold, a structure in the form of three dark holes appears in (a); the selection of a structure with

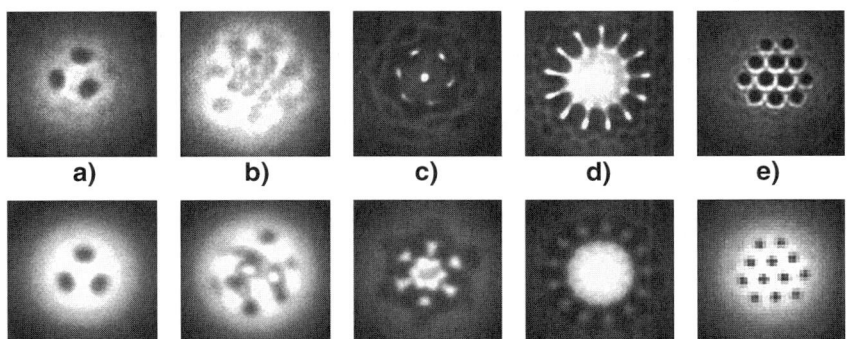

FIG. 12. Experimental (top) and corresponding simulated (bottom) patterns for increasing input intensities ((a)–(d)). (e) Well-developed pattern illustrating H^- hexagons. The frames have a size of 2.5 × 2.5 mm. (From Ackemann *et al.*, 1995)

dihedral symmetry in a situation in which the hexagonal structure is hindered by the boundary conditions agrees with the results of (Papoff *et al.*, 1993). Experiments with a larger aspect ratio show that a hexagonal lattice of dark holes (honeycomb hexagons) is indeed found in (e). In the smaller aspect ratio case, however, increasing the pump power leads to the formation of a pattern with five holes in a pentagonal arrangement, which gives way to the blurred, nonstationary pattern in (b). From this bright spots emerge; the number of spots (3 to 7) increase with power; (c) shows the hexagonal pattern. Further on, flowerlike patterns with a plateau in the beam center are observed in (d). The number of petals starts with 7 and increases up to 14; in some cases these patterns rotate. Flowerlike patterns in atomic vapor with single feedback mirror were first observed in (Grynberg *et al.*, 1994); the characteristics were however different in that case, in which the authors can interpret these structures as high-order Gauss-Laguerre modes.

In the Kerr slice system, the only effect of the feedback beam is to modulate the phase of the input beam; hence one can instead synthesize a Kerr effect through any kind of transducer that detects the intensity of the feedback beam and encodes the phase of the input beam in proportion. This function is performed by spatial phase modulators in general, and liquid crystal light valves (LCLV) in particular. There is a vast literature on pattern formation in feedback systems using LCLV. It displays a dazzling range of static and dynamic patterns (Akhmanov *et al.*, 1988; Vorontsov, 1993) and includes, within the repertoire of this system, a direct analogue of the Kerr slice with single mirror (Firth and Vorontsov, 1993), described by a closely similar model.

Using the same experimental setting introduced by the Moscow group of Akhmanov and Vorontsov (Akhmanov *et al.*, 1988; Vorontsov, 1993) and, in particular, utilizing an azimuthal rotation of the feedback wavefront, Arecchi and collaborators observed roll-hexagons competition (Pampaloni *et al.*, 1993), two dimensional crystals, and quasi-crystals (Ramazza *et al.*, 1995).

A model for pattern formation in a liquid crystal light valve with feedback has been extensively analyzed in (Neubecker *et al.*, 1995). The Darmstadt group performed a sequence of experiments with this system, observing hexagonals patterns (Neubecker *et al.*, 1994) and, most interesting, addressable solitary spots in the transverse plane (Kreuzer *et al.*, 1996; Schreiber *et al.*, 1997), see Section III.

The spatiotemporal dynamics of a thin-layer laser with feedback mirror has been considered in (Samson, 1996).

D. ANALOGY WITH HYDRODYNAMICS. VORTICES AND OTHER DEFECTS

The existence of general analogies between optical and hydrodynamical patterns has led to the use of terminologies as "laser hydrodynamics" (Brambilla, Lugiano, *et al.*, 1991) or "dry hydrodynamics" (Arecchi, 1991; Arecchi, 1994).

Let us illustrate from a formal viewpoint the analogy between the laser equations and the hydrodynamical ones, as formulated in (Brambilla, Lugiato *et al.*,

1991), see also (Akhmanov et al., 1972). In the limit $\kappa \ll \gamma_\perp, \gamma_\parallel$, in which one can eliminate adiabatically the atomic variables from Eqs. (II.25–II.27), one obtains in the laser case $A_I = 0$, $\Delta = 0$, assuming for simplicity that also $\theta = 0$

$$\frac{\partial A}{\partial \tau'} = -\left[1 - i\bar{a}\left(\frac{1}{4}\overline{\nabla}^2_\perp - \rho^2 + 1\right)\right]A + 2C\frac{A}{1 + |A|^2}\chi(\rho) \quad \text{(II.50)}$$

where we set $\tau' = \kappa t$, $\rho = r/w_0$, $\bar{a} = \eta/\kappa$, $\overline{\nabla}^2_\perp = w_0^2 \nabla^2_\perp$.

In order to reformulate (II.50) in the form of "hydrodynamical equations," we set $F = |F|e^{i\Phi}$ and we introduce a "mass density" $\sigma = |A|^2$ and a "velocity" $\mathbf{v} = \frac{\bar{a}}{2}\overline{\nabla}\Phi$, where $\overline{\nabla} = w_0 \nabla$, and ∇ is the gradient in the transverse plane. With some algebraic manipulations, Eq. (II.50) and its complex conjugate can be cast in the form

$$\frac{\partial \sigma}{\partial \tau'} + \overline{\nabla} \cdot (\sigma \mathbf{v}) = 2\sigma\left[\frac{2C}{1+\sigma}\chi(\rho) - 1\right] \quad \text{(II.51)}$$

$$\frac{\partial \Phi}{\partial \tau'} = -\frac{\bar{a}}{4}|\overline{\nabla}\Phi|^2 + \bar{a}\left[\frac{\overline{\nabla}^2 \sqrt{\sigma}}{4\sqrt{\sigma}} + 1 - \rho^2\right] \quad \text{(II.52)}$$

We can compare (II.51) and (II.52) with the fundamental equations of hydrodynamics, namely the equation of continuity for the mass density σ

$$\frac{\partial \sigma}{\partial \tau'} + \nabla \cdot (\sigma \mathbf{v}) = 0 \quad \text{(II.53)}$$

and the Bernoulli equation for the velocity potential Φ (such that $\mathbf{v} = Q\nabla\Phi$, where the constant Q is introduced because Φ is assumed dimensionless and $\mathbf{v}$ in this case has the dimensions of a velocity)

$$\frac{\partial \Phi}{\partial \tau'} = -\frac{Q}{4\pi}|\nabla\Phi|^2 + p\frac{2\pi}{Q} \quad \text{(II.54)}$$

which involves the pressure term p. From the comparison between these two pairs of equations we can conclude that in the optical case the total "mass" is not conserved because of the presence of the two dissipative terms that appear in the r.h.s. of (II.51), whereas the analogy with the Bernoulli equation is complete if we put $Q = \pi\bar{a}$ and define a "pressure"

$$p = \frac{\bar{a}^2}{2}\left[\frac{\overline{\nabla}^2 \sqrt{\sigma}}{4\sqrt{\sigma}} + 1 - \rho^2\right] \quad \text{(II.55)}$$

In this way a formal analogy between lasers and hydrodynamics is established, in the sense that all the relevant physical quantities that describe a fluid—mass density, velocity and pressure—can be derived from the modulus and the phase of the electric field emitted by a laser.

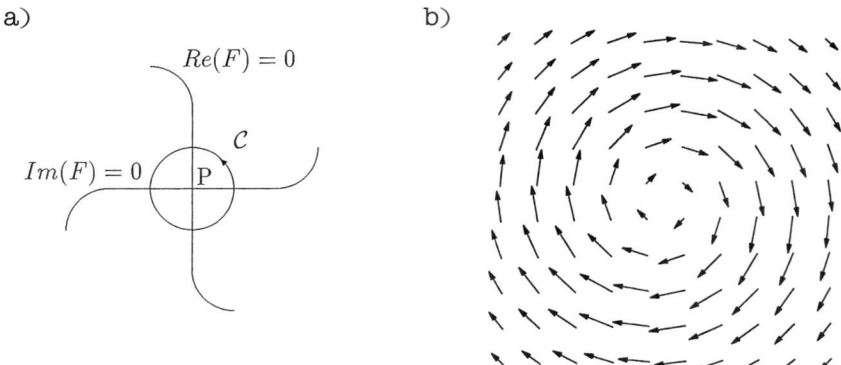

FIG. 13. (a) A phase singularity is a point P in the transverse plane where both the real and the imaginary part of the electric field vanish and the circulation of the electric field's phase over a loop $\mathscr{C}$ enclosing P is equal to a multiple of 2π. (b) Behavior of the transverse component of the Poynting vector around the phase singularity in a doughnut mode.

One of the main consequences of this analogy is the occurrence of phase singularities which play the same role as vortices in hydrodynamics and therefore are often called "optical vortices." These vortices are centered at isolated points where both the real and the imaginary part of the electric field vanish. In these points the intensity of the radiation is exactly zero and they appear as dark spots. Because A is a single-valued complex function, the variation of its phase along a circuit $\mathscr{C}$ surrounding one of these points P (Fig. 13a) is equal to an integer multiple of 2π

$$\Delta\Phi = \oint_{\mathscr{C}} \nabla\Phi \cdot d\vec{l} = 2m\pi \qquad (\text{II}.56)$$

The value of the integral does not change if $\mathscr{C}$ is shrunk to a loop of infinitesimal length enclosing P. Therefore the gradient diverges in P and this point is a phase singularity for the electric field. The smoothness of A implies that a phase singularity can exist only where $A = 0$. The integer m is called "topological charge" of the phase singularity. It is interesting to analyze the behavior of the Poynting vector around a phase singularity. First of all it must be noted that, even in the paraxial approximation, the electric and the magnetic fields associated with the laser radiation possess a longitudinal component. If the electric field is polarized along x and the magnetic field is polarized along y, their slowly varying envelopes are

$$\vec{A} = \left(A,\ 0,\ i\frac{1}{k_z}\frac{\partial A}{\partial x}\right) \qquad (\text{II}.57)$$

and

$$\vec{H} = \left(0,\ A,\ i\frac{1}{k_z}\frac{\partial A}{\partial y}\right) \tag{II.58}$$

The Poynting vector $\vec{S}$ for harmonic fields is defined as (Jackson, 1975).

$$\vec{S} = \frac{c}{8\pi}(\vec{F}\times\vec{H}^*) \tag{II.59}$$

and the real part of this vector represents the time average of the irradiated power per unit area. Taking into account Eqs. (II.57) and (II.58), one has

$$\text{Re}\ \vec{S} = \frac{c}{8\pi}|F|^2\left(\frac{\nabla\Phi}{k_z},\ 1\right) \tag{II.60}$$

The longitudinal components of the electric and magnetic fields produce the transverse component of the real part of the Poynting vector, which is proportional to the intensity of the electric field and directed as the gradient of its phase. Fig. 13b shows the behavior of Re $\vec{S}$ in the transverse plane for a doughnut mode with positive helicity. The presence of the phase singularity is associated with an angular momentum of the radiation. The effects of the exchange of angular momentum between radiation and matter have been recently investigated (Allen et al., 1992); a neutral atom in a field like that shown in Fig. 13b is subjected to the action of mechanical dipole forces arising from intensity and phase gradients (Brambilla, Cattaneo, et al., 1992).

The vortex structure in the field does not appear if one observes the intensity directly. However, the interference with another field can make the vortex visible. For example, the other field may be a plane wave with the same frequency propagating in the longitudinal direction. If, instead, the plane wave is slightly tilted with respect to the z-axis, the superposition transforms the vortices into other kinds of defects, that is, dislocations (Weiss, 1992) (see Fig. 14). This technique was used to analyze the statistical distribution of vortices in the transverse plane in a cavity containing a photorefractive material, as a function of the Fresnel number (Arecchi et al., 1991).

Defects in the radiation field were first discussed by Berry and collaborators (Berry, 1981; Berry et al., 1979). The authors considered the linear wave equation and identified appropriate solutions that display patterns with rich and complex structures, sometimes on different spatial scales simultaneously, including subwavelength structures. Vortices in nonlinear optical systems were first predicted by Coullet and collaborators (1989), who derived from the model (II.12–II.14) a Ginzburg-Landau equation valid in the neighborhood of the laser threshold; this equation turns out to include also a small diffusive contribution coefficient term (i.e., the coefficient of $\nabla^2_\perp$ is no longer purely imaginary). Figure 15 shows an

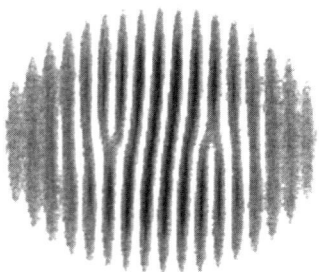

FIG. 14. Interferogram of a wave containing vortices of charge +1 and −1 with an inclined plane wave. The merging of interference fringes shows the vortices. The direction gives the sign of the charge and the number of fringes disappearing at the vortex location gives the magnitude of the vortex charge. Reprinted from *Phys. Rep. 219*. In C. O. Weiss, Spatio-temporal structures II. (p. 328) (1992), with kind permission from Elsevier Science - NL, Sara Burgerhartstraat 25, 1055 KV Amsterdam, The Netherlands.

example of the results obtained in these numerical calculations. It displays a main vortex in the center with the form of a spiral (see also Gil, 1992). In this model, the vortex arises from the presence of two distinct stationary solutions, the trivial solution $A = 0$ and the homogeneous nontrivial solution above threshold. Solutions that display vortices have an intensity distribution that connects these two basic configurations in the sense that it vanishes at the center of the vortex and presents a narrow core that rises steeply to the homogeneous nontrivial solution (Fig. 16) of which the vortex is a defect. Systematic experimental investigations

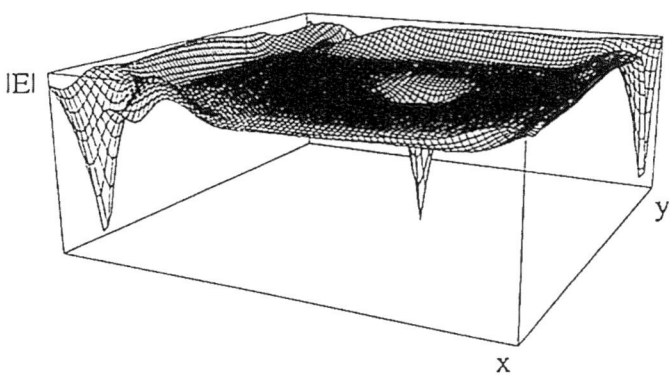

FIG. 15. Equiphase lines of the electric field. Four vortices are clearly visible. The phase difference between lines is $\pi/4$. Reprinted from *Opt. Comm. 73*. In P. Coullet, L. Gill, and F. Rocca, Optical vortices (p. 403) (1989), with kind permission from Elsevier Science - NL, Sara Burgerhartstraat 25, 1055 KV Amsterdam, The Netherlands.

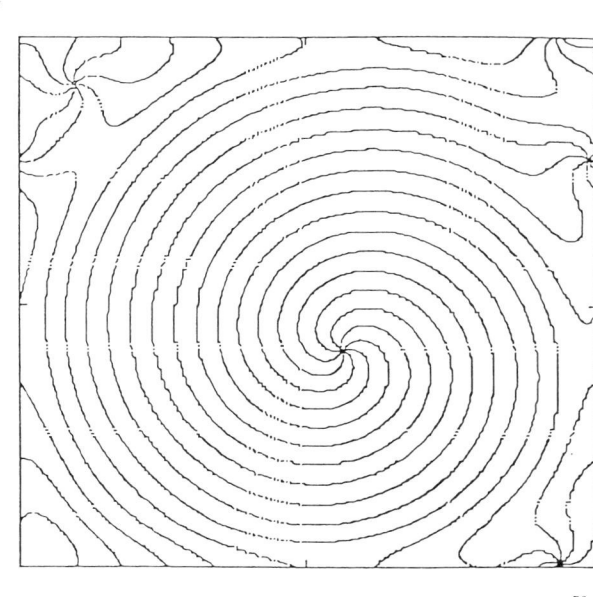

FIG. 16. The field amplitude corresponding to Fig. 15. The minima show the existence of optical vortices. Reprinted from *Opt. Comm. 73.* In P. Coullet, L. Gil, and F. Rocca, Optical vortices (p. 403) (1989), with kind permission from Elsevier Science - NL, Sara Burgerhartstraat 25, 1055 KV Amsterdam, The Netherlands.

on lasers with large Fresnel number show, however, a different behavior, with dynamical (chaotic) instead of steady-state patterns and, above all, with no presence of vortices (Huyet et al., 1995; Lovergnaux et al., 1996) (see Fig. 17). This difference is presumably due to the fact that the laser operates with more than one longitudinal cavity mode.

On the other hand, phase singularities are commonplace in class-A lasers (for class-B lasers, using the terminology introduced by Arecchi, see Khamin, 1995) with small Fresnel number, operating with one or few families of Gauss-Laguerre modes (Brambilla, Battipede, et al., 1991; Brambilla, Cattaneo, et al., 1994; Crates, et al., 1994) as will be shown in the following section, and in photorefractive materials (Arecchi et al., 1991; Arecchi et al., 1990). Vortices have been also observed, for example in phase-conjugate resonators (Liu and Indebetouw, 1992; Indebetouw and Liu, 1992), and in free propagation in connection with dark solitons (Swartzlander and Law, 1992; Baristiy et al., 1993).

A recent work (Yu et al., 1996) reports on the numerical observation of spiral wave formation in 2- and 3-level lasers, interpreted as a mechanism of excitable optical systems.

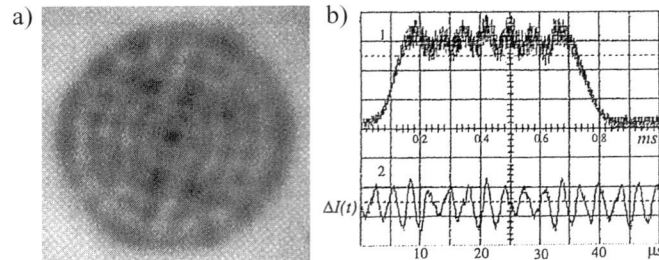

FIG. 17. (a) Average intensity pattern observed in a CO_2 laser on an infrared image plate. (b1) Intensity versus transverse coordinate measured with a rotating mirror, which deflects the beam onto a fast detector. The pattern is composed of four concentric rings. (b2) Ac component of the intensity at a fixed point. (From Huyet et al., 1995)

The fact that phase singularities also appear as solutions of the linear wave equation, for example in the Gauss-Hermite or Gauss-Laguerre modes of an empty cavity, has originated a strong discussion intended to distinguish between linear and nonlinear vortices and even between interesting and less interesting results. Therefore let us try to summarize here what we consider the only solid and objective points in this connection.

First, the distinction between linear and nonlinear has nothing to do with the level of interest. Defects in linear media arise exclusively because of the presence of spatial inhomogeneities in the material, in combination with diffraction and interference. They can be induced, for example, by the presence of lenses or mirrors and this explains, in particular, the phase singularities in the modes of an empty cavity.

In nonlinear media the nonlinearity of the dynamics plays the dominant role. For a given system setting, a certain number of cavity modes can contribute to the spatial structure. A priori, all linear combinations of these modes are possible; however, the nonlinearity is essential to select the one (or the ones in case of multistability) that the system realizes for a choice of the system parameters. For instance, the boundary-dependent structures, which arise from the competition of the modes of a frequency-degenerate family of Gauss-Laguerre modes in a laser, display phase singularities and correspond to the minima of an appropriate potential function (see the following section). A discussion of linear and nonlinear effects in optical vortices can be found in (Lippi et al., 1993).

As for the matter of interesting versus less interesting results, let us consider an experiment that displays a spatial pattern or structure. If one reproduces the observed configurations by an appropriate linear combination of modes, this is of course legitimate but of very limited interest. If, instead, one shows that the solutions of an appropriate nonlinear model reproduce correctly the observed structures and their variation with the parameters, this is a result of solid interest, independently of whether the nonlinear model is formulated in terms of PDEs or of ODEs for modal amplitudes.

In the patterns discussed previously for systems with translational symmetry one often meets defects, for example dislocations or penta-hepta defects in hexagonal patterns. Sometimes these defects are spontaneously eliminated in the long run, whereas sometimes they seem stabilized by the periodic boundary conditions even for long times.

E. Cavities with Spherical Mirrors

Let us start from Eqs. (II.25–II.27) and introduce the expansion in Gauss-Laguerre modes

$$A(r, \varphi, t) = \sum_{p,l=0}^{\infty} \sum_{i=1}^{2} a_{pli}(t) f_{pli}^0 \varphi(r, \varphi) \tag{II.61}$$

with $f_{pli}^0(r, \varphi) = f_{pli}(r, \varphi, z = 0)$, with modal amplitudes $a_{pli}(t)$. Taking into account the orthonormality relations

$$\int_0^\infty r\,dr \int_0^{2\pi} d\varphi\, f_{pli}^0(r, \varphi) f_{p'l'i'}^0(r, \varphi) = \delta_{p,p'} \delta_{l,l'} \delta_{i,i'} \tag{II.62}$$

and the eigenvalue equation

$$\left[\frac{w_0^2}{4} \nabla_\perp^2 - \frac{r^2}{w_0^2} \right] f_{pli}^0 \varphi(r, \varphi) = -(2p + l + 1) f_{pli}^0(r, \varphi) \tag{II.63}$$

one obtains from Eq. (II.25) a set of time-evolution equations for the mode amplitudes

$$\frac{d}{dt} a_{pli}(t) = -k \bigg\{ (1 + i\bar{\theta}_{pl}) a_{pli}$$

$$- \int_0^\infty r\,dr \int_0^{2\pi} d\varphi\, f_{pli}^0(r, \varphi) [A_I(r) + 2C\, P(r, \varphi, t)] \bigg\} \tag{II.64}$$

with

$$\bar{\theta}_{pl} = \theta + \frac{\eta}{k}(2p + l) = \frac{w_{pl} - w_0}{\kappa} \tag{II.65}$$

where we have taken into account Eqs. (II.16) and (II.23). This constitutes a set of ordinary differential equations (actually integro-differential equations), which must be coupled with the PDEs (II.26) and (II.27) with $A(r, \varphi, t)$ replaced by the expansion (II.61). We have most often solved this set of equations numerically; alternatively, one can also expand P and D over the Gauss-Laguerre mode basis, and one obtains a set of ODEs (Tredicce *et al.*, 1989).

Let us now focus on the case of a laser (that is, $A_I = 0$, $\Delta = 0$), and consider first the case that the gain line is so narrow that it excites only one frequency-degenerate family of modes, so that the sum in Eq. (II.61) can be restricted to the modes of that family. If, in addition, we assume for simplicity that the atomic frequency $\omega_a = \omega_0$ coincides with that of the frequency-degenerate family, we can set $\bar{\theta}_{pl} = 0$ in Eqs. (II.64). A detailed discussion of this problem can be found in (Prati et al., 1994; Brambilla, Battipede, et al., 1991). At steady state Eqs. (II.64), (II.26), (II.27), and (II.61) give

$$0 = a_{pli} - 2C \int_0^\infty r dr \int_0^{2\pi} d\varphi \, \chi(r) f^0_{pli}(r, \varphi) \frac{\sum'_{p'l'i'} f^0_{p'l'i'}(r, \varphi) a_{p'l'i'}}{1 + \left| \sum'_{p'l'i'} f^0_{p'l'i'}(r, \varphi) a_{p'l'i'} \right|^2}$$

(II.66)

where Σ' means that the sum is extended only to the modes of the frequency-degenerate family. Hence we introduce the function

$$V(\{a_{pli}\}, \{a^*_{pli}\}) = \int_0^\infty r dr \int_0^{2\pi} d\varphi \, [|A(r, \varphi)|^2 - 2C\chi(r)\ln(1 + |F(r, \varphi)|^2)]$$

(II.67)

with

$$A(r, \varphi) = \sum'_{pli} a_{pli} f^0_{pli}(r, \varphi)$$ (II.68)

The stationary equation (II.66) can be cast in the compact form

$$0 = \frac{\partial V}{\partial a^*_{pli}}$$ (II.69)

which shows that the stationary states coincide with the stationary points of the "potential" function V, which plays the role of a generalized free energy.

When $k \ll \gamma_\perp, \gamma_\parallel$ (class-A lasers), the atomic variables can be adiabatically eliminated and one can easily show (Brambilla, Lugiato, et al., 1991) that V is a monotonically decreasing function of time, so that the long-time solutions correspond to the local minima of V in the space of the amplitudes a_{pli}, a^*_{pli}. In order to find the stable stationary states, it is essential to allow the amplitude a_{pli} to be complex. Even if the patterns that arise from the interaction and competition of the modes of the family emerge through a spatial hole burning (gain) mechanism, the process is strongly affected by the phase.

As an example, let us consider the case of the family $2p + l = 2$. The minima of V correspond to four distinct kinds of stationary solutions, shown in Fig. 18. The first is the pure single mode solution $p = 2, l = 0$ (a). The fourth is a combination of the two modes $p = 0, l = 2$ with $\cos 2\varphi$ and $\sin 2\varphi$, which gives a doughnut configuration $\exp 2i\varphi$ and $\exp -2i\varphi$ (d). The other two are combinations of the three modes of the family, which are usually called the "leopard"

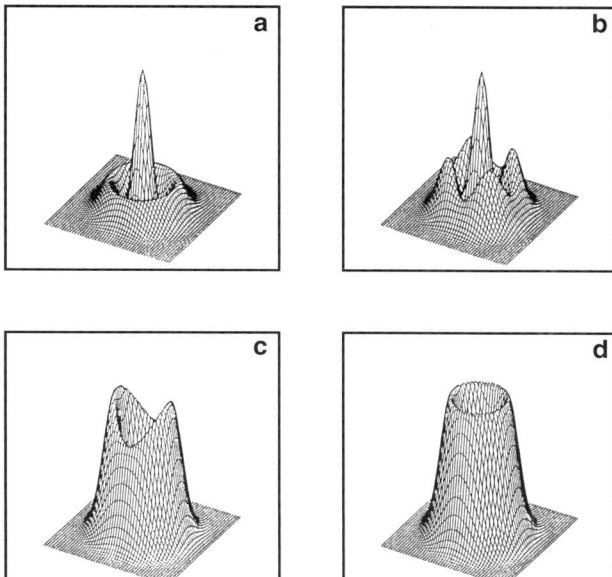

FIG. 18. Transverse intensity distribution for the stable pattern in the case $2p + l = 2$; (a) Gauss-Laguerre mode $p = 1, l = 0$; (b) the "leopard" configuration; (c) the "oval" pattern; (d) the doughnut configuration.

(b) and the "oval" (c) configuration. An experiment with sodium dimer laser (Brambilla, Battipede, *et al.*, 1991) shows exactly these configurations (Fig. 19). In the experiment one varies two control parameters, the pump parameter C and the width ψ of the pumped region. The mapping of the states in the plane (ψ, C) shows qualitatively good agreement between experiment and theory. In particular, two phenomena of spontaneous breaking of the rotational symmetry were observed: (1) the continuous transition from (a) to (b) in Fig. 18 with formation of four phase singularities; and (2) the continuous transition from (d) to (c), with the splitting of a phase singularity of charge 2 into two phase singularities of charge 1. As shown in (Colet *et al.*, 1991) the rotational symmetry can be restored by the slow random rotations of the pattern around the axis of the system.

In the case of class-B lasers the scenario is quite different. There is again a region of the plane (ψ, C) where the single mode stationary solution $p = 2, l = 0$ is stable. Elsewhere, however, one meets only dynamical solutions; basically, one finds a pattern similar to the leopard, which performs a rotational motion (Prati *et al.*, 1994; Prati *et al.*, 1995). This phenomenon has been observed in a CO_2 laser (Boscolo *et al.*, 1997).

Coming back to the case of class-A lasers, for families of order higher than 2, one finds an increasing number of phase singularities, arranged in regular arrays;

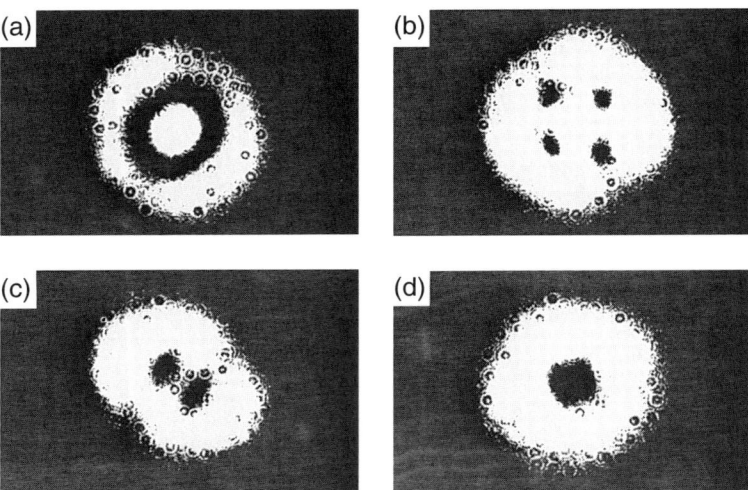

FIG. 19. Output intensity patterns observed in a sodium dimer laser. The four configurations a, b, c, d correspond to those of Fig. 18. The small-scale structures in the photographs are diffraction patterns caused by dust particles.

an example of these "phase singularity crystals" is shown in Fig. 20. In this case, it does not seem appropriate to call these phase singularities "defects." However, interference with a tilted plane wave leads to the appearance of dislocations like those of Fig. 14.

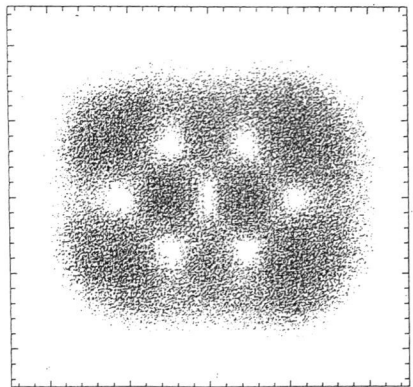

FIG. 20. Crystal of seven optical vortices formed by the frequency-degenerate family $2p + l = 3$. Reprinted from *Phys. Rep. 219.* In L. A. Lugiato, Spatio-temporal structures I. (p. 302) (1992), with kind permission from Elsevier Science - NL, Sara Burgerhartstraat 25, 1055 KV Amsterdam, The Netherlands.

Next, let us turn to the case that the gain line is able to excite a number of frequency-degenerate families. When the frequency difference between adjacent active modes is smaller than the cavity linewidth, one meets the phenomenon of *cooperative frequency locking* (Lugiato, Oldano, et al., 1988; Lugiato, Oppo, et al., 1988) in which the modes do not maintain their own frequency but instead lock to a common frequency value and oscillate in a synchronous way. The locking concerns also the relative phase of the modes, so that the output intensity has a stationary value and transverse configuration, corresponding to a sort of "supermode," where the common oscillation frequency ω corresponds to the average of the modal resonances, weighted by the intensity distribution of the modes in the stationary state

$$\omega = \frac{\sum_n \omega_n |a_n|^2}{\sum_n |a_n|} \tag{II.70}$$

where the index n labels the modes in play, and ω_n are the mode frequencies pulled by the atomic line. This phenomenon has been observed in CO_2 lasers (Tredicce et al., 1989) with good qualitative agreement with the prediction of theory. Additional experimental observations are described in (Tamm, 1988; Lovergnaux et al., in press). Recently, the same formula has been shown to be valid for the phase locking of the elements in an array of semiconductor lasers (Prati, Vecchione, and Vendramin, 1997). When frequency locking does not take place, the frequency competition among the modes gives rise to dynamical patterns. First, it must be noted that dynamical multimode self-oscillatory behaviors of transverse modes occur at much lower values of the gain than those required to obtain multimode self-pulsation in longitudinal modes of homogeneously broadened lasers (Lugiato, Prati, et al., 1988).

Let us assume that the Fresnel number of the cavity is such that only a small number of low-order mode families are involved. The frequencies that characterize the pattern dynamics correspond, in part, to the beat notes among the (mode-pulled) frequencies of the families in play. However, the nonlinear dynamics can create additional frequencies; an example is the frequency of the "rotating leopard" mentioned before in connection with class-B lasers. The simplest configuration arises from the superposition of the fundamental Gaussian mode and of the two modes of the adjacent family $2p + l = 1$

$$A(r, \varphi, t) = \exp\left[-\left(\frac{r}{w_0}\right)^2\right] \{a + br \exp[i(\varphi - \Delta\omega t)]\} \tag{II.71}$$

where $\Delta\omega$ is the frequency difference between adjacent families. The corresponding intensity pattern exhibits a regular rotation around the laser axis with frequency $\Delta\omega$. The configuration (II.71) is realized spontaneously by the laser as a consequence of an instability, which simultaneously breaks the time translational and the rotational symmetry of the Gaussian mode (Brambilla, Cattaneo, et al.,

1994); an experimental observation of a rotating pattern in Na_2 lasers (Coates *et al.*, 1994) confirms this.

The rotating structure (II.71) is the first of a sequence of dynamical configurations of increasing complexity that appear when the Fresnel number is gradually increased. As shown by Tredicce *et al.* (Green *et al.*, 1990), their appearance follows a bifurcation tree governed by $O(2)$ symmetry arguments. A variety of structures obtained theoretically and experimentally are discussed in (Prati *et al.*, 1994; Brambilla, Cattaneo, *et al.*, 1994; Coates *et al.*, 1994). Spectacular dynamical scenarios were found by Arecchi and collaborators in a photorefractive material pumped by a laser and contained in a resonant cavity (Arecchi, 1994; Arecchi *et al.*, 1990) at increasing transverse aspect ratio, the Fresnel number F. They first observed simple patterns alternating in the course of time (several seconds in photorefractive), either regularly (periodic alternance) or irregularly (chaotic itineracy). In the first case the system realizes in sequence a certain number of transverse patterns (Fig. 21); the sequence repeats itself again and again. The theory of alternation of pure patterns is based on a normal form model of the nonlinearities and on the role of heteroclinic connections (Arecchi, Boccaletti, *et al.*, 1992). For higher Fresnel number, one has an incoherent coexistence of many patterns (space–time chaos). This behavior, which can arise only when a very large number of cavity modes interact, is characterized by the fact that at each point of the transverse plane the time evolution of the field intensity is chaotic and the time

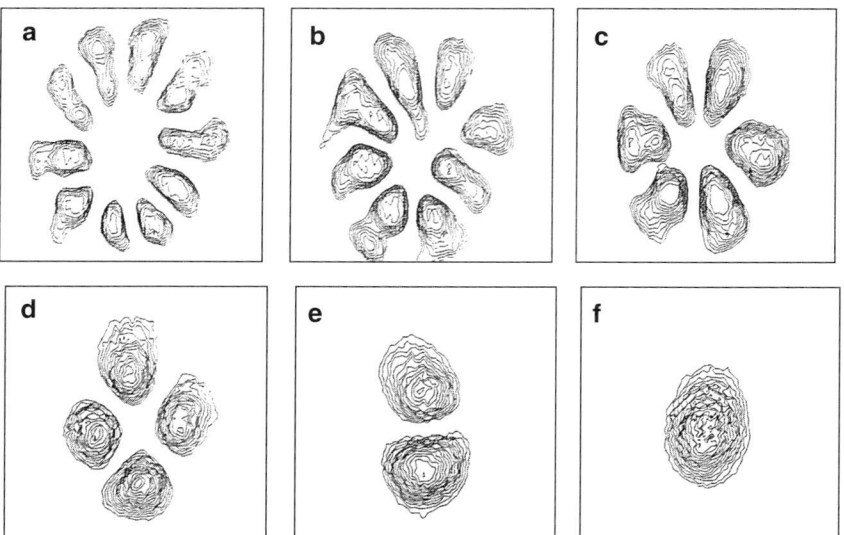

FIG. 21. Photorefractive oscillator: intensity patterns of the pure modes in their order of consecutive appearance in a cycle of periodic alternation at $F = 5$. (From Arecchi *et al.*, 1990)

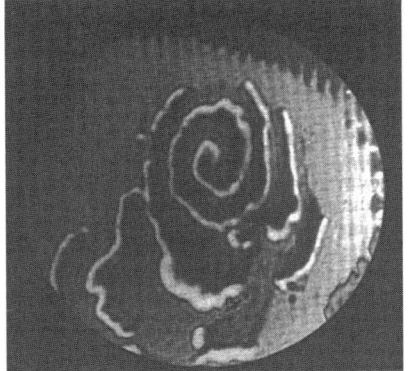 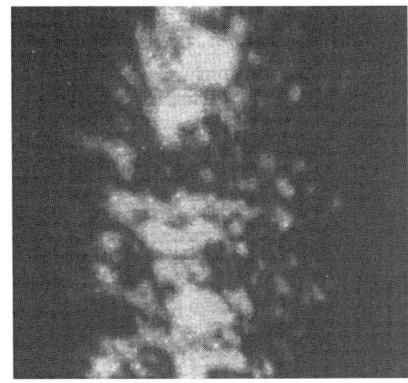

FIG. 22. Space–time chaos in a cavity with large Fresnel number containing a photorefractive material (right, courtesy F. T. Arecchi). Spiral pattern observed in a liquid crystal system with optical feedback (left, courtesy M. A. Vorontsov). Reprinted from *Phys. Rep. 219.* In L. A. Lugiato, Spatio-temporal structures I. (p. 305) (1992), with kind permission from Elsevier Science - NL, Sara Burgerhartstraat 25, 1055 KV Amsterdam, The Netherlands.

evolutions at different points are completely uncorrelated, provided that the distance exceeds a minimum value that is much smaller than the transverse dimension of the system. The authors (Arecchi, 1994; Arecchi *et al.,* 1991) characterized the statistical features of this behavior and observed continuous processes of creation and annihilation of optical vortices in agreement with the general picture of "defect mediated turbulence" described by Coullet, Gil, and Lega (1989) (an example is shown in Fig. 22 left). Optical vortices may decorrelate adjacent regions of the transverse plane (Harkness *et al.,* 1994). Strikingly complex and beautiful patterns, which include switching waves, spiral waves, domain walls, and turbulent behavior were observed by the group of Akhmanov and Vorontsov (see for example Akhmanov *et al.,* 1992) using a liquid crystal film in a configuration that includes a ring feedback similar to that used in video feedback experiments; Fig. 22 (right) shows a case of spiral wave.

F. APPLICATIVE ASPECTS

Let us first fix two concepts that are relevant in the perspective of encoding information in the transverse structure of the electric field and processing it in a completely optical and parallel way.

(a) Spatial multistability.
This is the coexistence of several distinct stable states in the phase space of the system for the same values of the parameters. It is quite different from the standard optical multistability, because in this case the difference

among the coexisting states does not lie so much in the total output intensity, but rather, in the spatial configuration of the electric field.
(b) Control.
By using injected external fields or pulses, or by inserting appropriate elements in a feedback loop, one can control the behavior of the system in the direction of the desired application.

We will discuss separately the case in which the system operates with few modes and the other extreme, in which the number of significant modes in play is extremely large. In the first case, one can achieve situations of spatial multistability, for example in the case of a laser operating with a frequency-degenerate family of modes (see the previous section). One has strong stability of each global pattern as a (spatially extended) fixed point in the phase space of the system. By using laser pulses it is possible to achieve low energy switching from an emission profile to another, as has been shown both theoretically and experimentally (Smith *et al.*, 1993). Schemes for optical associative memories based on these principles, utilizing the laser as the nonlinear discriminator element in the architecture, have been studied theoretically and experimentally (Smith *et al.*, 1993; Brambilla, Lugiato, *et al.*, 1992); the limitation of this approach lies in the order of spatial multistability that can be achieved in this way.

With respect to the goal of encoding information, patterns arising from combinations of few modes present an intrinsic limitation, because their various parts are strongly correlated with one another. Therefore any local modification induced by an external control in order to encode information, also affects other parts of the pattern, or, alternatively, the correlation leads to spontaneous elimination of the modification itself. For example, one might consider a lattice of intensity peaks, which arise from the superposition of few plane waves generated by a spatial instability as described in Section II.B (see Fig. 23a, in which we consider a square lattice). If the peaks were independent of one another, they could be addressed individually as pixels and one could encode information in binary form (Fig. 23b). However this is made impossible by the strong correlation between the points in the lattice.

In the opposite case of many modes, one opens the door to the world of complexity. However, one must avoid the situation that the behavior of the system becomes irregular; hence it is necessary to introduce control. In order to ensure the possibility of a reasonable implementation in practice, the control procedure must be simple.

It has been shown recently that by generating spatial solitons in a nonlinear cavity one can realize a set of intensity peaks that can be addressed (i.e., written and erased) individually and pinned down to precise locations, so that one deals indeed with an array of independent pixels as in Fig. 23b. A set of $N \times N$ solitons constitute a memory (or, in general, an optical processor) with $2^{N \times N}$ distinct states. This approach was pioneered by the work of Rosanov (1990, 1996) on

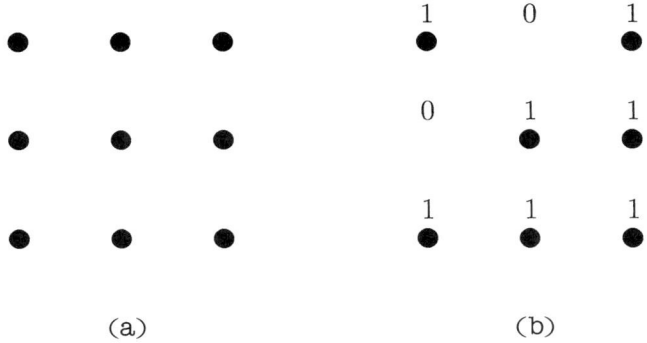

FIG. 23. (a) Square array of points. (b) If the points are independent of one another, they behave as binary pixels and one can encode a number in binary units.

"diffractive autosolitons" and of Moloney and collaborators (McLaughlin, et al., 1983). In the case of (Rosanov, 1996), the spatial solitons arise from the coexistence of two homogeneous stationary states.

In the cases that have been analyzed in the more recent literature, however, the spatial solitons are connected with the presence of a modulational, pattern-forming instability. They are related to the results obtained in the one-dimensional case in (McDonald and Firth, 1990, 1993). Recent theoretical/numerical predictions in two dimensions indicate the possibility of realizing these arrays using semiconductor large-area microcavities, as it is interesting for applications. The control of solitons seems reasonably simple and robust, and they behave as a self-organizing set, capable of accommodating errors in the addressing of writing/erasing beam. A detailed description of results for spatial solitons, together with the related literature, will be given in Section III.

A clear-cut description in optics of such concepts as spatio-temporal (s-t) chaos, weak turbulence, and s-t complexity is still to be gained thoroughly (Farjas et al., 1997; Arecchi et al., 1993; Huyet et al., 1995; Huyet and Tredicce, 1996; Harkness et al., 1996).

As a natural evolution of the topic of control of chaos there has been recent noteworthy contribution to the topic of control in extended systems in general (He and Gang, 1996; Lu and Harrison, 1994), and in optics in particular (Lu et al., 1996; Hess, forthcoming 1998; Hochheiser et al., 1997; Martin, Kent, et al., 1996; Martin, Scroggie, et al., 1996).

Of course, this subject is also interesting per se, independent of possible future applications. The aim is to stabilize a single unstable pattern out of the variety of such configurations visited by the system in a chaotic regime. Especially interesting for the simplicity of the control procedure (specifically, control of the spatial Fourier transform obtained by two lenses and a mask in a Fourier loop) is the scheme devised in (Martin, Scroggie *et al.*, 1996). This also allows control of the pattern formation process, by inducing the system to realize a certain pattern, for example a hexagonal pattern rather than a roll or a square pattern. This is intimately related to the mechanism of pattern formation illustrated in Section II.B.

An adaptive recognition and control technique introduced for low-dimensional chaos has been applied to pattern control in (Arecchi, Basti, *et al.*, 1994; Boccaletti and Arecchi, 1995; Boccaletti *et al.*, 1997).

The LCLV experiments are also suggestive of routes to applications. The Moscow group has already demonstrated optical logic and adaptive optics in this systems, and there are clear links to neural networks (Akhmanov, 1995). Some of these ideas may also be workable in faster optical materials such as semiconductors.

G. FURTHER ISSUES IN OPF

In this section we have gathered some of the topics that have not been treated in the previous sections.

First we note that several other classes of optical systems have been analyzed in the literature, in addition to those outlined in Section II.B. For example, second harmonic generating systems, based on the same parametric conversion as OPOs, have been studied in (Etrich *et al.*, 1997). Four-wave-mixing processes lead to OPF described by means of order parameter equations (deValcarcel *et al.*, 1996). The mere injection of a coherent homogeneous field in a laser system can account for different pattern-forming mechanisms (Georgiou and Mandel, 1994; Longhi, 1997; Brambilla, Brambilla, and Lugiato, forthcoming), while the free-lasing operation has been studied in three-level atomic systems, where two fields coherently interact and reciprocally influence the emerging structures in the respective transverse profile in the common frame of Swift-Hohenberg or Cross-Newell-Whitehead equations (Garcia-Ojalvo *et al.*, forthcoming). Issues in OPF in lasers with saturable absorbers have been studied in (Barsella *et al.*, 1994; Williowski *et al.*, 1994).

OPF in photorefractive materials has been extensively studied and the slow time scales typical of their electro-optical nonlinearity has allowed easy experimental detection of pattern formation and dynamics. Recent research, following the pioneering works described in Section II.E (Arecchi, Giacomelli, *et al.*, 1991; Arecchi, Giacomelli, *et al.*, 1992; Arecchi, Boccaletti, *et al.*, 1992), have considered both planar and nonplanar resonators; in the first case a few-mode dynamics can be predicted and observed, with creation and annihilation of phase singularities (Hannefuin *et al.*, 1994; Malos *et al.*, 1996; Heckenburg *et al.*, 1996; Indebetouw and Korwan, 1994) whose behavior, as the Fresnel number is increased, can be a flag for the onset of bulk effects in the dynamics (Arecchi, Boccaletti, *et al.*, 1995; Indebetouw and Korwan, 1996). In the planar resonator configuration or in the single plane mirror case, the formation of patterns has been theoretically and experimentally reported in different operating conditions (counterpropagating beams, anisotropic media, phase conjugation, four-wave mixing, two-wave mixing, etc.), showing similarities to structures reported in other optical systems (Chen and Abraham, 1995; Mamaev and Saffman, 1996a, 1996b; Honda *et al.,* 1997; Rehn and Kowarschik, 1996). Fast mapping techniques to reconstruct two-dimensional phase portraits have been reported in (Esselbach *et al.,* 1997). The onset of traveling waves in one transverse dimensional models showed conditions where collisions of the structures led to onset of spatio-temporal chaos (Leonardy *et al.,* 1996).

Throughout the present work the simplifying assumption of linearly polarized fields has been maintained, but polarization coupling mechanisms are relevant in several optical systems. The competition, for example, between two circular or orthogonal components of the field then mixes with the diffractive spatial coupling and can generate polarization instabilities that lead to pattern formation. In the case of vectorial Kerr (Geddes *et al.,* 1994) or four-level systems (Scroggie and Firth, 1996) such patterns closely resemble convective structures. Polarization spatial effects in vertical-cavity surface-emitting lasers (VCSEL) have been analyzed in (Martin-Regalado *et al.,* 1997; Prati *et al.,* 1997). Cavityless systems display polarized beam separation or mirroring during propagation (Roehricht *et al.,* 1995; Dangel and Holzner, 1997).

Finally, we mention the increasing interest in OPF in semiconductor devices. A model derived from microscopic modeling of carrier dynamics based on a density matrix approach has been presented in (Hess, 1994; Hess and Kahn, 1996) where complex patterns due to spatial and spectral holeburning are predicted by simulations. The complexity of some regimes has been studied by cross-correlation function analysis in (Fischer *et al.,* 1996). There is an extensive literature on transverse modes in VCSELs (Chang-Hasnain *et al.,* 1991; Tai *et al.,* 1993; Li *et al.,* 1994; Colstoun *et al.,* 1994; Huffaker *et al.,* 1994; Valle *et al.,* 1995; Michalzik and Ebeling, 1995; Zhao and McInerney, 1996). Phase locking and supermode formation in VCSEL arrays are also a subject of intense study (Orenstein *et al.,* 1991; Gourley *et al.,* 1991; Orenstein *et al.,* 1992; Catchmark *et al.,* 1996; Yao *et al.,* 1990; Prati, Vecchione, and Vendramin, 1997). Flowerlike patterns in VCSEL have been found recently (Woerdman).

III. OPF and Solitary Structures in Cavities

In this section we address the issue of the so-called localized structures or spatial solitons (SS) as previously introduced. Theoretical predictions and observations of SS in nonlinear cavities will be presented in different configurations and models, although a broader exposition will be devoted to SS in semiconductor devices, as they are always the most appealing systems for applications.

For a general paper on solitons in dissipative system see Christov and Velarole (1995).

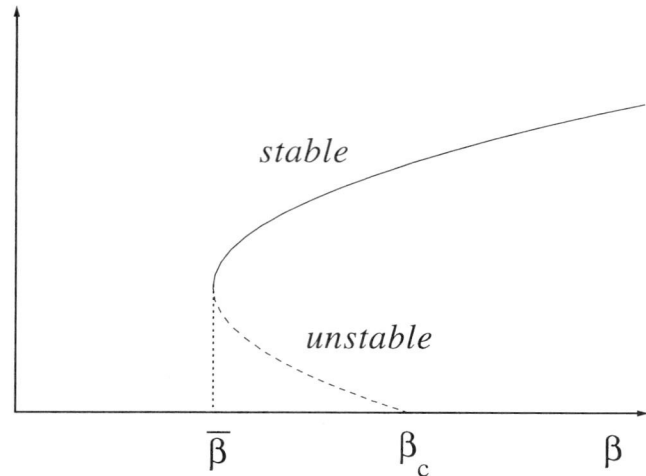

FIG. 24. The curve qualitatively shows the maximum intensity of the bifurcated pattern as a function of the control parameter β.

A. Localized Structures in Optics

In the case of systems with translational symmetry, one often meets the phenomenon of coexistence between a homogeneous stationary state and a modulated stationary solution, in the sense that in a certain interval of the control parameter β (say, the input field A_I) the two solutions are both stable and, according to the initial conditions, the system evolves to one or the other of the two. Referring to the bifurcation of a stationary pattern discussed in Section II.B (boundary for $\nu = 0$), this situation arises when the modulated solution bifurcates subcritically at the critical point β_c, that is, it emerges for $\beta < \beta_c$. In this case the bifurcated solution is unstable and, if one plots the maximum intensity of the bifurcated pattern as a function of β, one obtains a curve with negative slope (Fig. 24). However, usually this unstable branch bends for $\beta = \bar{\beta} < \beta_c$, its slope becomes positive and the solution becomes stable, so that there is coexistence for $\bar{\beta} < \beta < \beta_c$. An example of this behavior is given by the absorptive model (II.29), and with Fig. 9 there is coexistence between the solution with coexistence positive hexagons and the homogeneous state.

Stemming from this coexistence, we meet here another phenomenon that links the field of optics to hydrodynamics and nonlinear chemical reactions, namely the existence of the so-called "localized structures" (LS), which have been described in Ginzburg-Landau models (Fauve and Thual, 1988) and Swift-Hohenberg models (Glebsky and Lerman, 1995), and observed in fluids, granulates, and chemical systems (Gorshkov et al., 1994; Tsimring and Aranson, 1997; Dewel et al, 1995).

In the case of optics a LS is a solution for the field intensity where the two coexisting solutions appear simultaneously in different regions of the transverse plane. For example one observes a stable, stationary island containing a portion of the hexagonal pattern, embedded in a homogeneous background corresponding to the homogeneous solution. The existence of this kind of LS in nonlinear optical systems has been predicted by Tlidi, Mandel, and Lefever (Tlidi *et al.,* 1994; Tlidi and Mandel, 1994). Also, the case of chaotic oscillatory LS has been reported (Berre *et al.,* 1997).

LS arising from coexistence of homogeneous profiles and structures in the presence of a modulational instability are not the sole representatives in optics: a mainstream of researches (Rosanov, 1996) indicate that in 2-level systems, either passive or active, the well-known coexistence between low- and high-transmissivity branches of the plane wave steady-state curve may be a sufficient condition for realizing LS in the transverse plane when diffractive effects are accounted for. In this particular case the confinement of the portion of the solution corresponding to the high transmissivity branch occurs via a locking phenomenon between the transverse switching waves under conditions where their velocity in the transverse plane is vanishingly small. The reader will find in (Rosanov, 1996) how this phenomenon also allows one to realize bright and dark "autosolitons," to control them, and to study their interactions. The bright spots described in (Kreuzer *et al.,* 1991) might correspond to experimental observation of such structures.

We will concentrate on LS in the presence of modulational instabilities, and in the next sections will introduce in this context the concepts of formation and control of solitonic structures.

B. Exciting LS and Spatial Solitons (SS)

The location, shape, and size of LS constitute, so to say, free parameters spanning an infinitely degenerate manifold of stable solutions for the global profile of the emitted field. In this section we will illustrate how control can be gained on such qualifying parameters; for the sake of simplicity, we will consider the model for the saturable absorber (II.29). The same techniques have been successfully applied to other models, specified in the following.

A control on the LS location can be achieved by properly acting on the external field A_I. The idea is to maintain a homogeneous part whose amplitude locates the system's operating point somewhere in the region of coexistence with the stable branch of the hexagonal pattern, below the instability threshold A_{thr}, and to start from an initial condition corresponding to the homogeneous stationary state. Then we superimpose an input Gaussian pulse of duration t_p so that we can represent the total input field as (Brambilla, Lugiato, and Stefani, 1996; Firth and Scroggie, 1996).

$$A_I(x, y, t)$$
$$= \begin{cases} A_{I,hom} + \xi \exp\left\{-\frac{1}{\sigma^2}[(x - x_0)^2 + (y - y_0)^2]\right\}\exp i\varphi & t \le t_p \\ A_{I,hom} & t > t_p \end{cases}$$
(III.1)

We thus locally perturb the systems, by an amount ξ in an area centered around (x_0, y_0), whose size is measured by σ, in general much smaller than the medium section. If $\xi > A_{thr} - A_{I,hom}$ the action of this perturbation is to locally bring the system above the instability threshold, so that in that region the homogeneous solution loses stability and the field can realize a portion of the hexagonal lattice.

When the pulse is switched off, assuming t_p large enough in comparison to the most unstable modes' dynamics, the system persists on the modulated branch as its operating point lies in the coexistence region. A temporal sequence of the transverse intensity profile is shown in Fig. 25, where the process of local pattern formation can be followed, including a further boundary-effect-dominated dynamics that shapes the contour of the LS.

It is intuitive that the pulse width σ and its amplitude ξ determine the size of the LS while its location coincides with the location where the LS will approximatively be centered. The interesting aspect is that by reducing σ the LS shrinks, until the portion of the hexagonal structure is reduced (and increasingly influenced by boundary effects) and eventually coincides with a single intensity peak when the pulse size is (approximately) smaller than twice the critical wavelength associated to the MI. This single peak is generally indicated as a "spatial soliton" (SS).

There is an extensive literature on spatial solitons, that is, stationary, bright, self-confined intensity peaks that arise from the balance of diffraction and self-focusing (see for example Mamyshev et al., 1994; Crosignani et al., 1993); they correspond basically to the phenomenon of Kerr self-trapping described in (Akhmanov et al., 1972). The phenomenon described here, however, arises in nonlinear optical cavities; the cavity feedback is essential so that a stationary SS persists after the passage of the pulse. In this case the SS arises from the interplay of diffraction and nonlinearity, but also the cavity detuning plays an essential role as will be discussed in the following. In the case of the absorptive model, the laser pulse locally creates a bleached area, which allows the local intensity to rise. Thus the SS are here "optical bullet holes" (Firth and Scroggie, 1996) generated by the laser pulse.

In contrast to propagative solitons (in the NLS equation) no exact analytical expression for the SS solution is available up to date. Being a structure related to the global bifurcation, at instability threshold its size is on the order of the critical wavevector, although it is not strictly a subelement of the hexagonal lattice, due

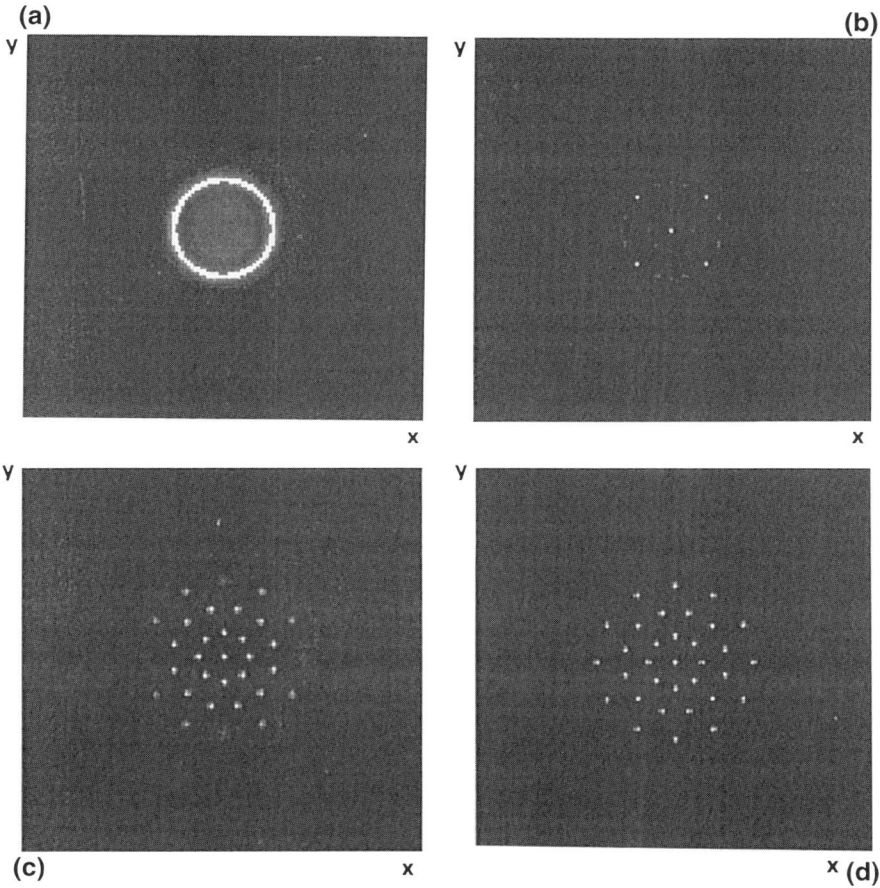

FIG. 25. The figure shows the evolution of the output field intensity profile when a broad Gaussian is injected. A stationary localized structure is visible at regime in panel (d).

to the boundary conditions imposing its connection to the homogeneous background, causing a certain deformation in its phase and intensity profile with respect to the single peak of the hexagons lattice (Firth and Scroggie, 1996). Thus the SS size scales as $\sqrt{\lambda \mathcal{L}/\mathcal{T}}$ (see the parameter a in Eq. (II.17), which scales spatial patterns), with a coefficient depending in particular on the detuning θ. When the nonlinearity is mainly absorptive, the SS exhibits a typical "solitonic" structure (Fig. 26a), whereas when the refractive nonlinearity dominates the SS develops rings around its centers (Fig. 26b).

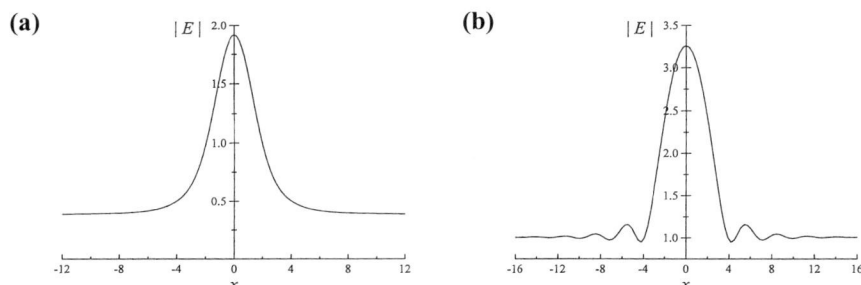

FIG. 26. Diametral section of the SS transverse intensity profile. (a) self-focusing active case: $C = 0.45$, $\theta = -2$, $\alpha = 5$, $\mu = 1$, $\eta = 0$, $\beta = 0$, $d = 1$; (b) self-defocusing passive case: $C = 50.5$, $\theta = -0.716$, $\Delta = 10$, $\mu = -1$, $\eta = 0$, $\beta = 0$, $d = 1$. The parameters refer to Eqs. (II.2, II.3).

1. Spatial Soliton Control

Several SS can be excited in the transverse field profile by using the procedure just described. The question arises now about the maximum density of independent SS that can be achieved, and what happens when the SS interact. It is thus essential to evaluate the minimum distance from an existing SS at which a second one can be created, without interacting with the former (Brambilla, Lugiato, and Stefani, 1996).

In the light of the previous comments on SS characters, it is intuitive to assume that the two subelements of the global lattice will interact when their distance will be on the order of or smaller than the hexagonal lattice transverse wavelength. Though this is a simplification, the idea is fundamentally substantiated by results.

Simulations in which the second SS is excited at location progressively closer to the first one, by using the same form (III.1) with $\varphi = 0$, indicate that two critical distances can be defined: D_{cr} and d_{cr} (with $D_{cr} > d_{cr}$). Let d be the distance between the existing SS and the transverse plane point where the Gaussian in the input field profile is centered, in order to excite a second structure; then we find the following:

1. If $d > D_{cr}$, a second independent SS is created.
2. If $D_{cr} > d > d_{cr}$, the two SS interact; the result of the interaction is that they move apart, until they reach a distance D_{cr}.
3. If $d < d_{cr}$, the existing SS may be erased and in this case a fully homogeneous profile is left. Alternatively, the two SS merge into one.

It turns out that D_{cr} is slightly larger (5–10%) than the hexagonal lattice wavelength $\lambda = 2\pi/k_c$. The value of d_{cr} in general is somewhat smaller (70–90%) than λ. It is remarkable that for $d < d_{cr}$ the existing SS may be canceled; this process allows *turning off* any soliton without influencing the others, locally restoring the homogeneous profile. However this erasing procedure does not seem useful in practice, because it works only under a rather critical control of the parameters t_p and ξ. A much better procedure is identified by taking $\varphi = \pi$ in Eq. (III.1), so that the inhomogeneous contribution is subtracted from the homogeneous background (see also (McDonald and Firth, 1990; McDonald and Firth, 1993)). When the "spot" is exactly superimposed to the existing SS, it locally creates the conditions to erase it. Of course there exists a minimum value of t_p, which causes the erasure of the LS for a given value of ξ and, as a rule of thumb, one observes that the product ξt_p is approximately constant. The advantage of this procedure is its robustness relative to the choices of σ, t_p and also with respect to the Gaussian phase φ: A broad set of values for φ exists where the cancellation takes place. Furthermore, some error in the superposition of the dark hole with respect to the SS is accommodated by the system.

Cavity SS of the kind described here have been experimentally observed recently with an organic material (Weiss, 1996), under conditions close to the absorptive model (II.29). The technique to write and erase SS, previously described, has been used in LCLV experiments (Thuering *et al.*, 1996). On the applicative point of view, SS can be intended as bistable elements, realizable in a two-dimensional plane with a very high degeneracy degree associated to the number and location of the intensity peaks and limited by the interaction distance of the SS. As in all systems exploiting a number of independently addressable bistable pixels, information can be stored, manipulated, and processed. To this purpose, it is desirable to trap a SS in its original switch-on location, avoiding the random motion induced by noise sources such as incoherent emission, pump or field fluctuations, thermally induced fluctuations, and so on. A remarkable property of SS is the possibility of moving them across the transverse plane without affecting them, by introducing suitable phase gradients in the field profile. On one hand, this allows pinning the SS in desired locations against random motion, and thus to associate a binary digit to each location we define in the transverse plane. This is obtained, for example, by introducing an appropriate phase modulation in the driving field; whenever they are created, the SS move spontaneously to the troughs of the phase profile (Firth and Scroggie, 1996). On the other hand, SS either can be rearranged by dynamically changing the phase profile or can be selectively brought to interaction in an externally controlled manner, which opens up perspectives to optical processing.

This concept of optical information storage and control is a mainstream of research for semiconductor arrays of optical elements (Koppa *et al.*, 1997). The peculiarity of the SS lies in the self-organization of the elements, which does not force one to physically isolate the bistable elements (by surface etching, for example).

C. SPATIAL SOLITONS IN SEMICONDUCTOR MODELS

We adopt a model suited to describing the basic features of broad-area semiconductor heterostructures in a microresonator. In order to preserve the complete homogeneity of the system we assume that the field radiates orthogonally with respect to the heterostructure's and resonator's layers. Moreover we account for two different settings of the modelized device: in one case we consider a purely passive system, such as multiple quantum wells in a Bragg microcavity, where free carriers are created by the purely optical excitation provided by an external plane wave beam. We call this setting the passive vertical cavity resonator (PVCR). In another case we assume a device structurally similar to a broad-area VCSEL, where a current is provided to achieve a free carrier population above the transparency value, but its intensity is kept some 5–10% below the lasing threshold. This setting operates as an optical amplifier where one benefits from the electrical/optical energy transfer, without running into the traveling wave and polarization instabilities of the semiconductor lasers. The latter setting will be designated as the active vertical cavity resonator (AVCR).

As it turns out, the model for both settings, in the paraxial and mean-field approximations, can be compactly cast as (Brambilla *et al.*, 1997).

$$\frac{\partial E}{\partial t} = -[(1 + \eta + i\theta) + 2Ci\Theta(N - 1) - i\nabla_\perp^2]E + E_I \qquad \text{(III.2)}$$

$$\frac{\partial N}{\partial t} = -\gamma[(1 + |E|^2)(N - 1) - \mu - \beta N^2 - d\nabla_\perp^2 N] \qquad \text{(III.3)}$$

where E and E_I are the normalized slowly varying envelopes of the intracavity and input (homogeneous) field, respectively. We introduce $N = (\tilde{N}/N_0)$, where $\tilde{N}$ is the carrier density and N_0 is its transparency value. Time t is scaled to the cavity decay rate and x, y are scaled to the diffraction length $\sqrt{a}$ (see Eq. II.17) whereas d is the scaled square of the diffusion length. The scaled nonradiative carrier recombination rate γ rules the carriers' dynamics. The parameter η accounts for parasitic linear absorption from Bragg reflectors, buffer layers, and so on, whereas β describes radiative recombination. In the PVCR case the nonlinearity is modeled via a Lorenzian excitonic resonance at ω_e with halfwidth γ_e, so the nonlinear coupling is expressed as $\Theta = (1 - i\Delta)/(1 + \Delta^2)$ where $\Delta = (\omega_e - \omega_0)/\gamma_e$. As no pump is supplied, $\mu = 0$. One should note that Θ accounts for both absorptive and dispersive coupling.

In the AVCR case, a current J is supplied, larger than the transparency value J_0, so that $\mu = J/J_0 - 1$ is the pump parameter; here $\Theta = i\alpha - 1$, where α is the linewidth enhancement factor of semiconductor lasers (Henry, 1982).

The physical reason for considering both passive and active configurations is twofold: on the one hand, active systems are more likely to undergo unwanted instabilities, though they require lower input field powers and operate essentially in the self-focusing regime, which is a bonus in stabilizing SS. On the other hand, passive devices are more stable, but they can be operated basically only on the self-defocusing side of the excitonic resonance.

1. *OPF in Semiconductor Models*

A schematic overview of the model's features shows that it differs from the absorber model (II.29) in two main respects: it accounts for the medium dynamics, and it includes diffusion, which is a strongly anti-patterning phenomenon that can suppress the modulational instability (Vitrant and Danckaert, 1994). We have found though that either diffusion or even the self-defocusing character of the nonlinear coupling to some extent, does not destroy the SS stability altogether. This result has a general valence in describing the role of diffusive processes not only in semiconductors but also in other classes of optical systems.

Figure 27 plots the steady-state curve associated with (III.2, III.3) for both the PVCR and AVCR cases, with a choice of parameters adapted to our model's scalings from experimental activities at CNET in Bagneux. The dotted portion of these curves indicates the domains where plane wave emission is unstable, and in particular the positive slope unstable branches are a consequence of a modulational instability that causes pattern formation. For both systems the character of the instability is stationary ($\nu = 0$ boundary in Section II.B).

Extended numerical simulations have been run (Brambilla *et al.*, 1997), evidencing a scenario where rolls and honeycomb lattices are found in the intensity transverse profile and sometimes they compete, forming more irregular and complex structures.

In particular the arrows in Fig. 27 mark the input intensity interval where stable

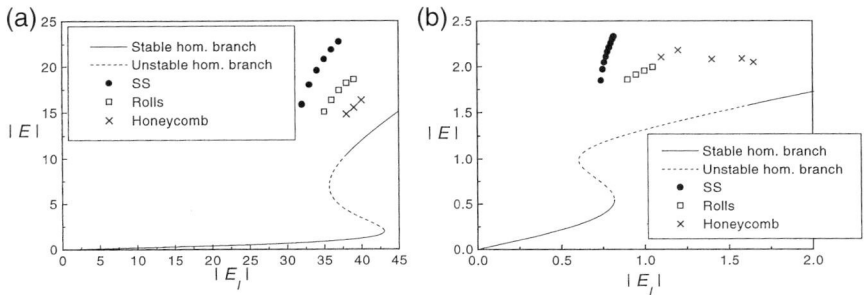

FIG. 27. Steady state of the homogeneous steady-state solutions of (III.2, III.3). Results from numerical simulations are superimposed. Each kind of pattern is indicated by a different symbol, positioned at the value of its maximum intensity. (a) Passive, self-defocusing case: parameters are $C = 40$, $\Delta = 1$, $\theta = -2$, $\beta = 1.6$, $\eta = 0.25$, $\mu = -1$, $d = 0.2$. (b) Active, self-focusing case: parameters are $C = 0.45$, $\alpha = 5$, $\theta = -2$, $\beta = 0$, $\eta = 0$, $\mu = 1$, $d = 0.052$.

SS are found. Note that in the PVCR case the parameter Δ is positive, indicating a self-defocusing character of the nonlinearity. For increasing values of d, the SS tend to broaden, but they do not disappear by diffusive spread for any of the experimentally meaningful values.

The profiles for the SS obtained numerically have been checked (Brambilla *et al.*, 1997) with those obtained by an appropriate generalization of the radial integration method described in (Firth and Scroggie, 1996).

A recent paper (Michaelis *et al.*, 1997) analyzes numerically a model similar to (III.2, III.3) but without absorptive terms, and predicts the formation of stable SS under conditions of large detuning Δ on the self-defocusing side of the nonlinearity. Preliminary numerical simulations of ours obtained with Eqs. (III.2, III.3) seem to confirm this result. Under these conditions, the guiding effect arising from the absorption bleaching is absent, the nonlinearity has an antiguiding role, and both field diffraction and carrier diffusion have a spreading effect. Hence in this case the mechanism that forms out cavity solitons would be quite different from the usual one that holds for standard SS, namely the balance between a spreading mechanism and nonlinearity. The picture proposed by Firth (in a private communication) is that the cavity SS arise from "the desire of the system to resonate with the cavity." As a matter of fact, in the absorptive model (II.29) the cavity detuning θ is exactly compensated by the Laplacian contribution at the SS center (Firth and Scroggie, 1996). This mechanism plays certainly a key role, but further investigations are necessary to obtain a complete picture.

A recent paper reports on the experimental observation of SS in a laser including a concentric cavity and a saturable absorber (Taranenko *et al.*, 1997).

Cavity SS containing a medium with a quadratic nonlinearity that gives rise to second harmonic generation (Etrich *et al.*, 1997) or to parametric downconversion (Staliunas and Sanchez-Morcillo, 1996) have been found numerically.

A recent paper (Samson and Vorontsov, 1997) predicts LS in an optical system containing a thin film of nonlinear medium having binary type refractive nonlinear response and a feedback mirror, and shows that the properties of these solutions can be used for nonlinear image processing and edge detection.

IV. Quantum Fluctuations and Optical Pattern Formation

Before the late 1980s, the literature on spatial aspects of quantum optical phenomena was quite limited (Drummond and Eberly, 1982; Heidmann *et al.*, 1984; Granger *et al.*, 1986; LaPorta and Slusher, 1991). The interest toward squeezed states of light relied mainly on the possiblity of producing (by optical homodyning) photon beams with a statistic in *time* more regular than the statistics of random events.

Sokolov and Kolobov addressed the issue of generating states of light with a statistical distribution of photons regular not only in time but also in space, that is, in the cross section of the beam (Kolobov and Sokolov, 1989, 1991; Kolobov, 1991, 1994; Kolobov and Kumar, 1993). Their investigations concerned mainly the squeezed light generated by the down-conversion process in a χ^2 crystal in the cavityless configuration.

Systematic investigations of quantum aspects of spatial patterns generated by nonlinear optical cavities, and, conversely, of spatial aspects of nonclassical states of the radiation field, were activated by our group in Milano (Lugiato and Castelli, 1992; Lugiato and Gatti, 1993; Gatti and Lugiato, 1995; Lugiato and Marzoli, 1995; Lugiato and Grynberg, 1995; Lugiato, Gatti, and Weidemann, 1997; Gatti et al., 1997; Marzoli et al., 1997; Lugiato, Gatti, Ritsch, et al., 1997; Gatti, Lugiato, Oppo, et al., 1997; Lugiato and Grangier, 1997; Castelli and Lugiato, 1997; Grynberg and Lugiato, 1993). Under the suggestion of Peter Knight, we coined the name "quantum image" to designate this extreme example of noisy image, where the noise is of quantum origin. Under a low-frequency observation, which averages over the fast dynamics of fluctuations, all or part of the image is washed out. However, appropriate spatial correlation functions are able to display a regular spatial modulation, even when average quantities display none. Even more important, correlation functions are able to give clear spatial evidence of the quantum nature of correlation existing between photons.

In the following, we illustrate some of the main concepts on the basis of a quantum model for a quadratic medium both with and without cavity.

A. QUANTUM MODELS FOR χ^2 MEDIA INCLUDING DIFFRACTION

Let us first consider the case of degenerate parametric down-conversion taking place in an optical resonator (Lugiato and Gatti, 1993; Gatti and Lugiato, 1995; Lugiato and Marzoli, 1995; Lugiato and Grynberg, 1995; Lugiato, Gatti, and Weidmann, 1997; Gatti et al., 1997; Marzoli et al., 1997; Lugiato, Gatti, Ritsch, et al., 1997; Gatti, Lugiato, Oppo, et al., 1997; Lugiato and Grangier, 1997; Castelli and Lugiato, 1997; Grynberg and Lugiato, 1993). We consider type I, degenerate, collinear phase-matching conditions in the nonlinear crystal (Boyd, 1992). The medium is enclosed in a single port cavity, on which a coherent, monochromatic field of frequency $2\omega_s$ is incident, and it is able to partially convert power from the pump frequency to signal frequency ω_s. The classical dynamics of such a system has been described in Section II.B, in the framework of mean field limit, single longitudinal mode, and paraxial approximation, and is formulated in terms of the couple of time evolution equations (II.31), (II.32).

In a quantum description, the intracavity pump and signal field are described by two operators $B_0(\vec{x})$, $B_1(\vec{x})$. More precisely, by B_0 and B_1 we denote the envelope operators for the two cavity longitudinal modes, which are closest to resonance with the pump frequency $2\omega_s$ and the signal frequency ω_s, respectively.

Dimensions are such that $\langle B_i^\dagger(\vec{x})B_i(\vec{x})\rangle$, where ($i = 0, 1$) is the mean number of photons per unit area in a transverse plane inside the cavity. Heisenberg operators for fields out of the cavity are linked to the intracavity ones by standard input/output relations (Gardiner, 1991):

$$B_i^{out}(\vec{x}, t) = \sqrt{2\kappa_i} B_i(\vec{x}, t) - B_i^{in}(\vec{x}, t) \qquad (i = 0, 1) \qquad (IV.1)$$

where κ_i are the cavity damping rates for the two fields, and B_i^{in} are envelope operators associated with the input fields. For the signal the input is in the vacuum state. Accordingly, $\langle B_i^{\dagger out}(\vec{x}, t)B_i^{out}(\vec{x}, t)\rangle$, ($i = 0, 1$) represents the expectation value of photon flux density (number of photons crossing unit area per unit time) immediately out of the cavity.

As is appropriate for an open system the model is formulated in terms of a master equation for the reduced density operator ρ of the two intracavity fields:

$$\frac{d\rho}{dt} = \frac{1}{\hbar}[H, \rho] + \Lambda_0 \rho + \Lambda_1 \rho \qquad (IV.2)$$

Here the Liouvillian terms:

$$\Lambda_i \rho = \int d^2x \, k_i [2B_i(\vec{x})\rho B(\vec{x})^\dagger - \rho B_i^\dagger(x) B_i(\vec{x})$$
$$- B_i^\dagger(x) B_i(\vec{x})\rho] \qquad (i = 0, 1) \qquad (IV.3)$$

describe the escape of photons from the cavity through the coupling mirror.

The reversible part of the dynamics is governed by three Hamiltonian terms:

$$H = H_{FREE} + H_{EXT} + H_{INT} \qquad (IV.4)$$

The free propagation of fields in a cavity with plane mirrors, including the effect of diffraction, is described by:

$$H_{FREE} = \hbar \kappa_0 \int d^2x [B_0^\dagger(\vec{x})(\theta_0 - ia_o \nabla_\perp^2) B_0(\vec{x})]$$
$$+ \hbar \kappa_1 \int d^2x [B_1^\dagger(\vec{x})(\theta_1 - ia_1 \nabla_\perp^2) B_1(\vec{x})] \qquad (IV.5)$$

where we have adopted an interaction picture in which the fast time evolution at frequency ω_s is eliminated for the signal and that at frequency $2\omega_s$ is eliminated for the pump; θ_i are the cavity detuning parameters, defined by (II.34), and the characteristic areas a_0, a_1 are given by (II.33).

For a resonator with spherical mirrors, the additional propagation phase shift due to mirror curvature is taken into account by modifying the Laplacian term in the following way:

$$a_i \nabla_\perp^2 \rightarrow \frac{\eta}{k_i}\left(\frac{w_i^2}{4}\nabla_\perp^2 - \frac{r^2}{w_i^2} + 1\right) \qquad (IV.6)$$

where η is the frequency separation between adjacent Gauss-Laguerre mode families, and w_0, w_1 ($w_0 = w_1/\sqrt{2}$) are the beam waists at the cavity center for pump and signal beam, respectively.

Coherent, monocromatic pumping is modeled by the Hamiltonian:

$$H_{EXT} = i\hbar \int d^2x \, [\mathcal{E}_I(\vec{x})B_0^\dagger(\vec{x}) - \mathcal{E}_I^*(\vec{x})B_0(\vec{x})] \tag{IV.7}$$

where $\mathcal{E}_I$ is the amplitude of the input field. Finally, the standard parametric Hamiltonian (Drummond *et al.*, 1981) is generalized to a spatially extended system as specified in (II.42).

By linearizing the model around the classical stationary solution below the OPO threshold, a much simpler model is obtained, where pump fluctuations and signal dynamics are decoupled. If, in addition, we consider a plane wave input, and, in the case of cavity with spherical mirrors we assume that the cavity mirrors do not reflect the pump, the full interaction Hamiltonian (II.42) reduces to:

$$H_{INT} = i\hbar\kappa_1 \frac{A_P}{2} \int d^2x \, [[B_1^\dagger(\vec{x})]^2 - B_1^2(\vec{x})] \tag{IV.8}$$

where A_P is the pump field amplitude in the classical steady state, multiplied by g/κ_1. Note that this linearization procedure amounts to neglecting the pump depletion due to parametric down-conversion in the nonlinear medium.

A model for optical parametric amplification (OPA) in cavityless configuration, including the effect of diffraction, was derived by Kolobov and Sokolov (Kolobov and Sokolov, 1989, 1991; Kolobov, 1991, 1994; Kolobov and Kumar, 1993). They considered a thin slab of nonlinear crystal of length L, ideally infinite in the transverse direction, on which a plane wave, monochromatic pump field is incident. They took into account nearly degenerate and nearly collinear phase-matching conditions into the crystal and neglected the pump depletion along the material. In this way an elementary frequency down-conversion process corresponds to the generation inside the crystal of a couple of signal photons, at frequencies $\omega_s + \Omega$ and $\omega_s - \Omega$, with transverse wave vectors $\vec{k}$ and $-\vec{k}$, which propagate in symmetrical direction with respect to the pump propagation direction (z direction).

By solving the field propagation equations along the crystal slab, these authors obtained linear input/output relations that connect signal operators at the exit of the crystal with those at the entrance:

$$e_{out}(\vec{k}, \Omega) = U(\vec{k}, \Omega)e_{in}(\vec{k}, \Omega) + V(\vec{k}, \Omega)e_{in}^\dagger(-\vec{k}, -\Omega) \tag{IV.9}$$

The coefficients $U(\vec{k}, \Omega)$ and $V(\vec{k}, \Omega)$, which can be found in (Kolobov and Sokolov, 1989), depend on the pump amplitude, on the parametric gain and on the phase mismatch along the propagation direction z

$$\Delta(\vec{k}, \Omega) = [k_{zP} - k_z(\vec{k}, \Omega) - k_z(-\vec{k}, -\Omega)] L \qquad (IV.10)$$

where k_z and k_{zP} are the z-components of the wave vectors for signal and pump, respectively (Kolobov and Sokolov, 1989). A quadratic approximation for the phase mismatch Δ is considered, that is, paraxial diffraction and quadratic dispersion relations for $k = k(\Omega)$ are taken into account. Note that in the context of OPA the phase mismatch Δ plays the same role as the cavity detuning θ_1 for the OPO. The input signal field is assumed to be in the vacuum state. This model has also been generalized to type II phase matching (Kolobov, 1991).

B. A Langevin Approach to Quantum Fluctuation Dynamics

In order to give a description of the fast dynamics of quantum fluctuations, the master equation model for the OPO with plane mirrors has been turned into a set of stochastic differential equations, for c-number fields in the Wigner representation.

In the case of linearized models, the quantum to classical correspondence (Haken, 1970; Charmichael, 1993), generalized to spatially extended system (Gatti, Weidemann, et al., 1997), allows one to transform the master equation into a Fokker-Planck equation for the Wigner functional, with positive definite diffusion matrix. In the case of the full Hamiltonian (II.42) the same procedure, gives, instead, an equation with third-order derivatives. By dropping these terms a Fokker-Planck equation is obtained, which can be interpreted in terms of a classical stochastic process. We believe that the nonlinear Langevin equations obtained in this way still reproduce the quantum dynamics of a macroscopic system, with the exception of a negligible neighborhood of the critical point.

By introducing appropriate scaled versions of the fields, the Langevin equations in the Wigner representation take the form of classical nonlinear equations (see Eqs (II.31, II.32) with a noise term added, which models vacuum fluctuations (Gatti, Weidemann, et al., 1997; Lugiato, Gatti, Ritsch, et al., 1997):

$$\frac{\partial}{\partial t} A_0(\vec{x}, t) = \kappa_0 \left[-\left(1 + i\theta_0 - i\frac{\kappa_1}{2\kappa_0}\nabla_\perp^2\right) A_0(\vec{x}, t) \right.$$
$$\left. + E - A_1^2(\vec{x}, t) + \sqrt{\frac{2}{\kappa_0 n_{th}}} \xi_0(\vec{x}, t) \right] \qquad (IV.11)$$

$$\frac{\partial}{\partial t} A_1(\vec{x}, t) = \kappa_1 \left[-(1 + i\theta_1 - i\nabla_\perp^2) A_1(\vec{x}, t) \right.$$
$$\left. + A_0(\vec{x}, t) A_1^*(\vec{x}, t) + \frac{1}{\sqrt{\kappa_0 n_{th}}} \xi_1(\vec{x}, t) \right] \qquad (IV.12)$$

In Eqs. (IV.11, IV.12) coordinates are scaled to the diffraction length $l_d = (a_1)^{1/2}$, E is the scaled amplitude of the input field $\mathcal{E}_I$, and $\xi_1(\vec{x}, t), \xi_0(\vec{x}, t)$ are Gaussian stationary stochastic processes, with zero average, and nonvanishing correlations:

$$\langle \xi_1^*(\vec{x}, t)\xi_1(\vec{x}', t') \rangle = \langle \xi_0^*(\vec{x}, t)\xi_0(\vec{x}', t') \rangle = \frac{1}{2}\delta(\vec{x} - \vec{x}')\delta(t - t') \quad (IV.13)$$

so that noise is white both in space and in time. The parameter

$$n_{th} = \frac{\kappa_1^2}{g^2} l_d^2 \quad (IV.14)$$

represents the number of pump photons in the characteristic area l_d^2 that are needed to trigger the threshold for signal generation, and it is hence a measure of the system size (Haken, 1970; Charmichael, 1993).

Besides providing symmetrically ordered expectation values, it is well known that in the Wigner representation the marginals of the quasi-probability distribution in classical phase space are the correct quantum-mechanical probability distributions. Hence we can interpret a single realization of the stochastic equations (IV.11, IV.12) as the result of a homodyne measurement, where a single quadrature component of the field is detected.

Figure 28 shows an example of stochastic realization of the system dynamics, obtained by numerically simulating Eqs. (IV.11, IV.12) in a square with periodic boundary conditions. The vertical left sequence shows snapshots of the near field distribution of the most amplified quadrature of the signal field, for increasing values of the input field E across the threshold region; the right vertical sequence displays the corresponding far field intensity distribution. Parameters are such that above threshold the signal emerges in the form of a roll pattern.

From a classical viewpoint the signal field below threshold is zero everywhere; in a quantum description, however, even below threshold there are signal photons, and the field vanishes only on average. More important, it is able to show a noteworthy level of self-organization, which anticipates the onset of an ordered structure above threshold (Gatti, Weidemann, *et al.*, 1997; Lugiato, Gatti, Ritsch, *et al.*, 1997).

Approaching threshold from below (the left frames of Fig. 28.1a, b, c, d) a spot pattern forms in the near field. The pattern is irregular and the spots perform a slow random motion in the transverse plane; as a result the time average of the field vanishes and the translational symmetry is recovered. However, in a single snapshot the distribution of spots is not random, but instead the probability of finding two spots at a distance r apart has maxima when r is an integer multiple of the critical wavelength $\lambda_c = 2\pi/k_c$. The far field distribution is concentrated on the critical circle, indicating the existence of a privileged angle of emission for

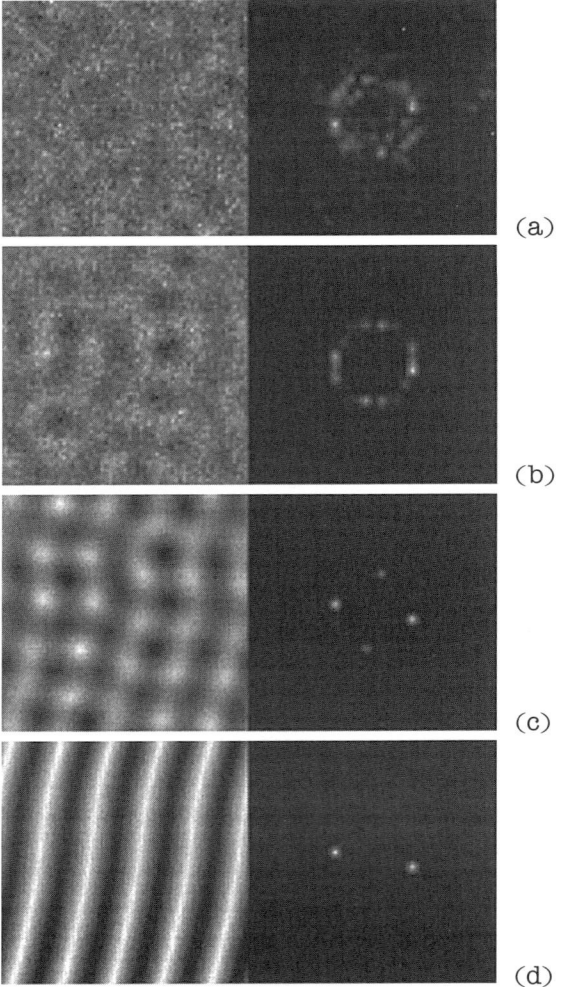

FIG. 28. Scan of the threshold region of the OPO. Snapshots of the spatial distribution of the most amplified quadrature component in the near field (left frames) and of the far field intensity distribution (right frames). (a) 4.2% below threshold, (b) 1% below threshold, (c) 3% above threshold, (d) 6% above threshold. Parameters are $\theta_1 = -1$, $\theta_0 = 0.7$, $\kappa_1 = \kappa_0$, $n_{th} = 10^6$.

the photon pair. In a single snapshot bright spots always appear in symmetrical pairs, which provides clear spatial evidence of the emission of correlated twin photons. Below threshold the position of the pairs of spots moves randomly over the critical circle, in such a way that the mean intensity distribution forms a annulus around the critical circle. Due to critical slowing down of fluctuations, this motion, as well as the motion of spots in the near field, becomes slower and slower as threshold is approached. Above threshold the rotational symmetry is broken by fixing the position of the pair of spots in the far field. This correspond to selecting an orientation for the rolls in the near field. In this way the sequence of Fig. 28a–d shows the transition from a quantum image below threshold to a classical image (roll pattern) above threshold. The sequence of Fig. 28 can be found in the form of an animation in (Gatti, Lugiato, Oppo, et al., 1997).

C. QUANTUM IMAGES IN THE NEAR FIELD

Stochastic simulations, such as the ones shown in Fig. 28, are able to reproduce the dynamics of the quantum image as it would appear under an observation that is fast enough to resolve the dynamics of fluctuations. Under a low-frequency observation, spatial correlation functions represent the main tool to highlight spatial properties of a field that on average appears structureless.

We consider here the spatial correlation function of field quadratures, in a plane immediately out of the cavity. This quantity can be measured in a homodyne detection scheme, by using for example two correlated CCD cameras (Gatti and Lugiato, 1995; Lugiato and Marzoli, 1995), and is defined as:

$$\Gamma(\vec{x}, \vec{x}', t) = \langle : \delta \mathscr{E}_H(\vec{x}, t) \mathscr{E}_H(\vec{x}', 0) : \rangle \quad (IV.15)$$

$$\mathscr{E}_H(\vec{x}, t) = B_1^{out}(\vec{x}, t) e^{-i\phi_L} + B_1(\vec{x}', t)^{out\dagger} e^{i\phi_L} \quad (IV.16)$$

where $\langle : \; : \rangle$ indicates chronologically and normally ordered expectation in the steady state, and ϕ_L is the phase of the local oscillator field (LOF) used in the detection.

For a resonator with plane mirrors, when input field is a plane wave, analytical calculations can be performed in the framework of the linearized model (IV.8) (see for example, Gatti and Lugiato, 1995; Gatti, Weidemann, et al., 1977). Fig. 29 shows examples of the results obtained for zero time delay $t = 0$, for increasing values of the input field below the threshold. As it must be for a system with translational and rotational symmetry, the function $\Gamma(\vec{x}, \vec{x}', t)$ depends only on the distance between the two points. When plotted as a function of this distance, the correlation shows a regular modulation, which occurs at the same wavelength λ_c of the roll pattern immediately above threshold. The function is exponentially damped, but the correlation length turns out to diverge at threshold (Gatti and Lugiato, 1995). Hence the function is able to reveal the presence of a certain degree of spatial order in the OPO field below threshold; moreover the spatial order becomes long-ranged as the critical point is approached, anticipating the onset of a regular spatial structure. It is worth mentioning that these phenomena are still present when an integral over time delay t is performed, that is, if a slow detector is used. Moreover these features of correlation functions are robust

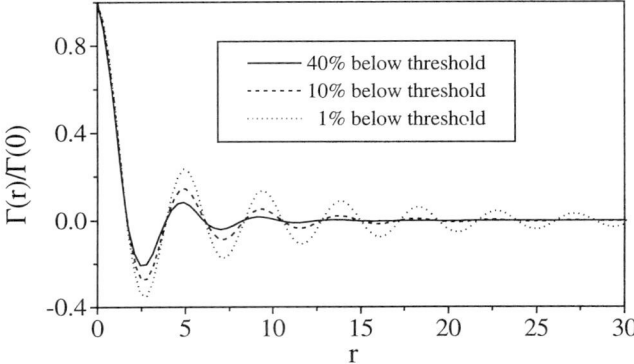

FIG. 29. OPO with plane mirror. Analytical results for the correlation function of field quadratures. The ratio $\Gamma(\vec{x}, \vec{x}', 0)/\Gamma(\vec{x}, \vec{x}, 0)$ is plotted versus the scaled distance $r = |\vec{x} - \vec{x}'|/l_d$, for $\phi_L = 0$ or $\pi/2$, $\theta_1 = -2$, and 40% below threshold (solid line), 5% below threshold (dashed line), 1% below threshold (dotted line).

toward the injection of pump beam with finite diameter, instead of a plane wave beam (Gatti, Weidemann, et al., 1997).

For a resonator with spherical mirror the symmetry broken at the threshold for parametric oscillation is that for rotation around the z axis. When a frequency degenerate family of Gauss-Laguerre modes, say that with $2p + l = q_c$, is resonant with the signal field, that is $\omega_{pl} = \omega_s$ (see Eq. II.23)), the signal field immediately above threshold emerges as a combination of modes belonging to this critical family, and in general the intensity distribution does not have cylindrical symmetry. Below threshold, where cylindrical symmetry is preserved, the correlation function (IV.15) is again able to anticipate the onset of the spatial structure (Lugiato and Marzoli, 1995). Figure 30 illustrates the situation when the critical family is $q_c = 2$. The function is plotted versus the angular separation between the two points, and it clearly reproduces the dependance on the angular variable of the modes $p = 0$, $l = 2$. Moreover on approaching threshold the angular correlation extends to the whole interval $(0, 2\pi)$, which is analogous to the divergence of the correlation length for a system with translational symmetry.

All these features of the correlation functions substantiate nicely the traditional analogy between the laser threshold region and second-order phase transitions (Graham and Haken, 1970; DeGiorgio and Scully, 1970), by adding the spatial aspect that was absent in the previous treatments.

The phenomena here described are "classical," in the sense that the same functions could describe spatial correlations of thermal fluctuations. In order to find signatures of the quantum nature of fluctuations, we have to consider the correlation function of the field quadrature $\phi_L = \pi/2$ (phases are measured with respect to the pump phase), which corresponds to the most deamplified (i.e., squeezed) quadrature component. It turns out that at small distance $r = |\vec{x} - \vec{x}'|$ this function takes on negative values. In the case of planar resonator, for example:

$$\lim_{|\vec{x}-\vec{x}'|\to 0, t\to 0} \Gamma(\vec{x}, \vec{x}', t) = -A_P \left(\frac{1 - A_P}{1 + A_P}\right)^{1/2} \qquad \text{(IV.17)}$$

This behavior has no classical counterpart, where we take the coherent states as a reference for classical states of radiation. To elucidate this point, let us consider the fluctuations δi of the photocurrent, which result from a homodyne detection of the signal performed over a small portion ΔS of the beam, and over a short time interval Δt:

$$\langle \delta i^2 \rangle \propto \frac{1}{\Delta t} \int_0^{\Delta t} dt \int_{\Delta S} d^2x \int_{\Delta S} d^2x' \, \langle \delta \mathscr{E}_H(\vec{x}, t) \delta \mathscr{E}_H(\vec{x}', t) \rangle \qquad \text{(IV.18)}$$
$$\approx \Delta S + \Delta t \Delta S^2 \Gamma(\vec{x}, \vec{x}, 0)$$

The first term at rhs of (IV.18) represents the level of noise of the coherent state (shot noise), which corresponds to a Poisson distribution of photocounts; the second may assume negative values for suitable choice of the local oscillator phase ϕ_L, which shows that photocurrent fluctuations may have statistics more regular than Poisson statistics.

This result is related to what was obtained by Kolobov and Sokolov (1991) for the cavityless configuration. Figure 31 shows an example of the correlation function (IV.15) calculated at the exit face of the crystal in a optical parametric amplifier. Such a configuration of the correlation function led the authors to introduce the idea of *antibunching in space–time*, when referring to the photocurrent density measured by homodyne detection: the system is able to produce a photocount

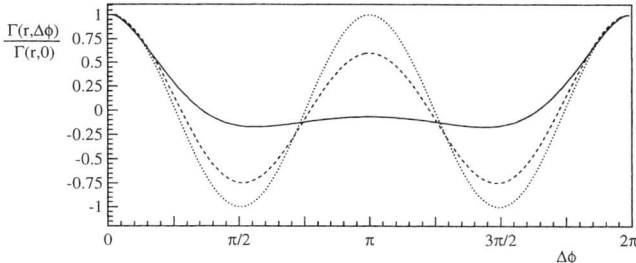

FIG. 30. OPO with spherical mirror. Plot of the ratio $\Gamma(\vec{x}, \vec{x}', 0)/\Gamma(\vec{x}, \vec{x}, 0)$ as a function of the angular separation between the two points, along a circle of radius $r = w_1/\sqrt{2}$, when the critical family is $q_c = 2$, for $\phi_L = 0$ or $\pi/2$ and 50% below threshold (solid line), 1% below threshold (dashed line), limiting behavior at threshold (dotted line).

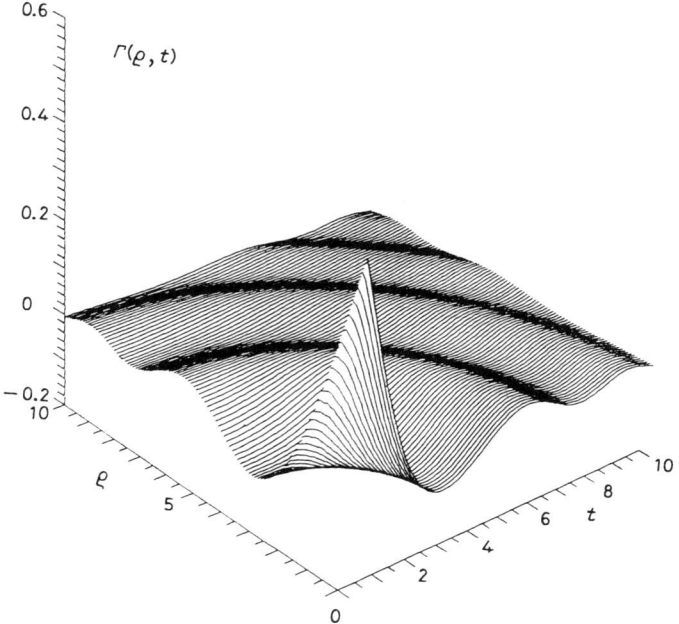

FIG. 31. OPA. Dimensionless spatio-temporal correlation function of field quadratures at the crystal exit is plotted versus the distance ρ between the two points and the time delay t, for $\phi_L = \pi/2$, and for a phase mismatch $\Delta < 0$. (From Kolbov and Sokolov, 1991)

distribution more ordered than that of random events (Poisson distribution) not only in time but also in space. It can be noted that the rigorous definition of sub-Poissonian photon distribution in space–time would correspond to the situation $\Gamma(r = 0, t = 0) < 0$, as it is in the cavity case, whereas in Fig. 31 there is a small region around $r = 0, t = 0$ where the function has a positive peak. However, while averaging in space–time this region gives a negligible contribution, because the volume element in two-dimensional space grows as rdr. A configuration where rigorous sub-Poissonian character in space–time is obtained in the optical parametric amplifier requires careful control of the phase of the squeezed light by means of a lens (Kolobov and Sokolov, 1991).

For this kind of states of light, one can envisage applications to optical image processing (Kolobov, 1994; Kolobov and Kumar, 1993). The advantage with respect to traditional squeezed light consists in having several independent regions in the cross section of the beam, where light is squeezed; that is, the source has several "quantum" channels, each of which produces an object image with fluctuations below shot-noise limit.

1. Spatial Structure of Squeezed States

With respect to applications, the situation seems more promising in the cavityless configuration. From inspection of Eqs. (IV.17) and (IV.18), it is clear that the amount of noise reduction when detecting a small portion of the beam becomes negligible on approaching threshold for the cavity case. As a matter of fact the cavity acts as a filter, which selects a small bandwidth of spatial modes that close to threshold reduces only to those modes exactly resonant with the cavity (critical circle). A natural question arises then as to whether it is possible in a measurement to probe the level of quantum fluctuations in a single spatial mode, rather than in a region of space. To answer this question we have to come back to the very definition of spectrum of squeezing through the homodyne detection, as was done in (Lugiato and Gatti, 1993; Gatti and Lugiato, 1995; Lugiato and Marzoli, 1995).

In a balanced homodyne detection scheme, the field is made to interfere with a strong local oscillator field (LOF). When the whole beam is detected, the measured quantity is proportional to the observable:

$$E_H = \left[\int d^2x \, \rho_L^2(\vec{x}) \right]^{-1/2} \int d^2x \, \rho_L(\vec{x}) \, [B_1^{out}(\vec{x}, t)e^{-i\phi_L} + B_1(\vec{x}, t)^{out\dagger}e^{i\phi_L}]$$

(IV.19)

where we have considered a LOF of the form $\alpha_L(\vec{x}) = \rho_L(\vec{x})\exp(-i\phi_L)$. From formula (IV.19) it is clear that the homodyne detection on the one hand selects the quadrature of the field through the phase of the LOF, and on the other performs the projection of the field onto the LOF spatial distribution $\rho_L(\vec{x})$. Hence the detected level of quantum noise reduction below shot noise (i.e., squeezing) depends on the spatial configuration of the LOF.

We are interested here in the spectrum of fluctuations of the homodyne field around the stationary mean value $\delta E_H = E_H - \langle E_H \rangle$:

$$V(\omega) = \int_{-\infty}^{\infty} dt \, e^{-i\omega t} \langle \delta E_H(t) \delta E_H(0) \rangle \quad \text{(IV.20)}$$
$$= 1 + S(\omega) \quad \text{(IV.21)}$$

where we have indicated by $S(\omega)$ the normally ordered part of the spectrum $V(\omega)$. In this way $S(\omega) = 0$ corresponds to the shot noise level of fluctuations, whereas $S(\omega) = -1$ indicates complete suppression of quantum noise at frequency ω (that is, perfect squeezing), for the quadrature component ϕ_L.

Let us consider an orthonormal set of functions in the transverse plane $\{f_l\}_l$. By varying the spatial distribution of the LOF among the functions of the set, one is able to explore the level of fluctuations in the single-mode spatial components of the field, that is, the whole *spatial structure of the squeezed states* (Lugiato and Gatti, 1993; Gatti and Lugiato, 1995). In the case of the OPO with spherical

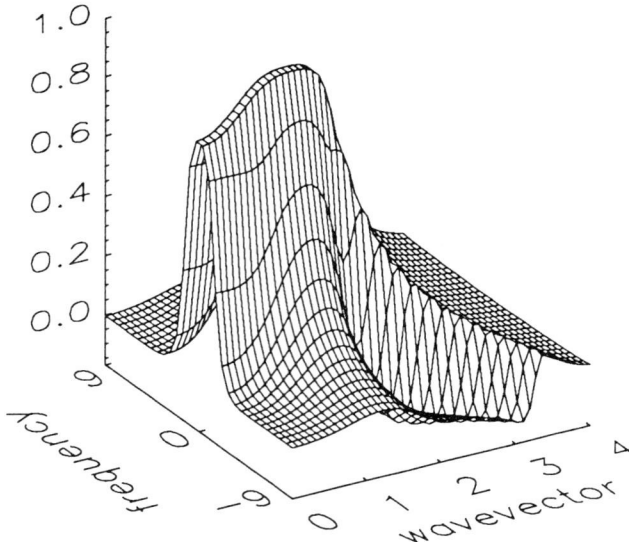

FIG. 32. OPO. The spectrum of squeezing S, changed of sign, is plotted as a function of the scaled frequency ω/κ_1, and of the modulus of the scaled transverse wavevector $l_d|\vec{k}|$, which identifies the angle of emission of the twin photons, at the instability threshold. $\theta_1 = -1.5$.

mirrors, the set appropriate to describe the spatial structure of the squeezed vacuum generated below threshold is the Gauss-Laguerre basis.

For a resonator with planar mirrors, a choice of a LOF of the form $\rho_L(\vec{x}) \propto \cos(i\vec{k} \cdot \vec{x})$ corresponds to probing the level of squeezing in two propagation directions, symmetrically tilted at angle $\delta_k \approx |k|/k_z$ with the cavity axis.

In Figure 32 the spectrum S, changed of sign, is plotted as a function of the frequency ω and the modulus of the wavevector $\vec{k}$. Here signal detuning is $\theta_1 = -1.5$, and hence critical modes correspond to $l_d k_c = 1.5$. For each propagation direction δ_k (that is, for each value of the wavevector), the phase ϕ_L has been chosen to optimize the level of squeezing. A significant reduction of quantum noise below shot-noise level is shown by a large region of wave vectors around $k = k_c$, for zero frequency (Lugiato and Gatti, 1993; Gatti and Lugiato, 1995). The same plot can be read as describing the level of fluctuations on Gauss-Laguerre mode basis, by substituting the k vector variable with the variable $\eta/\kappa_1(2p + l)$ (Lugiato and Gatti, 1993).

From an experimental point of view, the issue of exactly mode-matching the LOF to the transverse modes of the cavity can be a very delicate one (LaPorta and Slusher, 1991). In this connection, two remarks are in order:

(i) All the information about the spatial structure of squeezed states is encoded in the correlation function (IV.15), and can be extracted by making a suitable functional trasformation (Gatti and Lugiato, 1995).

(ii) There is a special resonator geometry, where the level of achievable squeezing does not depend critically on the intensity distribution of the LOF that probes the field, namely the case of confocal geometry (Lugiato and Grangier, 1997). For a confocal resonator, all the Gauss-Laguerre modes with l even (or alternatively odd) are frequency degenerate. In the approximation of thin medium and plane-wave pump, it has been shown (Lugiato and Grangier, 1997) that by using an appropriate matching lens, the degree of squeezing that can be obtained is largely independent from the shape of the LOF, provided its spatial distribution is even (odd) for exchange $\vec{x} \to -\vec{x}$.

D. QUANTUM IMAGES IN THE FAR FIELD

The fundamental character of the twin photon emission is well known (see for example Hong *et al.*, 1987; Kwait *et al.*, 1995). In the case of degenerate parametric down-conversion of type I, the two photons of the pair are degenerate both in polarization and in frequency. However, recent analyses (Marzoli *et al.*, 1997; Lugiato, Gatti, Ritsch, *et al.*, 1997) have shown that the far field distribution is able to provide a clear spatial evidence of the emission of twin photons, and carries a spatial signature of the quantum nature of the correlation between the photons of the pair.

The discussion will be limited to the case of an OPO with spherical mirrors, in the *quasi-planar* configuration, when separation between adjacent Gauss-Laguerre modes η is much smaller than separation between longitudinal modes. Similar results hold also for an OPO with plane mirrors.

We start by considering the direct detection of photons in a plane at great distance z from the cavity. Our calculations show that the average intensity distribution $\langle B_1^{\dagger out}(\vec{x}) B_1^{out}(\vec{x}) \rangle$ has cylindrical symmetry around the cavity axis, and, well below threshold, corresponds to a broad peak, which does not show any privileged angle of emission for the signal photons.

More interesting features are displayed by the correlation function of the intensity fluctuations:

$$G(\vec{x}, \vec{x}', t) = \langle : \delta I(\vec{x}, t) \delta I(\vec{x}', 0) : \rangle \qquad (IV.22)$$

with

$$\delta I(\vec{x}, t) = B_1^{\dagger out}(\vec{x}, t) B_1^{out}(\vec{x}, t) - \langle B_1^{\dagger out}(\vec{x}, t) B_1^{out}(\vec{x}, t) \rangle \qquad (IV.23)$$

Figure 33 shows two plots of the function, obtained by fixing point $\vec{x}$ and varying point $\vec{x}'$, for zero time delay. In the near field (Fig. 33a) the correlation has a

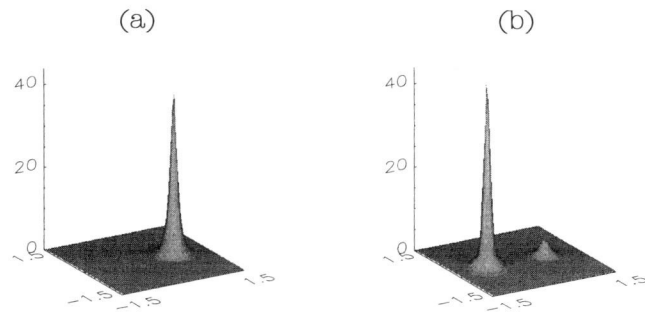

FIG. 33. Spatial correlation function of the intensity fluctuations $G(\vec{x}, \vec{x}', t = 0)$, plotted as a function of the second point $\vec{x}'$. (a) near field, (b) far field ($z = 200z_0$). 50% below threshold, and $\eta/\kappa_1 = 0.1$.

single high peak for $\vec{x}' = \vec{x}$, much narrower than the intensity distribution. Moving to the far field (Fig. 33b) G undergoes a dramatic transformation: the peak at $\vec{x}' = \vec{x}$ drops, while another peak grows on the other side with respect to the system axis, that is for $\vec{x}' = -\vec{x}$. This result gives clear evidence of the emission of twin photons, from a point of view of spatial fluctuations. Twin photons are produced with no special angle of emission; but if one of the two is emitted in some direction, its twin will propagate in a direction symmetrical with respect to the system axis, due to conservation of transverse momentum. Propagation at large distances separates the pair, and then in the far field, it is natural to find the maximum correlation between two points symmetrically located with respect to the z axis.

Even more important, it can be shown that this correlation is of quantum nature, in the sense that no classical field can produce a correlation function $G_z(\vec{x}, \vec{x}', 0)$ such that $G_z(\vec{x}, -\vec{x}, 0) > G_z(\vec{x}, \vec{x}, 0)$. The simple proof, based on the Cauchy-Schwartz inequality, can be found in (Lugiato, Gatti, Ritsch, et al., 1997), and it is the spatial counterpart of the proof that a classical field cannot show antibunching in time.

A relevant consequence of this shape of the correlation function in the far field is the existence of an observable with sub-shot-noise fluctuations (Marzoli et al., 1997). Let us consider two regions in the cross section of the beam, say S_1 and S_2, symmetrical with respect to the axis of the system. Let us focus on the total intensities detected in these two regions, and on their difference. We calculated the spectrum of fluctuations in the intensity difference, and showed that for zero frequency it is well below the shot-noise level. Provided that the system operates not too close to threshold, and for appropriate choices of the regions $S_{1,2}$, we found a quantum noise reduction up to 85% (Marzoli et al., 1997).

This result recalls the behavior of the spectrum of fluctuations in the signal and idler intensity difference (Reynaud *et al.*, 1987; Heidmann *et al.*, 1987) when considering a nondegenerate OPO. Also, in this case the reduction of fluctuations below shot noise indicates the high correlation in the process of parametric downconversion. Here, however, the correlation concerns not global (i.e., the total signal and idler intensities), but rather local observables, that is, the intensity of light collected from symmetrical regions in the transverse plane.

A similar phenomenon occurs in the case of a homodyne detection of the far field fluctuations (Lugiato, Gatti, Ritsch, *et al.*, 1997). Let us consider two pairs of noncommuting observables: X_1, Y_1, corresponding to two orthogonal quadratures of the signal field, detected in a region S_1 of the beam cross-section; and X_2, Y_2, corresponding to field quadratures detected in the symmetric and separated region S_2. For a suitable choice of the phase of the local oscillator field that probes the signal, it can be shown that fluctuations in the difference $X_1 - X_2$ and in the sum $Y_1 + Y_2$ can be reduced below shot-noise limit. The amount of noise reduction can range up to 80%. This result is not very sensitive either to the size of the detection regions, or to the waist of the Gaussian profile of the LOF, provided that the LOF phase profile matches the phase profile of Gauss-Laguerre modes in the detection plane (Lugiato, Gatti, Ritsch, *et al.*, 1997; Gatti *et al.*, in preparation).

Moreover, it can be shown that the quantum correlation existing between pairs of symmetrical regions in the transverse plane at large distance from the cavity is deeper than that implied by squeezing in the appropriate observables. Actually, variables measured in region S_1 are correlated to the ones measured in S_2 also in the Einstein-Podolsky-Rosen (1935) (EPR) sense; from a measurement of X_1, Y_1 in region S_1, it is possible to infer the values for variables X_2, Y_2 of the spatially separated region S_2 with a precision apparently violating Heisenberg rules (Gatti, Petsas, *et al.*, in preparation).

This result is again reminiscent of the EPR aspects demonstrated, theoretically (Reid and Drummond, 1988; Reid, 1988) and experimentally (Ou *et al.*, 1992), for a nondegenerate OPO below threshold. In that case, the two pair of conjugate variables were two orthogonal quadratures for a signal and an idler beam, distinguished by their polarization. In our case of degenerate, spatially extended OPO the "signal" and the "idler" beams correspond, respectively, to the field measured in regions S_1 and S_2.

From an applicative viewpoint, the analysis of the spatial aspects of squeezing and of the spatial properties of entangled photon pairs can lead to the exploration of the ultimate limit imposed by quantum noise on the visibility of images, or on the discernibility of edges. Research in this direction is pursued by Saleh (1997), and Fabre (in preparation). Another relevant topic is the "noiseless" amplification of images, that is, amplification that does not degrade the signal-to-noise ratio (Kolobov and Lugiato, 1995).

A recent paper discusses spatial quantum noise in laser diodes (Poizat *et al.*, submitted).

E. FINAL CONSIDERATIONS ON QUANTUM ASPECTS

In Sections IV.B–IV.D we have analyzed some quantum phenomena arising in the OPO below the threshold. However, interesting above threshold quantum phenomena also appear.

In the case of the roll pattern formation, the two signal beams in the far field, corresponding to the two spots in Figure 28d, are highly quantum correlated, and exhibit sub-shot-noise fluctuations in the intensity difference (Lugiato and Castelli, 1992), and EPR aspects (Castelli and Lugiato, 1997).

As below the threshold, these features arise from the perfect quantum correlation existing between the twin photons emitted in the down-conversion process. Above threshold, however, the mean number of signal photons is large, and we are in the presence of quantum effects in a macroscopic system. By introducing a breaking of the translational symmetry, one meets aspects related to the wave-particle duality (Lugiato and Grynberg, 1995). The presence of quantum correlation between the beams that form the far field has also been confirmed in the case of more complex structures, as the hexagonal patterns that form in $\chi^{(3)}$ media (Grynberg and Lugiato, 1993). A general description of these results can be found in (Lugiato, Gatti, and Weidemann, 1997).

Related results concerning propagation in cavityless configurations have been obtained in (Sibilia et al., 1994) for $\chi^{(2)}$ media and in (Nagasako et al., 1997) for $\chi^{(3)}$ media.

A recent experiment performed in the Laboratories of the Milano University at Como (Gatti, Lugiato, Oppo, et al., 1997) exhibits features closely related to the quantum images described in Section IV.B. In this experimental setting, picosecond pump pulses propagate in a sample of lithium triborate, where nondegenerate frequency down-conversion takes place. Despite the differences from the OPO case described in Section IV.A, the configuration of the signal field in the near field (Fig. 34b) and in the far field (Fig. 34a) are closely similar to those of

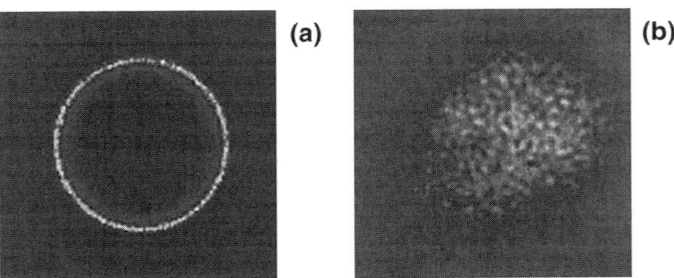

FIG. 34. Experimental results at Como Laboratories. (a) Single-shot far field intensity profile of the signal field, (b) corresponding near-field intensity distribution.

Figs. 28a or b. This similarity concerns only the classical aspects of quantum images; in order to substantiate the quantum nature of these phenomena it will be necessary to analyze the spatial correlation between signal and idler fields, along the lines of the discussion in Section IV.D.

A recent paper (Santagiustina *et al.*, in press) predicts very interesting noise-sustained patterns in a passive optical cavity containing a Kerr medium, when the input beam is slightly tilted with respect to the cavity axis. These patterns arise as a macroscopic manifestation of dynamically amplified quantum noise, with amplification factors up to 10^5.

V. Acknowledgments

We thank the ESPRIT Basic Research Program and the TMR Program of the European Union for their support to our project. We are grateful to all the colleagues who provided us with a wealth of information and/or contributed with figures included in this chapter.

VI. References

Ackemann, T. (1997). Drift instability and locking behavior of two-dimensional optical patterns. *Chaos Solitons and Fractals* (forthcoming).

Ackemann, T., Logvin, Yu. A., Heuer, A., and Lange, W. (1995). Transition between positive and negative hexagons in optical pattern formation. *Phys. Rev. Lett.* 75, 3450.

Ackhmanov, S., Vorontsov, M. A., and Ivanov, Y. (1988). Large-scale transverse nonlinear interaction in laser beams; new types of nonlinear waves; onset of "optical turbulence. *JETP Lett.* 47, 707.

Akhmanov, S. A. (1995). Information processing and nonlinear physics: From video pulses to waves and structures. In M. A. Vorontsov and W. B. Miller (Eds.), *Self-organization in optical systems and applications in information technology*. Springer-Verlag (Berlin); Vorontsov, M. A. (1990). Problems of large neurodynamics system modelling: Optical synergetics and neural networks. *SPIE Proc.* 1402, 116.

Akhmanov, S. A., Vorontsov, M. A., Ivanov, V. Yu., Larichev, A. V., and Zheleznykh, N. I. (1992). Controlling transverse-wave interactions in nonlinear optics: Generation and interaction of spatiotemporal structures. *J. Opt. Soc. Am. B* 9, 78–90.

Akhmanov, S. A., Khoklov, R. V., and Suchkorukov, A. P. (1972). Self-focusing, self-defocusing and self-modulation of laser beams. In F. T. Arecchi and E. O. Schultz-DuBois (Eds.), *Laser handbook* (p. 1151 ff.). North Holland (Amsterdam).

Allen, L., Beijenbergen, M. W., Spreeuw, R. J., and Woerdman, J. P. (1992). Orbital angular momentum of light and the transformation of Laguerre-Gaussian laser modes. *Phys. Rev. A* 45, 8185.

Arecchi, F. T. (1991). Space-time complexity in nonlinear optics. *Physics D* 51, 450–464.

Arecchi, F. T. (1994). Optical morphogenesis: Pattern formation and competition in nonlinear optics. *Il Nuovo Cimento A* 107, 1111.

Arecchi, F. T., Basti, G., Boccaletti, S., and Perrone, A. L. (1994). Adaptive recognition of a chaotic dynamics. *Europhys. Lett.* 26, 327.

Arecchi, F. T., Boccaletti, S., Giacomelli, G., Puccioni, G. P., Ramazza, P. L., and Residori, S. (1995). Boundary dominated versus bulk dominated regime in optical space time complexity. *Int. Journ. of Bifurc. and Chaos* 4, 1281.

Arecchi, F. T., Boccaletti, S., Mindlin, G. B., and Perez-Garcia, C. (1992). Periodic and chaotic alternation in systems with imperfect O(2) symmetry. *Phys. Rev. Lett.* 69, 3723–3726.

Arecchi, F. T., Boccaletti, S., Ramazza, P. L., Residori, S. (1993). Transition from boundary controlled to bulk controlled regimes in optical pattern formation. *Phys. Rev. Lett.* 70, 2277.

Arecchi, F. T., Giacomelli, G., Ramazza, P. L., and Residori, S. (1990). Experimental evidence of chaotic itinerancy and spatiotemporal chaos in optics. *Phys. Rev. Lett.* 65, 2531–2534.

Arecchi, F. T., Giacomelli, G., Ramazza, P. L., and Residori, S. (1991). Vortices and defect statistics in optical chaos. *Phys. Rev. Lett.* 67, 3749.

Baranova, N. B., Zel'dovich, B. Ya., Mamaev, A. V., Pilipetskii, N. F., and Shkukov, V. V. (1981). Dislocations of the wavefront of a speckle-inhomogeneous field (theory and experiment). *Sov. Phys. JETP Lett.* 33, 196.

Baristiy, I. V., Bazhenov, V. Yu., Soskin, M. S., and Vasnetsov, M. V. (1993). Optics of light beams with screw dislocations. *Opt. Comm.* 103, 422.

Barsella, A., Alcantara, P., Jr., Arimondo, E., Brambilla, M., and Prati, F. (1994). Dynamics of transverse patterns in a laser with saturable absorber. *Chaos, Solitons and Fractals* 4, 1665.

Berry, M. V. (1981). Singularities in waves and rays. In R. Balian *et al.*, (Eds.), *Physics of defects*. Les Houches Session XXXV (p. 453). North Holland, and references therein.

Berry, M. V., Nye, J. F., and Wright, F. J. (1979). The elliptic umbilic diffraction catastrophe. *Phil. Trans. Roy. Soc. Lond.* 291, 453.

Boccaletti, S. and Arecchi, F. T. (1995). Adaptive control of chaos. *Europhys. Lett.* 31, 127.

Boccaletti, S., Giaquinta, A., and Arecchi, F. T. (1997). Adaptive recognition and filtering of noise using wavelets. *Phys. Rev. E* 55, 5393.

Bonifacio, R. and Lugiato, L. A. (1978). Bistable absorption in a ring cavity. *Lett. Nuovo Cimento* 21, 505.

Boscolo, I., Bramati, A., Malvezzi, M., and Prati, F. (1997). Three-mode rotating pattern in a CO_2 laser with a high cylindrical symmetry. *Phys. Rev. A* 55, 738.

Boyd, R. W. (1992). *Nonlinear Optics*. Academic Press (Boston).

Brambilla, E., Brambilla, M., and Lugiato, L. A. (in preparation). Spatial and spatio-temporal instabilities in lasers with injected signals.

Brambilla, M., Battipede, F., Lugiato, L. A., Penna, V., Prati, F., Tamm, C., and Weiss, C. O. (1991). Transverse laser patterns. I. Phase singularity crystals. *Phys. Rev. A* 43, 5090.

Brambilla, M., Cattaneo, M., Lugiato, L. A., Pirovano, R., Pitzen, C., and Prati, F. (1992). Spatio temporal instabilities in nonlinear optical system. In R. Friedrich and A. Wunderlin (Eds.), *Evolution of dynamical structures in complex systems* (p. 83). Springer-Verlag (Berlin).

Brambilla, M., Cattaneo, M., Lugiato, L. A., Pirovano, R., Prati, F., Kent, A. J., Oppo, G.-L., Coates, A. B., Weiss, C. O., Green, C., D'Angelo, E. J., and Tredicce, J. R. (1994). Dynamical transverse laser patterns. I. Theory. *Phys. Rev. A* 49, 1427–1451.

Brambilla, M., Lugiato, L. A., and Stefani, M. (1996). Interaction and control of optical localised structures. *Europhys. Lett.* 34, 109; *Chaos* 6, 368.

Brambilla, M., Lugiato, L. A., Pinna, M. V., Prati, F., Pagani, P., Vanotti, P., Li, M. Y., and Weiss, C. O. (1992). The laser as nonlinear element for an optical associative memory. *Opt. Comm.* 92, 145–164.

Brambilla, M., Lugiato, L. A., Penna, V., Prati, F., Tamm, C., and Weiss, C. O. (1991). Transverse laser patterns II. Variational principle for pattern selection, spatial multistability and laser hydrodynamics. *Phys. Rev. A* 43, 5114–5120.

Brambilla, M., Lugiato, L. A., Prati, F., Spinelli, L., and Firth, W. J. (1997). Spatial soliton pixels in semiconductor devices. *Phys. Rev. Lett.* 79, 2042.

Carmichael, H. (1993). An open system approach to quantum optics. *Lecture Notes in Physics* m18. Springer (Berlin).

Castelli, F. and Lugiato, L. A. (1997). Realization of the Einstein-Podolsky-Rosen paradox in the far-field of the optical parametric oscillator above threshold. *J. Mod. Opt.* 44, 765.

Catchmark, J. M., Rogers, L. E., Morgan, R. A., Asom, M. T., Guth, G. D., and Christodoulides, D. N. (1996). Optical characteristics of multitransverse-mode two-dimensional vertical-cavity top surface-emitting laser arrays. *IEEE JQE* 32, 986–995.

Chan, J. and Lefever, R. (Eds.). (1995). Inhomogeneous phases and pattern formation. *Physica A* 213, 1–276.

Chang-Hasnain, C. J., Harbison, J. P., Hasnain, G., Von Lehmen, A. C., Florez, L. T., and Stoffel, N. G. (1991). Dynamics, polarization, and transverse mode characteristics of vertical cavity surface emitting lasers. *IEEE J. Quantum Electron.* 27, 1402–1409.

Chen, Z. and Abraham, N. B. (1995). Pattern dynamics in a bidirectional photorefractive ring resonator, *Appl. Phys. B* 60, 183.

Christov, C. I. and Velarde, M. G. (1995). Dissipative solitons. *Physica D* 86, 323.

Coates, A. B., Weiss, C. O., Green, C., D'Angelo, E. J., Tredicce, J. R., Brambilla, M., Cattaneo, M., Lugiato, L. A., Pirovano, R., Prati, F., Kent, A. J., and Oppo, G.-L. (1994). Dynamical transverse laser patterns. II. Experiments. *Phys. Rev. A* 49, 1452–1466.

Colet, P., San Miguel, M., Brambilla, M., and Lugiato, L. A. (1991). Fluctuations in transverse laser patterns. *Phys. Rev. A* 43, 3862.

de Colstoun, F. B., Khitrova, G., Fedorov, V. A., Nelson, T. R., Lowry, C., Brennan, T. M., Hammons, B. G., and Maker, P. D. (1994). Transverse modes, vortices and vertica-cavity surface-emitting lasers. *Chaos Solitons and Fractals* 4, 1575–1596.

Coullet, P., Gil, L., and Lega, J. (1989). Defect-mediated turbulence. *Phys. Rev. Lett.* 62, 1619.

Coullet, P., Gil, L., and Rocca, F. (1989). Optical vortices. *Opt. Comm.* 73, 403–407.

Crosignani, B., Segev, M., Engin, D., diPorto, P., Yariv, A., and Salamo, G. (1993). Self-trapping of optical beams in photorefractive media. *J. Opt. Soc. Am. B* 10, 446.

D'Alessandro, G. P. and Firth, W. J. (1992). Hexagonal spatial patterns for a Kerr slice with a feedback mirror. *Phys. Rev. A* 46, 537.

Dangel, S. and Holzner, R. (1997). Dynamics of light pattern formation for polarized laser beams in sodium vapor. *JOSA B* (forthcoming).

Dawling, T. P., Scully, M. O., and De Martini, F. (1991). Radiative patterns of a classical dipole in a cavity. *Opt. Comm.* 82, 415.

Degiorgio, V. and Scully, M. O. (1970). Analogy between the laser threshold region and a second order phase transition. *Phys. Rev. A* 2, 1170–1177.

Dewel, J., Borckmans, P., DeWit, A., Rudovics, B., Perraud, J.-J., Dulos, E., Boissonade, J., and DeKepper, P. (1995). Pattern selection and localized structures in reaction-diffusion systems. *Physics A* 213, 181 and references quoted therein.

Drummond, P. D. and Eberly, J. H. (1982). Transverse coherence and scaling in four-dimensional simulations of superfluorescence. *Phys. Rev. A* 25, 3446–3448.

Drummond, P. D., McNeil, K. J., and Walls, D. F. (1981). Non-equilibrium transitions in sub/second harmonic generation. II. Quantum theory. *Opt. Acta* 28, 211–225.

Drummond, P. D., McNeil, K. J., and Walls, D. F. (1980). Non-equilibrium transitions in sub/second harmonic generation. I. Semiclassical theory. *Opt. Acta* 27, 321.

Einstein, A., Podolsky, B., and Rosen, N. (1935). Can quantum-mechanical description of physical reality be considered complete? *Phys. Rev* 47, 777.

Esselbach, M., Kiessling, A., and Rehn, H. (1997). Transient phase measurement using a self-pumped phase conjugate mirror as an optical novelty filter. *JOSA B* 14, 846.

Etrich, C., Michaelis, W., Peschel, U., and Lederer, F. (1997). Nonlinear waves in intracavity vectorial SHG. *Chaos Solitons and Fractals* (forthcoming).

Etrich, C., Peschel, U., and Lederer, F. (1997). Solitary waves in quadratically nonlinear resonators. *Phys. Rev. Lett.* 79, 2454.

Fabre, C., in preparation.

Farjas, J., Hannequin, D., Dangoisse, D., and Glorieux, P. (1997). Symmetries in the transition to turbulence in optics. *Phys. Rev A,* forthcoming.

Fauve, S. and Thual, O. (1988). Localized structures generated by subcritical instability. *J. Phys. France* 49, 1829.

Firth, W. J. (1990). Spatial instabilities in a Kerr medium with a single feedback mirror. *J. Mod. Opt.* 37, 151.

Firth, W. J. (1991). Spontaneous spatial patterns in nonlinear optics. In M. Bertolotti and E. R. Pike (Eds.), *ECOOSA '90—quantum optics* (p. 173 ff.). Institute of Physics (Bristol).

Firth, W. J. (1995). Pattern formation in passive nonlinear optical systems. In M. Vorontsov and W. B. Miller (Eds.), *Self-organization in optical systems and application to information technology.* Springer-Verlag (Berlin).

Firth, W. J. and Lord, A. (1996). Two-dimensional solitons in a Kerr cavity. *J. Mod. Opt.* 43, 1071.

Firth, W. J. and Parè, C. (1988). Transverse modulational instabilities for counterpropagating beams in Kerr media. *Opt. Lett.* 13, 1096–1098.

Firth, W. J. and Scroggie, A. C. (1996). Optical bullet holes: Robust controllable localised states of a nonlinear cavity. *Phys. Rev. Lett.* 76, 1623.

Firth, W. J. and Scroggie, A. J. (1994). Spontaneous pattern formation in an absorbtive system. *Europhys. Lett.* 26, 521.

Firth, W. J. and Vorontsov, M. A. (1983). Adaptive phase distortion suppression in a nonlinear system with feedback mirrors. *J. of Mod. Opt.* 40, 1841.

Firth, W. J. Private communication.

Firth, W. J., Scroggie, A. J., McDonald, G. S., and Lugiato, L. A. (1992). Hexagonal patterns in optical bistability. *Phys. Rev. A* 46, 3609.

Fischer, I., Hess, O., Elsaesser, W., and Goebel, E. (1996). Complex spatio-temporal dynamics on a picosecond timescale in the near field of a broad area semiconductor laser. *Europhys. Lett.* 35, 579.

Garcia-Ojalvo, J., Vilaseca, R., and Torrent, M. C. (forthcoming). Coupled pattern formation near threshold in a broad-area cascade laser. *Phys. Rev. A.*

Gardiner, C. W. (1991). *Quantum noise.* Springer-Verlag (Berlin).

Gatti, A. and Lugiato, L. A. (1995). Quantum images and critical fluctuations in the optical parametric oscillator below threshold, *Phys. Rev. A* 52, 1675.

Gatti, A., Lugiato, L. A., Oppo, G.-L., Martin, R., DiTrapani, P., and Berzanskis, A. (1997). From quantum to classical images. *Optics Express* 1, 21.

Gatti, A., Petsas, K., Lugiato, L. A., and Marzoli, I., in preparation.

Gatti, A., Wiedemann, H., Lugiato, L. A., Marzoli, I., Oppo, G.-L., and Barnett, S. M. (1997). Langevin treatment of quantum fluctuations and optical patterns in optical parametric oscillators below threshold. *Phys. Rev. A* 56, 877–897.

Geddes, J. B., Moloney, J. V., Wright, E. M., and Firth, W. J. (1994). Polarization patterns in a nonlinear cavity. *Opt. Comm.* 111, 623.

Georgiou, M. and Mandel, P. (1994). Transverse effects in a laser with an injected signal. *Chaos, Solitons and Fractals* 4, 1657.

Gil, L., Emilsson, K., and Oppo, G.-L. (1992). Dynamics of spiral waves in a spatially inhomogeneous Hopf bifurcation. *Phys. Rev. A* 45, R567–R570.

Giusfredi, G., Valley, J. F., Pon, R., Khitrova, G., and Gibbs, H. M. (1988). Optical instabilities in sodium vapor. *J. Opt. Soc. Am. B* 5, 1181–1191.

Glebsky, L. Y. and Lerman, L. M. (1995). On small stationary localized solutions for the generalized Swift-Hohenberg equation. *Chaos* 5, 424.

Gorshkov, K. A., Korzinov, L. N., Rabinovich, M. I., and Tsimring, L. S. (1994). Random pinning of localized states and the birth of deterministic disorder within gradient models. *J. Stat. Phys.* 74, 1033.

Gourley, P. L., Warren, M. E., Hadley, G. R., Vawter, G. A., Brennan, T. M., and Hammons, B. E. (1991). Coherent beams from high efficiency two-dimensional surface-emitting semiconductor laser arrays. *Appl. Phys. Lett.* 58, 890–892.

Graham, R. and Haken, H. (1970). Laserlight—First example of a second order phase transition far away from thermal equilibrium. *Zeit. f. Phys.* 237, 31–46.

Grangier, P., Aspect, A., Heidmann, A., and Reynaud, S. (1986). Observation of photon antibunching in phase matched multiatom resonance fluorescence. *Phys. Rev. Lett.* 57, 687–690.

Green, C., Mindlin, G. B., D'Angelo, E. J., Solari, H. G., and Tredicce, J. R. (1990). Spontaneous symmetry breaking in laser: The experimental side. *Phys. Rev. Lett.* 65, 3124–3127.

Grynberg, G. and Lugiato, L. A. (1993). Quantum properties of hexagonal patterns. *Opt. Comm.* 101, 69–73.

Grynberg, G., Le Bihan, E., Verkerk, P., Simoneau, P., Leite, J. R. R., Bloch, D., Le Boiteux, S., and Ducloy, M. (1988). Observation of instabilities due to mirrorless four-wave mixing oscillation in sodium. *Opt. Comm.* 67, 363–366.

Grynberg, G., Maitre, A., and Petrossian, A. (1994). Flowerlike patterns generated by a laser beam transmitted through a rubidium cell with a single feedback mirror. *Phys. Rev. Lett.* 72, 2379.

Haelterman, M. and Sheppard, A. P. (1994). Polarization instability, multistability and transverse localized structures in Kerr media. *Chaos Solitons and Fractals* 4, 1731.

Haelterman, M. and Vitrant, G. (1992). Drift instability and spatiotemporal dissipative structures in a nonlinear Fabry-Perot resonator under oblique incidence. *J. Opt. Soc. Am. B* 9, 1563.

Haelterman, M., Trillo, S., and Wabnitz, S. (1991). Low dimensional modulational chaos in diffractive nonlinear cavities. *Opt. Comm.* 93, 343.

Haken, H. (1970). Laser theory. In L. Genzel (Ed.), *Enciclopaedia of Physics* XXV/2C. Springer-Verlag (Berlin).

Haken, H. (1977). *Synergetics, an introduction*. Springer-Verlag (Berlin).

Hannequin, D., Dambly, L., Dangoisse, D., and Glorieux, P. (1994). Weakly multimode dynamics of a photorefractive oscillator. *Journal de Physique III* 4, 2459.

Harkness, G. K., Lega, J., and Oppo, G-L. (1997). The effects of mirror curvature on the laser bifurcations. I. Amplitude equations close to threshold. II. Far from threshold transition to travelling waves. *Phys. Rev. A*, submitted.

Harkness, G. K., Lega, J., and Oppo, G.-L. (1996). Measuring disorder with correlation functions of averaged patterns. *Physica D*, 96, 26.

Harkness, G. K., Lega, J. C., and Oppo, G.-L. (1994). Correlation function in the presence of optical vortices. *Chaos Solitons and Fractals* 4, 1519.

He, K. F. and Gang, H. (1996). Feedback control of chaotic motions and unstable wave packets in a space-time dependent system. *Phys. Rev. E*, 53, 2271.

Heckenberg, N. R., Vaupel, M., Malos, J., and Weiss, C. O. (1996). Optical vortex pair creation and annihilation and helical astigmatism of a nonplanar ring resonator. *Phys. Rev. A* 54, 2369.

Heidman, A., Horowicz, R. J., Reynaud, S., Giacobino, E., Fabre, C., and Camy, G. (1987). Observation of quantum noise reduction on twin laser beams. *Phys. Rev. Lett.* 59, 2555.

Heidmann, A., Reynaud, S., and Cohen-Tannoudji, C. (1984). Photon noise reduction and coherence properties of squeezed fields. *Opt. Comm.* 52, 235.

Henry, C. H. (1992). Theory of the linewidth of semiconductor lasers. *IEEE J. Quantum Electron.* 18, 259.

Hess, O. (1994). Spatio-temporal complexity in multi-stripe and broad area semiconductor lasers. *Chaos, Solitons and Fractals* 4, 1597.

Hess, O. (1998). Controlling complex temporal and spatio-temporal dynamics in semiconductor lasers. *Chaos Solitons and Fractals* (forthcoming).

Hess, O. and Kahn, T. (1996). Maxwell Bloch equations for spatially inhomogeneous semiconductor lasers. *Phys. Rev. A* 54, 3347.

Hochheiser, D., Moloney, J. V., and Lega, J. (1997). Controlling optical turbulence. *Phys. Rev. A* 55, R4011.

Honda, T., Matsumoto, H., Sedlatschek, M., Denz, C., and Tschudi, T. (1997). Spontaneous formation of hexagons, squares and squeezed hexagons in a photorefractive phase conjugator with virtually internal feedback mirror. *Opt. Comm.* 133, 293.

Hong, C. K., Ou, Z. Y., and Mandel, L. (1987). Measurement of subpicosecond time intervals between two photons by interference. *Phys. Rev. Lett.* 59, 2044.

Horak, R., Bajer, J., Bertolotti, M., and Sibilia, C. (1995). Diffraction-free field in a planar nonlinear waveguide. *Phys. Rev. E,* 52, 4421.

Huffaker, D. L., Deppe, D. G., and Rogers, T. J. (1996). Transverse mode behavior in native-oxide-defined low threshold vertical-cavity lasers. *Appl. Phys. Lett.* 65, 1611–1613.

Huyet, G. and Rica, S. (1996). Spatio-temporal instabilities in the transverse patterns of lasers. *Physica D,* 215.

Huyet, G. and Tredicce, J. R. (1996). Spatio-temporal chaos in the transverse section of a laser. *Physica D,* 96, 209.

Huyet, G., Martinoni, C. M., Tredicce, J. R., and Abraham, N. B. (1995). Spatio-temporal dynamics of lasers with large Fresnel number. *Phys. Rev. Lett.* 75, 4027.

Huyet, G., Martinoni, M. C., Tredicce, J. R., and Rica, S. (1995). Spatiotemporal dynamics of lasers with large Fresnel number. *Phys. Rev. Lett.* 75, 4027.

Ikeda, K. I. (1979). Multiple-valued stationary state and its instability of the transmitted light by a ring cavity system. *Opt. Comm.* 30, 257.

Indebetouw, G. and Korwan, D. R. (1996). Off Bragg model for the photorefractive phase conjugate resonator. *Opt. Comm.* 129, 205.

Indebetouw, G. and Korwan, D. R. (1994). Model for vortices nucleation in a photorefractive phase conjugate resonator. *Journ. Mod. Opt.* 41, 941.

Indebetouw, G., and Liu, S. R. (1992). Defect-mediated spatial complexity and chaos in a phase-conjugate resonator. *Opt. Comm.* 91, 321–330.

Jackson, J. D. (1975). *Classical Electrodynamics.* Wiley (New York).

Jacobsen, P. K., Indik, R. A., Moloney, J. V., Newell, A. C., Winful, H. G., and Raman, L. (1991). Diode-laser array modes: discrete and continuous models and their stability. *J. Opt. Soc. Am. B* 8, 1674.

Jakobsen, P. K., Lega, J., Feng, Q., Staley, M., Moloney, J. V., and Wright, E. M. (1994). Nonlinear transverse modes of large aspect ratio, homogeneously broadened laser I. Analysis and numerical simulation. *Phys. Rev. A* 49, 4189–4200.

Jakobsen, P. K., Moloney, J. V., Newell, A. C., and Indik, R. (1992). Space-time dynamics of wide-gain-section lasers. *Phys. Rev. A* 45, 8129–8137.

Khanin, Ya. I. (1995). *Principles of laser dynamics.* Elsevier North-Holland (Amsterdam).

Kivshar, Y. S. and Yang, Xiaoping. (1994). Dynamics of dark solitons, *Chaos Solitons and Fractals* 4, 174.

Kolobov, M. (1991). Spatial-multimode broadband squeezing in both polarization components of an electromagnetic field. *Phys. Rev. A* 44, 1986–1994.

Kolobov, M. (1994). Quantum noise reduction in optical image processing. Role of the spatial coherence of the source. *Phys. Rev. A.*

Kolobov, M. and Kumar, P. (1993);. Sub-shot-noise microscopy: Imaging of paint phase objects with squeezed light. *Opt. Lett.* 18, 849–851.

Kolobov, M. and Lugiato, L. A. (1995). Noiseless amplification of optical images. *Phys. Rev. A* 52, 4930.

Kolobov, M. I. and Sokolov, I. V. (1989). Spatial behavior of squeezed states of light and quantum

noise in optical images. *Sov. Phys. JETP* 69, 1097; Squeezed states of light and quantum noise-free optical images. *Phys. Lett. A* 140, 101.
Kolobov, M. I. and Sokolov, I. V. (1991). Multimode squeezing, antibounching in space and noise-free optical images. *Europhys. Lett.* 15, 271.
Koppa, P., Chavel, P., Oudar, J. L., Kuszelewicz, R., Schnell, J. P., and Pocholle, J. P. (1997). Demonstration of optically controlled data routing with the use of multiple-quantum-well bistable elements and electro-optical devices. *Appl. Opt.* 36, 5706–5717.
Kreuzer, M., Gottschling, H., and Tschudi, T. (1991). *Mol. Cryst. Liq. Cryst.* 207, 219.
Kreuzer, M., Schreiber, A., and Thuering, B. (1996). Evolution and switching dynamics of solitary spots in nonlinear optical feedback systems. *Mol. Cryst. Liq. Cryst.*, 282.
Kwait, P. G., Mattle, K., Weinfurter, H., Zeilinger, A., Sergienko, A. V., and Shih, Y. (1995). New high-intensity source of polarization entangled photon pairs. *Phys. Rev. Lett.* 75, 4337.
La Penna, P. and Giusfredi, G. (1993). Spatiotemporal instabilities in a Fabry-Perot resonator filled with sodium vapor. *Phys. Rev. A* 48, 2299.
Lange, W., Logvin, Yu. A., and Ackemann, T. (1996). Spontaneous optical patterns in an atonic vapor: Observation and simulation. *Physica D* 96, 230.
LaPorta, A. and Slusher, R. E. (1991). Squeezing limit at high parametric gain. *Phys. Rev. A* 44, 2013–2022.
Law, C. T. and Swartzlander, G. A., Jr. (1994). Polarized optical vortex solitons: instabilities and dynamics in Kerr nonlinear media. *Chaos Solitons and Fractals* 4, 1759.
Le Berre, M., Leduc, D., Patrascu, S., Ressayre, E., and Tallet, A. (1997). Beyond the mean-field model of the ring cavity. *Chaos Solitons and Fractals,* forthcoming. Le Berre, M., Patrascu, S., Ressayre, E., and Tallet, A. (1997). Localised structures in chaotic patterns. From disorder to ordering. *Phys. Rev. A* 56, 3150–3160.
Le Berre, M., Patrascu, A. S., Ressayre, E., Tallet, A., and Zheleznyk, N. J. (1994). Spatial patterns in a passive ring cavity with atoms. *Chaos Solitons and Fractals* 4, 1389.
Lefever, R., Lugiato, L. A., Wang, K., Abraham, N. B., and Mandel, P. (1989). Phase dynamics of transverse diffraction patterns in the laser. *Phys. Lett. A* 135, 254–268.
Lega, J., Jakobsen, P. K., Moloney, J. V., and Newell, A. C. (1994). Nonlinear transverse modes of large aspect ratio homogeneously broadened laser. II: Pattern analysis near and beyond threshold. *Phys. Rev. A* 49, 4201–4212.
Leonardy, J., Kaiser, F., Belic, M. R., and Hess, O. (1996). Running transverse waves in optical phase conjugation. *Phys. Rev. A* 53, 4519.
Li, H., Lucas, T. L., McInerney, J., and Morgan, R. A. (1994). Transverse modes and patterns of electrically pumped vertical cavity surface emitting lasers. *Chaos Solitons and Fractals* 4, 1619–1636.
Lippi, G. L., Ackemann, T., Hoffer, L. M., Gahland, A., and Lange, W. (1993). Interplay of linear and nonlinear effects in the formation of optical vortices in a nonlinear resonator. *Phys. Rev. A* 48, R4043.
Liu, S. R. and Indebetouw, G. (1992). Periodic and chaotic spatiotemporal states in a phase-conjugate resonator using photorefractive $BaTiO_3$ phase-conjugate mirror. *J. Opt. Soc. Am. B* 9, 1507.
Logvin, Yu. A., Ackemann, T., and Lange, W. (1997). Subhexagons and ultrahexagons as a result of a secondary instability. *Phys. Rev. A* 55, 4538.
Logvin, Yu. A., Samson, B. A., Afanasèv, A. A., Samson, A. M., and Loiko, N. A. (1996). Triadic Hopf-static structures in two dimensional optical pattern formation. *Phys. Rev. A* 54, R4548.
Longhi, S. (1996). Travelling-wave states and secondary instabilities in optical parametric oscillators. *Phys. Rev. A* 53, 4488.
Longhi, S. (1997). Transverse patterns in a laser with injected signal. *Phys. Rev. A* 56, 2397.
Lovergnaux, E., Hennequin, D., Dangoisse, D., and Glorieux, P. (1996). Transverse mode competition in a CO_2 laser. *Phys. Rev. A* 53, 4435.

Lovergnaux, E., Slekys, G., Dangoisse, D., and Glorieux, P. (in press). Coupled longitudinal and transverse self organization in lasers induced by transverse mode locking. *Phys. Rev. A.*

Lu, W. and Harrison, R. G. (1994). Controlling chaos using continuous interference feedback: Proposal for all optical devices. *Opt. Comm.* 109, 457.

Lu, W., Yu, D., and Harrison, R. G. (1996). Control of patterns in spatio-temporal chaos in optics. *Phys. Rev. Lett.* 76, 3316.

Lugiato, L. A. (1992). Spatio-temporal structures. Part I. *Phys. Rep.* 219, 293–310.

Lugiato, L. A. (1994). Transverse nonlinear optics: Introduction and review. *Chaos Solitons and Fractals* 4, 1251–1258.

Lugiato, L. A. and Castelli, F. (1992). Quantum noise reduction in a spatial dissipative structure. *Phys. Rev. Lett.* 68, 3284.

Lugiato, L. A. and Gatti, A. (1993). Spatial structure of a squeezed vacuum. *Phys. Rev. Lett.* 70, 3868.

Lugiato, L. A. and Grangier, Ph. (1997). Improving quantum-noise reduction with spatially multimode squeezed light. *J. Opt. Soc. Am. B* 14, 225–231.

Lugiato, L. A. and Grynberg, G. (1995). Quantum picture of optical patterns: Complementarity and wave-particle aspects. *Europhys. Lett.* 29, 675–680.

Lugiato, L. A. and Lefever, R. (1987). Spatial dissipative structures in passive optical systems. *Phys. Rev. Lett.* 58, 2209–2211.

Lugiato, L. A. and Marzoli, I. (1995). Quantum spatial correlation in the optical parametric oscillator with spherical mirrors. *Phys. Rev. A* 52, 4886.

Lugiato, L. A. and Narducci, L. M. (1992). Multistability, chaos and spatiotemporal dynamics. In J. Dalibard, J. M. Raimond and J. Zinn-Justin (Eds.), *Fundamental systems in quantum optics* (p. 945). Les Houches Session LIII, Elsevier Science Publishers B.V.

Lugiato, L. A. and Oldano, C. (1988). Stationary spatial patterns in passive optical systems: Two level atoms. *Phys. Rev. A* 37, 3896.

Lugiato, L. A. and Strini, G. (1980). On the squeezing obtainable in parametric oscillators and bistable absorption. *Opt. Comm* 41, 67.

Lugiato, L. A., Gatti, A., and Wiedemann, H. (1997). Quantum fluctuations and nonlinear optical patterns. In S. Reynaud, E. Giacobino, and J. Zinn-Justin (Eds.), *Quantum fluctuations.* Elsevier North-Holland (Amsterdam).

Lugiato, L. A., Gatti, A., Ritsch, H., Marzoli, I., and Oppo, G.-L. (1997). Quantum images in nonlinear optics. *J. Mod. Opt.* 44 1899–1915.

Lugiato, L. A., Oldano, C., and Narducci, L. M. (1988). Cooperative frequency locking and stationary spatial structures in lasers. *J. Opt. Soc. Am. B* 5, 879–888.

Lugiato, L. A., Oppo, G.-L., Tredicce, J. R., Narducci, L. M., and Pernigo, M. A. (1990). Instabilities and spatial complexity in a laser. *J. Opt. Soc. Am.* 7, 1019–1033.

Lugiato, L. A., Oppo, G.-L., Pernigo, M. A., Tredicce, J. R., Narducci, L. M., and Bandy, D. K. (1988). Spontaneous spatial pattern formation in lasers and cooperative frequency locking. *Opt. Comm.* 68, 63–88.

Lugiato, L. A., Prati, F., Narducci, L. M., Ru, P., Tredicce, J. R., and Bandy, D. K. (1988). Role of transverse effects in laser instabilities. *Phys. Rev. A* 37, 3847–3866.

Lugiato, L. A., Wang, K., Abraham, N. B. (1994). Spatial pattern formation and instabilities in resonators with nonlinear dispersive media. *Phys. Rev. A* 49, 2049.

Macdonald, M. and Eichler, J. H. (1992). Spontaneous optical pattern formation in a nematic liquid crystal with feedback mirror. *Opt. Comm.* 89, 289.

McDonald, G. S. and Firth, W. J. (1993). Switching dynamics of spatial solitary wave pixels. *J. Opt. Soc. Am. B* 10, 1081–1089.

Malos, J., Vaupel, M., Staliunas, K., and Weiss, C. O. (1996). Dynamical structures in a photorefractive oscillator. *Phys. Rev. A* 53, 3559.

Mamaev, A. V. and Saffman, M. (1996a). Pattern formation in a linear photorefractive oscillator. *Opt. Comm.* 128, 281.

Mamaev, A. V. and Saffman, M. (1996b). Hexagonal optical patterns in anisotropic nonlinear media. *Europhys. Lett.* 34, 669.
Mamyshev, P. V., Villeneuve, A., Stegeman, G. I., and Aitchison, J. S. (1994). Steerable optical waveguides found by bright spatial solitons in AlGaAs. *Electron Lett.* 30, 726–727.
Mandel, P. (1997). *Theoretical problems in cavity nonlinear optics.* Cambridge University Press (Cambridge).
Mandel, P., Georghiou, M., and Erneaux, T. (1993). Transverse effects in coherently driven nonlinear cavities. *Phys. Rev. A* 47, 4277.
Manneville, P. (1990). *Dissipative structures and weak turbulence,* Academic Press (San Diego).
Martin, R., Kent, A. J., and Oppo, G.-L. (1996). Feasibility of controlling complex dynamics in multitransverse mode lasers. *Opt. Comm.* 130, 414.
Martin, R., Scroggie, A. J., Oppo, G.-L., and Firth, W. J. (1996). Stabilization, selection and tracking of unstable patterns by Fourier space techniques. *Phys. Rev. Lett.* 77, 4007.
Martin-Regalado, J., Balle, S., and San Miguel, M. (1997). Polarization and transverse mode dynamics of gain guided VCSELs. *Opt. Lett.* 22, 460–463.
Marzoli, I., Gatti, A., and Lugiato, L. A. (1997). Spatial quantum signatures in parametric downconversion. *Phys. Rev. Lett.* 78, 2092.
McDonald, G. S. and Firth, W. J. (1990). Spatial solitary-wave optical memory. *J. Opt. Soc. Am. B* 7, 1328.
McLaughlin, D. W., Moloney, J. V., and Newell, A. C. (1983). *Phys. Rev. Lett* 51, 75.
Mertens, S., Dewel, G., Borckmans, P., and Engelhardt, R. (1997). Pattern selection in bistable systems. *Europhys. Lett.* 37, 109.
Michaelis, D., Peschel, U., and Lederer, F. (1997). Multistable localized structures and superlattices in semiconductor optical resonators. *Phys. Rev. A* 56 R3366–R3369.
Michalzik, R., and Ebeling, K. J. (1995). Generalized BV diagrams for higher order transverse modes in planar vertical-cavity laser diodes. *IEEE JQE* 31, 1371–1379.
Moloney, J. V. and Gibbs, H. M. (1982). Role of diffractive coupling and self-focusing or defocusing in the dynamical switching of a bistable optical cavity. *Phys. Rev. Lett.* 48, 1607–1610.
Nagasako, E. M., Boyd, R. W., and Agarwal, G. S. (1997). Vacuum field induced filamentation in laser beam propagation. *Phys. Rev. A* 55 1412.
Neubecker, R., Oppo, G.-L., Thuering, B., and Tschudi, T. (1995). Pattern formation in a liquid-crystal valve with feedback, including polarization, saturation, and internal threshold effects. *Phys. Rev. A* 52, 791.
Neubecker, R., Thuering, B., and Tschudi, T. (1994). Formation and characterization of hexagonal patterns in a single feedback experiment. *Chaos Solitons and Fractals* 4, 1307.
Newell, A. C. and Moloney, J. V. (1992). *Nonlinear optics.* Addison-Wesley (Redwood City, CA).
Nicolis, G. (1995). *Introduction to nonlinear science.* Cambridge University Press (Cambridge).
Nicolis, G., and Prigogine, I. (1977). *Self-organization in nonequilibrium systems.* Wiley (New York).
Oppo, G.-L., Brambilla, M., and Lugiato, L. A. (1994). Formation and evolution of roll patterns in optical parametric oscillators. *Phys. Rev. A* 49, 2028.
Oppo, G.-L., D'Alessandro, G. P., and Firth, W. J. (1991). Spatiotemporal instabilities of lasers in models reduced via center manifold techniques. *Phys. Rev. A* 44, 4712–4720.
Orenstein, M., Kapon, E., Harbison, J. P., Florez, L. T., and Stoffel, N. G. (1992). Large two-dimensional arrays of phase-locked vertical cavity surface emitting lasers. *Appl. Phys. Lett.* 60, 1535–1537.
Orenstein, M., Kapon, E., Stoffel, N. G., Harbison, J. P., Florez, L. T., and Wullert, J. (1991). Two-dimensional phase-locked arrays of vertical-cavity semiconductor lasers by mirror reflectivity modulation. *Appl. Phys. Lett.* 58, 804–806.
Otsuka, K. (1990). Self-induced turbulence and chaotic itineracy in coupled laser systems. *Phys. Rev. Lett.* 65, 329.
Ott, E., Grebogi, C., and Yorke, J. A. (1990). Controlling chaos. *Phys. Rev. Lett.* 64, 1196.

Ou, Z. Y., Pereira, S. F., and Kimble, H. J. (1992). Realization of the Einstein-Podolsky-Rosen paradox for continuous variables in nondegenerate parametric amplification. *Appl. Phys. B* 55, 265–278.

Ouazzardini, A., Adachihara, H., Moloney, J. V., McLaughlin, D. W., and Newell, A. C. (1988). Spontaneous spatial symmetry breaking in passive optical feedback systems. *J. Phys.* Coll C2, Suppl. 6, 455.

Pampaloni, E., Residori, S., and Arecchi, F. T. (1993). Roll-hexagon transition in a Kerr-like experiment. *Europhys. Lett.* 24, 647.

Papoff, F., D'Alessandro, G., Oppo, G.-L., and Firth, W. J. (1993). Local and global effects of boundaries on optical-pattern formation in Kerr media. *Phys. Rev. A* 48, 634.

Petrossian, A., Dambly, L., and Grynberg, G. (1995). Drift instability for a laser beam transmitted through a rubidium cell with feedback mirror. *Europhys. Lett.* 29, 209.

Prati, F., Brambilla, M., and Lugiato, L. A. (1994). Pattern formation in lasers. *La Rivista del Nuovo Cimento* 17, 1.

Prati, F., Tissoni, G., San Miguel, M., and Abraham, N. (1997). Vector vortices and polarization state of low-order transverse modes in a VCSEL. *Opt. Commun.* 143, 133–146.

Prati, F., Vecchione, D., and Vendramin, G. (1977). Frequency locking of supermodes and stability of the out-of-phase locked state in one-dimensional and two-dimensional arrays of vertical-cavity surface-emitting lasers. *Opt. Lett.* 22, 1633.

Prati, F., Zucchetti, L., and Molteni, G. (1995). Rotating patterns in class-B lasers with cylindrical symmetry. *Phys. Rev. A* 51, 4093.

Ramazza, P. L., Bigazzi, B., Pampaloni, E., Residori, S., and Arcchi, F. T. (1995). One dimensional transport-induced instabilities in an optical system with nonlocal feedback. *Phys. Rev. E* 52, 5524.

Rehn, H. and Kowarschik, R. (1996). Transverse optical structures in a ring resonator with internal phase coupling. *Optics Lett.* 21, 1505.

Reid, M. D. (1988). Demonstration of the Einstein-Podolsky-Rosen paradox using nondegenerate parametric amplification. *Phys. Rev. A* 40, 913.

Reid, M. D. and Drummond, P. D. (1988). Quantum correlation of phase in nondegenerate parametric oscillations. *Phys. Rev. Lett.* 60, 2731–2733.

Reynaud, S., Fabre, C., and Giacobino, E. (1987). Quantum fluctuations in a two mode parametric oscillator. *J. Opt. Soc. Am. B* 4, 1520–1524.

Roehricht, B., McCord, A. W., Brambilla, M., Prati, F., Dangel, S., Eschle, P., and Holzner, R. (1995). Spatial separation of circularly polarized laser beams in sodium vapor. *Optics Comm.* 118, 601.

Rosanov, N. N. (1990). Diffractive autosolitons in nonlinear interferometers. *J. Opt. Soc. Am. B* 7, 1057.

Rosanov, N. N. (1996). Transverse patterns in wide-aperture nonlinear optical systems. In E. Wolf (Ed.), *Progress in Optics XXXV* (pp. 1–60).

Rosanov, N. N., Smirnov, V. A., and Vyssotina, N. V. (1994). Numerical simulations of interaction of bright spatial solitons in medium with saturable nonlinearity. *Chaos Solitons and Fractals* 4, 1767.

Saleh, B. (forthcoming). Quantum imaging. In *Proceedings of the fifth international conference on squeezed states and uncertainty relations.* Balatonfured, May 1997.

Samson, B. A. (1996). Spatiotemporal dynamics of a thin-layer laser. *Phys. Rev. A* 53, 517.

Samson, B. A. and Vorontsov, M. A. (1997). Localized states in nonlinear optical system with binary-phase slice and feedback mirror. *Phys. Rev. A* 56, 1621–1626.

Santagiustina, M., Colet, P., San Miguel, M., and Welgraev, D. (in press). Noise sustained convective structure. *Phys. Rev. Lett.*

Schreiber, A., Thuring, B., Kreuzer, M., and Tschudi, T. (1997). Experimental investigation of solitary structures in a nonlinear optical feedback system. *Opt. Comm.* 136, 415.

Scroggie, A. J. and Firth, W. J. (1996). Pattern formation in an alkali metal vapor with a feedback mirror. *Phys. Rev. A* 53, 2752.

Scroggie, A. J., Firth, W. J., McDonald, G. S., Tlidi, M., Lefever, R., and Lugiato, L. A. (1994). Pattern formation in a passive ring cavity. *Chaos Solitons and Fractals* 4, 1323.
Sibilia, C., Schiavone, V., Bertolotti, M., Horak, R., and Perina, I. (1994). Nonclassical spatial properties of light propagation in dissipative nonlinear waveguides. *J. Opt. Soc. Am. B* 11, 2175.
Smith, C. P., Dihardja, Y., Weiss, C. O., Lugiato, L. A., Prati, F., and Vanotti, P. (1993). Low energy switching of laser doughnut modes and pattern recognition. *Opt. Comm.* 102, 505–514.
Staliunas, K. (1993). Laser Ginzburg-Landau equation and laser hydrodynamics. *Phys. Rev. A* 48, 1573–1581.
Staliunas, K. and Sanchez-Morcillo, V. J. (1996). Localized structures in degenerate optical parametric oscillators. *Opt. Comm.* 139, 306.
Swartzlander, G. A., Jr. and Law, C. T. (1992). Optical vortex-solitons observed in Kerr nonlinear media. *Phys. Rev. Lett.* 69, 2503.
Tai, K., Lai, Y., Huang, K. F., Huang, T. C., Lee, T. D., and Wu, C. C. (1993). Transverse mode emission characteristics of gain-guided surface emitting lasers. *Appl. Phys. Lett.* 63, 2624–2626.
Tamburrini, M., Bonavita, M., Wabnitz, S., and Santamato, E. (1993). *Opt. Lett.* 18, 855.
Tamburrini, M., Ciaramella, F., and Santamato, E. (1994). Hexagonal beam filamentation in a liquid crystal film with single feedback mirror. *Chaos Solitons and Fractals* 4, 1355.
Tamm, C. H. (1988). Frequency locking of two transverse optical modes of a laser. *Phys. Rev. A* 38, 5960.
Taranenko, W. B., Staliunas, K., and Weiss, C. O. (1997). Spatial soliton laser: Localized structures in a laser with a saturable absorber in a self-imaging resonator. *Phys. Rev. E* 56, 1582.
Thuering, B., Kreutzer, M., Schreiber, A., and Tschudi, T. (1996). In *EQEC96, Proceedings of the 1996 European quantum electronic conference* (Hamburg). European Physical Society (Geneva).
Tlidi, M. and Mandel, P. (1994). Spatial patterns in nascent optical bistability. *Chaos Solitons and Fractals* 4, 1475.
Tlidi, M., Georghiou, M., and Mandel, P. (1993). Transverse patterns of nascent optical bistability. *Phys. Rev. A* 48, 4605.
Tlidi, M., Lefever, R., and Mandel, P. (1996). *Quantum Semicl. Opt.* 8, 931.
Tlidi, M., Mandel, P., and Lefever, R. (1994). *Phys. Rev. Lett.* 73, 64.
Tlidi, M., Mandel, P., and Lefever, R. (1994). Localized structures and localized patterns in optical bistability. *Phys. Rev. Lett.* 73, 640.
Tredicce, J. R., Quel, E. J., Ghazzawi, A. M., Green, C., Pernigo, M. A., Narducci, L. M., and Lugiato, L. A. (1989). Spatial and temporal instabilities in a CO_2 laser. *Phys. Rev. Lett.* 62, 1274–1277.
Tsimring, L. S. and Aranson, I. (1997). Cellular and localized structures in a vibrated granular layer. *Phys. Rev. Lett.* 79, 213.
Turing, A. M. (1952). *Phyl. Trans. R. Soc. London B* 237, 37.
de Valcarcel, G. J., Staliunas, K., Roldan, E., and Sanchez-Morcillo, V. J. (1996). Transverse patterns in degenerate optical parametric oscillation and degenerate four-wave mixing. *Phys. Rev. A* 54, 1609–1624.
Valle, A., Sarma, J., and Shore, K. A. (1995). Dynamics of transverse mode competition in vertical cavity surface emitting laser diodes. *Opt. Commun.* 115, 297–302.
Vitrant, G. and Danckaert, J. (1994). Suppression of modulational instabilities by optical transport in nonlinear optical resonators. *Chaos Solitons and Fractals* 4, 1369.
Vorontsov, M. A. (1993). Akhseals: New class of spatio-temporal instabilities of optical fields. *Quantum Electron.* 23, 269.
Vorontsov, M. A., Iroshnikov, N. G., and Abernathy, R. L. (1994). Diffractive patterns in a nonlinear optical two dimensional feedback system with field rotation. *Chaos Solitons and Fractals* 4, 1701.
Wang, K., Abraham, N. B., and Lugiato, L. A. (1993). Leading role of optical phase instabilities in the formation of certain laser transverse patterns. *Phys. Rev. A* 47, 1263.

Wang, S. S. and Winful, H. G. (1988). Dynamics of phase locked semiconductor laser arrays. *Appl. Phys. Lett.* 52, 1774–1776.

Weiss, C. O. (1992). Spatio-temporal structures. Part II. Vortices and defects in lasers. *Phys. Rep.* 219, 311–328.

Weiss, C. O. (1996). ESPRIT LTR project PASS 21.112 report.

Wilkowski, D., Hannequin, D., Dangoisse, D., and Glorieux, P. (1994). Multistability and periodic alternance in a multimode CO_2 laser with a saturable absorber. *Chaos, Solitons and Fractals* 4, 1683.

Woerdman, J. P., private communication.

Yariv, A. (1989). *Quantum electronics* (3rd ed., pp. 136–154). Wiley (New York).

Yoo, H.-J., Hayes, J. R., Paek, E. G., Scherer, A., and Kwon, Y.-S. (1990). Array mode analysis of two-dimensional phased arrays of vertical cavity surface emitting lasers. *IEEE JQE* 26, 1039–1051.

Yu, Dejin, Lu, W. and Harrison, R. G. (1996). Origin of spiral wave formation in excitable optical systems. *Phys. Rev. Lett.* 77, 5051.

Zhao, Y. and McInerney, J. (1996). Transverse-mode control of vertical-cavity surface-emitting lasers. *IEEEE JQE* 32, 1950–1958.

Zheleznikh, N. I., Le Berre, M., Ressayre, F., and Tallet, A. (1994). Rotating spiral waves in a nonlinear optical system with spatial interaction. *Chaos Solitons and Fractals* 4, 1717.

Index

Active vertical cavity resonator (AVCR), 276
ALEPH detector, 43
Annulus of unstable wave vectors, 241
Antibunching in space–time, 287–288
Anti-Stokes emission. *See* Laser cooling of solids, anti-Stokes
ASTRID ring, 115, 117
  bunched beam cooling, 140
  laser cooling experiments using $Li+$ ions at, 136
Atomic beam magnetic resonance method (Berkeley experiment), 26–31
Atomic effects caused by nuclear magnetic moment, 24–25
Atom trapping and cooling, 25–26
A–V (pseudovector–vector), 21

$Be+$ ions, laser cooling experiments using, 137–138
Berkeley experiment (atomic beam magnetic resonance method), 26–31
Bernoulli equation, 253
Bifurcation analysis, 244
Bohr mass shift corrections, 83–88
Bohr-Weisskopf effect, 147, 154
Breit-Schawlow correction, 147
Bunched beams, cooling of, 139–142

Cabbibo–Kobayashi–Maskawa (CKM) matrix, 8
Capture range, cooling with improved, 138–139
Cauchy-Schwartz inequality, 292
Clebsch-Gordan coefficient, 27
Coefficient of performance (COP), 169, 171
  refrigerator, 206, 208, 209–210, 212, 214
Cone of unstable wave vectors, 241
Configuration interaction (CI) method, 58, 62–63, 68
  core-excited doublet states and, 77, 79
  first ionization potential and, 90
  highly excited quartet states and, 79
  oscillator strengths and, 102
  relativistic corrections and, 80
Configuration interaction-Hylleraas (CI-HY) approach, 58, 63, 68–69
  core-excited doublet states and, 77
  first ionization potential and, 90
  lower bound estimates for E_{NR} and, 71
  oscillator strengths and, 102
Cooperative frequency locking, 263
CP violation, 2, 7
  left-right symmetric model and, 13
  multi-Higgs model and, 12
  standard model and, 8–9
  supersymmetric model and, 11
  weak dipole moment and tau leptons, 39–44
Critical circle, 241
Cross-Newell-Whitehead equations, 268
CRYRING, 115, 127–128
CRYSTAL storage rings, 133

Dirac EDM Hamiltonian, 7, 14
Dirac-Fock calculations, 154
Dirac wave-function, 16–17
Doppler effect
  capture range and, 138
  measurement of HFS splitting, 149–150
  transverse, 142–144
Doppler formula, 120–121
Doppler-free two-step laser-induced recombination, 131
Doppler limit, 132
Dye solutions, laser cooling and organic, 178–181

Einstein-Podolsky-Rosen (EPR), 293
Electric dipole moments (EDMs)
  magnetic and EDM of neutrinos, 44–48
  nucleon, 4
  research on, 1–6
  valence electron, 4

Electric dipole moments, electron
  atomic effects caused by nuclear magnetic moment, 24–25
  calculation of enhancement factor for paramagnetic atoms, 15–21
  detection of, 3–4
  experimental searches for, 25–31
  paramagnetic molecules and, 32–36
  P,T-odd electron–nucleon interaction, 21–24
  Schiff's theorem, 14–15
Electric dipole moments, lepton
  detection of, 5–6
  left-right symmetric models, 12–13
  multi-Higgs models, 12
  proper-Lorentz-invariant interaction, 6–7
  standard model, 7–10
  supersymmetric models, 10–12
  weak dipole moment and tau, 39–44
Electric dipole moments, muon
  practicality problems with muonic atoms, 36–37
  precession of muons in storage ring, 37–39
Electric dipole transition, oscillator strengths and, 101–102
Electroluminescence, thermodynamics of, 167–170
Electron-beam ion sources (EBIS), 117
Electron-beam ion trap (EBIT), 117, 152
Electron cooler laser experiments
  laser-induced recombination (LIREC) in, 122–125
  laser-induced recombination using a single laser step, 125–128
  two-step laser-induced recombination, 128–131
Electron-cyclotron-resonance (ECR), 117
Electrons. *See* Electric dipole moments, electron
Enhancement factor for paramagnetic atoms, calculation of, 15–21
ESR ring, 115, 129–131
  measurement of HFS splitting of hydrogen in, 149–152

Faraday rotation, 36
Field shift contribution, 84
First ionization potential, calculation of, 90–93

Flux temperatures, 206, 210
Fokker-Planck equation, 282
Franck-Condon factors, 176

Gallium arsenide, laser cooling and, 181–184
Gases, laser cooling and, 175–177
Gaugino, 11
Gaunt factor, 123, 124
Gauss-Hermite modes, 257
Gauss-Laguerre functions, 232, 252, 257
  cavities with spherical mirrors, 259–265
Ginzburg-Landau equations, 232, 255, 270
g-2 experiment, 3, 25

Hamiltonian
  See also Lithium (Li) atom, nonrelativistic Hamiltonian and
  Dirac EDM, 7, 14
  one-electron, 22
  P,T-odd, 21–22
  relativistic Breit-Pauli, 80–82
  scalar electronic, 23
Hartree-Fock method, 18, 19, 20
  See also Multiconfiguration Hartree-Fock (MCHF) methods
Heinsenberg operators, 280, 293
Higgs models, multi, 12
Hills-type differential equation, 119
Hund's case, 32
Hydrodynamics
  dry, 252
  laser, 252–253
  optical pattern formation and analogy with, 252–259
Hylleraas (HY) approach, 58–60, 63, 69, 72
  first ionization potential and, 90
  lower bound estimates for E_{NR} and, 71
  oscillator strengths and, 102
  relativistic corrections and, 80, 82
Hyperfine coupling constants, 93–100

Intrabeam scattering, 138
Ives-Stilwell experiment, 142–144

Kennedy-Thorndike experiment, 142
Kerr medium model, 244, 247–248
Kinematic effects in storage rings, 117–122

Kramers' doublets, 188
Kuramoto-Shivashinsky equations, 232

Lagrangian density, 6
Lamb shift correction, 88, 146
LAMPF, 144
Langevin approach, quantum fluctuation and, 282–285
Laser cooling
  bunched beams, cooling of, 139–142
  capture range cooling, 138–139
  experiments using $Be+$ and $Mg+$ ions, 137–138
  experiments using $Li+$ ions at ASTRID, 136
  experiments using $Li+$ ions at TSR, 133–136
  principles of, in storage ring, 131–133
  transverse velocity component cooling, 138
Laser cooling of solids, anti-Stokes
  basic concepts, 161–163
  comparison to laser Doppler cooling, 164–165
  experimental results, summary of, 163–164
  flux versus brightness temperatures, 173
  future for, 221–223
  gases and, 175–177
  maximum cooling power and optical refrigerators, 217–221
  minimum temperature, 215–217
  organic dye solutions and, 178–181
  rare-earth ions and, 186–188
  ruby and, 184–186
  semiconductors and, 181–184
  thermodynamics and the limits of, 204–214
  thermodynamics of electro-luminescence, 167–170
  thermodynamics of photoluminescence, 170–175
  of Ytterbium-doped ZBLANP glass, 174, 188–204
  viability of, 165–167
Laser Doppler cooling, 164–165
Laser-induced recombination (LIREC)
  in electron cooler, 122–125
  experiments with TSR, 125–127
  at reduced electron temperatures, 127–128
  two-step, 128–131
  using a single laser step, 125–128
Laser spectroscopy. See Storage ring laser spectroscopy
Left-right symmetric models, 12–13
Lepton electric dipole moments
  detection of, 5–6
  left-right symmetric models, 12–13
  multi-Higgs models, 12
  proper-Lorentz-invariant interaction, 6–7
  standard model, 7–10
  supersymmetric models, 10–12
  weak dipole moment and tau, 39–44
Lifetimes, lithium, 103
Linear wave equation, 255, 257–258
Liquid crystal light valves (LCLV), 252, 268, 275
Lithium (Li) atom, calculations of
  background research, 57–58
  computational approaches, 58–63
  convergence considerations, 62–63
  first ionization potential, 90–93
  future for, 103–105
  hyperfine coupling constants, 93–100
  lifetimes, 103
  mass shift corrections, 83–88
  mathematical issues, 63–67
  optimization algorithm, use of global, 61
  optimization procedures, 61
  orbital exponent parameters, selection of, 60–61
  oscillator strengths, 101–102
  polarizabilities, 100–101
  quantum electrodynamic corrections, 88–90
  relativistic Breit-Pauli Hamiltonian, 80–82
  term energies for lower-lying doublet states, 91–93
  transition isotope shifts, 84, 86–88
Lithium (Li) atom, nonrelativistic Hamiltonian and
  computational approaches used, 68–69
  core-excited doublet states, 76–79
  distribution functions, 72
  estimates of low-lying doublet states of, 70–71

Lithium (Li) atom (*continued*)
  excited states, 72–80
  formula, 67–68
  ground states, 68–72
  highly excited quartet states, 79–80
  lower bound estimates for E_{NR}, 69, 71–72
  low-lying excited doublet states, 72–74
  low-lying quartet states, 74–76
  $Li+$ ions, laser cooling experiments using
    at ASTRID, 136
    at TSR, 133–136
Long structures (LS). *See* Optical pattern formation, spatial solitons and
Lorentz-invariant interaction, proper, 6–7
Lorentz transformation, 120, 142–143

$Mg+$ ions, laser cooling experiments using, 137–138
Magnetic resonance spectroscopy, 138
Many-body perturbation theory (MBPT), 20, 58
  first ionization potential and, 90
  oscillator strengths and, 102
Mass polarization correction, 83
Mass shift corrections, lithium atoms and, 83–88
Maxwell-Bloch equations (MBE), 233–239
Michelson-Morley experiment, 142
Minimal supersymmetric standard model (MSSM), 11
Multiconfiguration Hartree-Fock (MCHF) methods, 58, 63
  first ionization potential and, 90
  oscillator strengths and, 102
Multi-Higgs models, 12
Multiphonon absorption, 215–216
Muon electric dipole moments
  practicality problems with muonic atoms, 36–37
  precession of muons in storage ring, 37–39

NAP-M proton storage ring, 133
Neutrinos, magnetic and EDMs of, 44–48
Newell-Whitehead equations, 232
Nonlinear Schrödinger (NLS) equation, 244
Nonlinear stability analysis, 244

OPAL detector, 42–43
Optical parametric amplification (OPA), 281
Optical parametric oscillator (OPO) example, 245–247
Optical pattern formation (OPF)
  analogy with hydrodynamics, 252–259
  applications of, 229–230, 232–233, 265–268
  cavities with spherical mirrors, 259–265
  configurations used in studying, 231
  cooperative frequency locking, 263
  far and near field patterns, 241–243
  Kerr medium example, 244, 247–248
  laser example, 244–245, 259–264
  Maxwell-Bloch equations (MBE), 233–239
  optical parametric oscillator (OPO) example, 245–247
  photorefractive materials and, 268–269
  polarization coupling mechanisms and, 269
  radiation field defects, 255–257
  research on, 230–233, 268–269
  semiconductors and, 269
  single feedback mirror systems, 249–252
  spatial multistability and control, 265–268
  systems with translational and rotational symmetry, 239–249
Optical pattern formation, quantum fluctuations and
  future of, 294–295
  images in the far field, 291–293
  images in the near field, 285–291
  Langevin approach to, 282–285
  models for X^2 media including diffraction, 279–282
  research on, 278–279
  spatial structure of squeezed states, 289–291
Optical pattern formation, spatial solitons and, 270
  control of spatial solitons, 274–275
  exciting long structures and, 271–275
  localized structures in optics, 270–271
  semiconductor models and, 276–278
Optical vortices, 254
Oscillator strengths, 101–102

Paramagnetic atoms, calculation of enhancement factor for, 15–21
Paramagnetic molecules, EDMs of, 32–36
Parity nonconservation (PNC), 22–23
Passive vertical cavity resonator (PVCR), 276
Pauli spin operator, 2, 6, 22
Phase singularity, 254, 262
Photoluminescence, thermodynamics of, 170–175
Photorefractive materials, optical pattern formation and, 268–269
Photo-refrigerating effect, 175–176
Photothermal deflection spectroscopy, 191–197
Planck scale, 11
Plane wave approximation, 230
Polarizabilities, lithium, 100–101
Polarization coupling mechanisms, optical pattern formation and, 269
Poynting vector, 254–255
Proper-Lorentz-invariant interaction, 6–7
PS–S (pseudoscalar–scalar), 21
P,T-odd electron–nucleon interaction, 21–24

Quantum electrodynamics (QEDs), corrections, 88–90, 115
Quantum electrodynamics (QEDs), ground-state hyperfine splitting of hydrogen-like ions and
  background information, 146–149
  experimental technique in ESR, 149–152
  nuclear contributions to, 154–155
  spectroscopy of infrared transition, 152–154
Quantum fluctuations. *See* Optical pattern formation, quantum fluctuations and
Quantum Monte Carlo (QMC) calculations, 102

Radial equations, coupled, 17
Raman scattering, lattice, 215, 217
Rare-earth ions, laser cooling and, 186–188
Refrigerators
  general, 205–206
  maximum cooling power and optical, 217–221

model three-level optical, 210–214
simplification of three-level optical, 208–210
three-level optical, 206–208
Resonant laser ionization, 114
RFQ ring, 133
Ring pattern, 241
Ruby, laser cooling and, 184–186
Rydberg states, 123, 125, 128, 129

Saturation spectroscopy, 115
Schiff's theorem, 4, 14–15
Schrödinger equation, 83, 231
  nonlinear (NLS), 244
Semiconductors
  laser cooling and, 181–184
  optical pattern formation and, 269
  spatial solitons and, 276–278
Spatial multistability and control, 265–268
Spatial solitons (SS). *See* Optical pattern formation, spatial solitons and
Spatial structure of squeezed states, 289–291
Special relativity, storage rings and
  experimental setup at TSR, 144–145
  results, 145–146
  test schemes for the theory of, 142
  transverse Doppler-effect analysis, 142–144
Spherical mirrors, cavities with, 259–265
S–PS (scalar–pseudoscalar), 21
Standard model, EDM 7–10
Stark effect, 101
  linear, 3
  quadratic, 3
Stefan-Boltzmann constant, 200
Sternheimer equation, 17
Stokes lines, 165
Stokes' rule or law, 165
Storage ring laser spectroscopy
  ASTRID ring, 115, 117
  CRYRING, 115, 127–128
  ESR ring, 115, 129–131, 149–152
  history of research on, 115
  kinematic effects in rings, 117–122
  laser cooling, 131–142
  laser-induced recombination (LIREC) in electron cooler, 122–125

Storage ring laser spectroscopy (*continued*)
  laser-induced recombination using a single laser step, 125–128
  outlook, 155–156
  properties of existing heavy-iron rings, 115–117
  quantum electrodynamics and, 146–155
  special relativity and, 142–146
  TSR ring, 115, 125–127, 129–131, 133–136, 140, 141, 144–146
  tune of ring, 119
  two-step laser-induced recombination, 128–131
Supersymmetric models (SUSY), EDM, 10–12
Swift-Hohenberg equations, 268, 270

Tau-Charm Factory, 44
Tau lepton electric dipole moments. *See* Electric dipole moments, lepton
Temperatures
  flux, 206, 210
  laser cooling and minimum, 215–217
Theory of relativity, 114–115
Thermodynamics of fluorescence cooling
  of electroluminescence, 167–170
  limits of, 204–214
  photoluminescence, 170–175
  viability of anti-stokes, 165–167
Thomas-Fermi equation, 19
Tietz potential, modified, 18–19
Time evolution equation, 249
T–PT (tensor–pseudotensor), 21
Transition isotope shifts (TIS), 84, 86–88
Transverse Doppler-effect analysis, 142–144
Transverse nonlinear optics. *See* Optical pattern formation (OPF)

Transverse velocity component cooling, 138
TSR ring, 115, 125–127, 129–131
  bunched beam cooling, 140, 141
  laser cooling experiments using $Li+$ ions at, 133–136
  test of special relativity at, 144–146
Turing instabilities, 241, 243
Two-step laser-induced recombination
  Doppler-free, 131
  experiments with ESR and TSR, 129–131
  motivation, 128–129

Urbach edge, 215–216

V–A (vector–pseudovector), 21
van der Waals molecules, use of, 36
Vavilov's Law, 163
Vertical-cavity surface-emitting lasers (VCSELs), 269, 276
Volume shift, 84
Vortices, 252–259

Weak dipole moment (WDM), EDM of tau lepton and, 39–44

Ytterbium-doped ZBLANP glass, laser cooling of, 174
  absorption and emission properties, 189–191
  bulk cooling experiments, 198–204
  Kramers' doublets, 188
  level structure of, 188, 189
  photothermal deflection spectroscopy, 191–197

ZBLANP glass. *See* Ytterbium-doped ZBLANP glass, laser cooling of

Contents of Volumes in This Serial

Volume 1

Molecular Orbital Theory of the Spin Properties of Conjugated Molecules, *G. G. Hall and A. T. Amos*

Electron Affinities of Atoms and Molecules, *B. L. Moiseiwitsch*

Atomic Rearrangement Collisions, *B. H. Bransden*

The Production of Rotational and Vibrational Transitions in Encounters between Molecules, *K. Takayanagi*

The Study of Intermolecular Potentials with Molecular Beams at Thermal Energies, *H. Pauly and J. P. Toennies*

High-Intensity and High-Energy Molecular Beams, *J. B. Anderson, R. P. Andres, and J. B. Fen*

Volume 2

The Calculation of van der Waals Interactions, *A. Dalgarno and W. D. Davison*

Thermal Diffusion in Gases, *E. A. Mason, R. J. Munn, and Francis J. Smith*

Spectroscopy in the Vacuum Ultraviolet, *W. R. S. Garton*

The Measurement of the Photoionization Cross Sections of the Atomic Gases, *James A. R. Samson*

The Theory of Electron-Atom Collisions, *R. Peterkop and V. Veldre*

Experimental Studies of Excitation in Collisions between Atomic and Ionic Systems, *F. J. de Heer*

Mass Spectrometry of Free Radicals, *S. N. Foner*

Volume 3

The Quantal Calculation of Photoionization Cross Sections, *A. L. Stewart*

Radiofrequency Spectroscopy of Stored Ions I: Storage, *H. G. Dehmelt*

Optical Pumping Methods in Atomic Spectroscopy, *B. Budick*

Energy Transfer in Organic Molecular Crystals: A Survey of Experiments, *H. C. Wolf*

Atomic and Molecular Scattering from Solid Surfaces, *Robert E. Stickney*

Quantum Mechanics in Gas Crystal-Surface van der Waals Scattering, *E. Chanoch Beder*

Reactive Collisions between Gas and Surface Atoms, *Henry Wise and Bernard J. Wood*

Volume 4

H. S. W. Massey—A Sixtieth Birthday Tribute, *E. H. S. Burhop*

Electronic Eigenenergies of the Hydrogen Molecular Ion, *D. R. Bates and R. H. G. Reid*

Applications of Quantum Theory to the Viscosity of Dilute Gases, *R. A. Buckingham and E. Gal*

Positrons and Positronium in Gases, *P. A. Fraser*

Classical Theory of Atomic Scattering, *A. Burgess and I. C. Percival*

Born Expansions, *A. R. Holt and B. L. Moiselwitsch*

Resonances in Electron Scattering by Atoms and Molecules, *P. G. Burke*

Relativistic Inner Shell Ionizations, *C. B. O. Mohr*

Recent Measurements on Charge Transfer, *J. B. Hasted*

Measurements of Electron Excitation Functions, *D. W. O. Heddle and R. G. W. Keesing*

Some New Experimental Methods in Collision Physics, *R. F. Stebbings*

Atomic Collision Processes in Gaseous Nebulae, *M. J. Seaton*

Collisions in the Ionosphere, A. Dalgarno

The Direct Study of Ionization in Space, *R. L. F. Boyd*

Volume 5

Flowing Afterglow Measurements of Ion-Neutral Reactions, *E. E. Ferguson, F. C. Fehsenfeld, and A. L. Schmeltekopf*

Experiments with Merging Beams, *Roy H. Neynaber*

Radiofrequency Spectroscopy of Stored Ions II: Spectroscopy, *H. G. Dehmelt*

The Spectra of Molecular Solids, *O. Schnepp*

The Meaning of Collision Broadening of Spectral Lines: The Classical Oscillator Analog, *A. Ben-Reuven*

The Calculation of Atomic Transition Probabilities, *R. J. S. Crossley*

Tables of One- and Two-Particle Coefficients of Fractional Parentage for Configurations $s_\sigma s'_u p_q$, *C. D. H. Chisholm, A. Dalgarno, and E R. Innes*

Relativistic Z-Dependent Corrections to Atomic Energy Levels, *Holly Thomis Doyle*

Volume 6

Dissociative Recombination, *J. N. Bardsley and M. A. Biondi*

Analysis of the Velocity Field in Plasmas from the Doppler Broadening of Spectral Emission Lines, *A. S. Kaufman*

The Rotational Excitation of Molecules by Slow Electrons, *Kazuo Takayanagi and Yukikazu Itikawa*

The Diffusion of Atoms and Molecules, *E. A. Mason and T. R. Marrero*

Theory and Application of Sturmian Functions, *Manuel Rotenberg*

Use of Classical mechanics in the Treatment of Collisions between Massive Systems, *D. R. Bates and A. E. Kingston*

Volume 7

Physics of the Hydrogen Master, *C. Audoin, J. P. Schermann, and P Grivet*

Molecular Wave Functions: Calculations and Use in Atomic and Molecular Processes, *J. C. Browne*

Localized Molecular Orbitals, *Harel Weinstein, Ruben Pauncz, and Maurice Cohen*

General Theory of Spin-Coupled Wave Functions for Atoms and Molecules, *J. Gerratt*

Diabatic States of Molecules—Quasi-Stationary Electronic States, *Thomas F. O'Malley*

Selection Rules within Atomic Shells, *B. R. Judd*

Green's Function Technique in Atomic and Molecular Physics, *Gy. Csanak, H. S. Taylor, and Robert Yaris*

A Review of Pseudo-Potentials with Emphasis on Their Application to Liquid Metals, *Nathan Wiser and A. J. Greenfield*

Volume 8

Interstellar Molecules: Their Formation and Destruction, *D. McNally*

Monte Carlo Trajectory Calculations of Atomic and Molecular Excitation in Thermal Systems, *James C. Keck*

Nonrelativistic Off-Shell Two-Body Coulomb Amplitudes, *Joseph C. Y. Chen and Augustine C. Chen*

Photoionization with Molecular Beams, *R. B. Cairns, Halstead Harrison, and R. I. Schoen*

The Auger Effect, *E. H. S. Burhop and W. N. Asaad*

Volume 9

Correlation in Excited States of Atoms, *A. W. Weiss*

The Calculation of Electron–Atom Excitation Cross Sections, *M. R. H. Rudge*

Collision-Induced Transitions between Rotational Levels, *Takeshi Oka*

The Differential Cross Section of Low-Energy Electron–Atom Collisions, *D. Andrick*

Molecular Beam Electric Resonance Spectroscopy, *Jens C. Zorn and Thomas C. English*

Atomic and Molecular Processes in the Martian Atmosphere, *Michael B. McElroy*

Volume 10

Relativistic Effects in the Many-Electron Atom, *Lloyd Annstrong, Jr. and Serge Feneuille*

The First Born Approximation, *K. L. Bell and A. E. Kingston*

Photoelectron Spectroscopy, *W. C. Price*

Dye Lasers in Atomic Spectroscopy, *W. Lange, J. Luther and A. Steudel*

Recent Progress in the Classification of the Spectra of Highly Ionized Atoms, *B. C. Fawcett*

A Review of Jovian Ionospheric Chemistry, *Wesley T. Huntress, Jr.*

Volume 11

The Theory of Collisions between Charged Particles and Highly Excited Atoms, *I. C. Percival and D. Richards*

Electron Impact Excitation of Positive Ions, *M. J. Seaton*

The R-Matrix Theory of Atomic Process, *P. G. Burke and W. D. Robb*

Role of Energy in Reactive Molecular Scattering: An Information-Theoretic Approach, *R. B. Bernstein and R. D. Levine*

Inner Shell Ionization by Incident Nuclei, *Johannes M. Hansteen*

Stark Broadening, *Hans R. Griem*

Chemiluminescence in Gases, *M. F. Golde and B. A. Thrush*

Volume 12

Nonadiabatic Transitions between Ionic and Covalent States, *R. K. Janev*

Recent Progress in the Theory of Atomic Isotope Shift, *J. Bauche and R.-J. Champeau*

Topics on Multiphoton Processes in Atoms, *P. Lambropoulos*

Optical Pumping of Molecules, *M. Broyer, G. Goudedard, J. C. Lehmann, and J. Vigué*

Highly Ionized Ions, *Ivan A. Sellin*

Time-of-Flight Scattering Spectroscopy, *Wilhelm Raith*

Ion Chemistry in the D Region, *George C. Reid*

Volume 13

Atomic and Molecular Polarizabilities—A Review of Recent Advances, *Thomas M. Miller and Benjamin Bederson*

Study of Collisions by Laser Spectroscopy, *Paul R. Berman*

Collision Experiments with Laser-Excited Atoms in Crossed Beams, *I. V. Hertel and W. Stoll*

Scattering Studies of Rotational and Vibrational Excitation of Molecules, *Manfred Faubel and J. Peter Toennies*

Low-Energy Electron Scattering by Complex Atoms: Theory and Calculations, *R. K. Nesbet*

Microwave Transitions of Interstellar Atoms and Molecules, *W. B. Somerville*

Volume 14

Resonances in Electron Atom and Molecule Scattering, *D. E. Golden*

The Accurate Calculation of Atomic Properties by Numerical Methods, *Brian C. Webster, Michael J. Jamieson, and Ronald E. Stewart*

(e, 2e) Collisions, *Erich Weigold and Ian E. McCarthy*

Forbidden Transitions in One- and TwoElectron Atoms, *Richard Marrus and Peter J. Mohr*

Semiclassical Effects in Heavy-Particle Collisions, *M. S. Child*

Atomic Physics Tests of the Basic Concepts in Quantum Mechanics, *Francis M. Pipkin*

Quasi-Molecular Interference Effects in Ion–Atom Collisions, *S. V. Bobashev*

Rydberg Atoms, *S. A. Edelstein and T. F. Gallagher*

UV and X-Ray Spectroscopy in Astrophysics, *A. K. Dupree*

Volume 15

Negative Ions, *H. S. W. Massey*
Atomic Physics from Atmospheric and Astrophysical Studies, *A. Dalgarno*
Collisions of Highly Excited Atoms. *R. F. Stebbings*
Theoretical Aspects of Positron Collisions in Gases, *J. W. Humberston*
Experimental Aspects of Positron Collisions in Gases, *T. C. Griffith*
Reactive Scattering: Recent Advances in Theory and Experiment, *Richard B. Bernstein*
Ion–Atom Charge Transfer Collisions at Low Energies, *J. B. Hasted*
Aspects of Recombination, *D. R. Bates*
The Theory of Fast Heavy Particle Collisions, *B. H. Bransden*
Atomic Collision Processes in Controlled Thermonuclear Fusion Research, *H. B. Gilbody*
Inner-Shell Ionization, *E. H. S. Burhop*
Excitation of Atoms by Electron Impact, *D. W. O. Heddle*
Coherence and Correlation in Atomic Collisions, *H. Kleinpoppen*
Theory of Low Energy Electron-Molecule Collisions, *P. G. Burke*

Volume 16

Atomic Hartree–Fock Theory, *M. Cohen and R. P. McEachran*
Experiments and Model Calculations to Determine Interatomic Potentials, *R. Düren*
Sources of Polarized Electrons, *R. J. Celotta and D. T. Pierce*
Theory of Atomic Processes in Strong Resonant Electromagnetic Fields, *S. Swain*
Spectroscopy of Laser-Produced Plasmas, *M. H. Key and R. J. Hutcheon*

Relativistic Effects in Atomic Collisions Theory, *B. L. Moiseiwitsch*
Parity Nonconservation in Atoms: Status of Theory and Experment, *E. N. Fortson and L. Wilets*

Volume 17

Collective Effects in Photoionization of Atoms, *M. Ya. Amusia*
Nonadiabatic Charge Transfer, *D. S. F. Crothers*
Atomic Rydberg States, *Serge Feneuille and Pierre Jacquinot*
Superfluorescence, *M. F. H. Schuurmans, Q. H. F. Vrehen, D. Polder, and H. M. Gibbs*
Applications of Resonance Ionization Spectroscopy in Atomic and Molecular Physics, *M. G. Payne, C. H. Chen, G. S. Hurst, and G. W. Foltz*
Inner-Shell Vacancy Production in Ion–Atom Collisions, *C. D. Lin and Patrick Richard*
Atomic Processes in the Sun, *P. L. Dufton and A. E. Kingston*

Volume 18

Theory of Electron–Atom Scattering in a Radiation Field, *Leonard Rosenberg*
Positron–Gas Scattering Experiments, *Talbert S. Stein and Walter E. Kauppila*
Nonresonant Multiphoton Ionization of Atoms, *J. Morellec, D. Normand, and G. Petite*
Classical and Semiclassical Methods in Inelastic Heavy-Particle Collisions, *A. S. Dickinson and D. Richards*
Recent Computational Developments in the Use of Complex Scaling in Resonance Phenomena, *B. R. Junker*
Direct Excitation in Atomic Collisions: Studies of Quasi-One-Electron Systems, *N. Anderson and S. E. Nielsen*
Model Potentials in Atomic Structure, *A. Hibbert*
Recent Developments in the Theory of Electron Scattering by Highly Polar Molecules, *D. W. Norcross and L. A. Collins*

Quantum Electrodynamic Effects in Few-Electron Atomic Systems, *G. W. F. Drake*

Volume 19

Electron Capture in Collisions of Hydrogen Atoms with Fully Stripped Ions, *B. H. Bransden and R. K. Janev*

Interactions of Simple Ion–Atom Systems, *J. T. Park*

High-Resolution Spectroscopy of Stored Ions, *D. J. Wineland, Wayne M. Itano, and R. S. Van Dyck, Jr.*

Spin-Dependent Phenomena in Inelastic Electron–Atom Collisions, *K. Blum and H. Kleinpoppen*

The Reduced Potential Curve Method for Diatonic Molecules and Its Applications, *F. Jenč*

The Vibrational Excitation of Molecules by Electron Impact, *D. G. Thompson*

Vibrational and Rotational Excitation in Molecular Collisions, *Manfred Faubel*

Spin Polarization of Atomic and Molecular Photoelectrons, *N. A. Cherepkov*

Volume 20

Ion–Ion Recombination in an Ambient Gas, *D. R. Bates*

Atomic Charges within Molecules, *G. G. Hall*

Experimental Studies on Cluster Ions, *T. D. Mark and A. W. Castleman, Jr.*

Nuclear Reaction Effects on Atomic Inner-Shell Ionization, *W. E. Meyerhof and J.-F. Chemin*

Numerical Calculations on Electron-Impact Ionization, *Christopher Bottcher*

Electron and Ion Mobilities, *Gordon R. Freeman and David A. Armstrong*

On the Problem of Extreme UV and X-Ray Lasers, *I. L Sobel'man and A. V. Vinogradov*

Radiative Properties of Rydberg State, in Resonant Cavities, *S. Haroche and J. M. Ralmond*

Rydberg Atoms: High-Resolution Spectroscopy and Radiation Interaction—Rydberg Molecules, *J. A. C. Gallas, G. Leuchs, H. Walther and H. Figger*

Volume 21

Subnatural Linewidths in Atomic Spectroscopy, *Dennis P. O'Brien, Pierre Meystre, and Herbert Walther*

Molecular Applications of Quantum Defect Theory, *Chris H. Greene and Ch. Jungen*

Theory of Dielectronic Recombination, *Yukap Hahn*

Recent Developments in Semiclassical Floquet Theories for Intense-Field Multiphoton Processes, *Shih-I Chu*

Scattering in Strong Magnetic Fields, *M. R. C. McDowell and M. Zarcone*

Pressure Ionization, Resonances, and the Continuity of Bound and Free States, *R. M. More*

Volume 22

Positronium—Its Formation and Interaction with Simple Systems, *J. W. Humberston*

Experimental Aspects of Positron and Positronium Physics, *T. C. Griffith*

Doubly Excited States, Including New Classification Schemes, *C. D. Lin*

Measurements of Charge Transfer and Ionization in Collisions Involving Hydrogen Atoms, *H. B. Gilbody*

Electron–Ion and Ion–Ion Collisions with Intersecting Beams, *K. Dolder and B. Pearl*

Electron Capture by Simple Ions, *Edward Pollack and Yukap Hahn*

Relativistic Heavy-Ion–Atom Collisions, *R. Anholt and Harvey Gould*

Continued-Fraction Methods in Atomic Physics, *S. Swain*

Volume 23

Vacuum Ultraviolet Laser Spectroscopy of Small Molecules, *C. R. Vidal*

Foundations of the Relativistic Theory of Atomic and Molecular Structure, *Ian P. Grant and Harry M. Quiney*

Point-Charge Models for Molecules Derived from Least-Squares Fitting of the Electric Potential, *D. E. Williams and Ji-Min Yan*

Transition Arrays in the Spectra of Ionized Atoms, *J. Bauche, C. Bauche-Arnoult, and M. Klapisch*

Photoionization and Collisional Ionization of Excited Atoms Using Synchroton and Laser Radiation, *E. J. Wuilleumier, D. L. Ederer, and J. L. Picqué*

Volume 24

The Selected Ion Flow Tube (SIDT): Studies of Ion–Neutral Reactions, *D. Smith and N. G. Adams*

Near-Threshold Electron–Molecule Scattering, *Michael A. Morrison*

Angular Correlation in Multiphoton Ionization of Atoms, *S. J. Smith and G. Leuchs*

Optical Pumping and Spin Exchange in Gas Cells, *R. J. Knize, Z Wu, and W. Happer*

Correlations in Electron–Atom Scattering, *A. Crowe*

Volume 25

Alexander Dalgarno: Life and Personality, *David R. Bates and George A. Victor*

Alexander Dalgarno: Contributions to Atomic and Molecular Physics, *Neal Lane*

Alexander Dalgarno: Contributions to Aeronomy, *Michael B. McElroy*

Alexander Dalgarno: Contributions to Astrophysics, *David A. Williams*

Dipole Polarizability Measurements, *Thomas M. Miller and Benjamin Bederson*

Flow Tube Studies of Ion–Molecule Reactions, *Eldon Ferguson*

Differential Scattering in He–He and He$^+$–He Collisions at KeV Energies, *R. F. Stebbings*

Atomic Excitation in Dense Plasmas, *Jon C. Weisheit*

Pressure Broadening and Laser-Induced Spectral Line Shapes, *Kenneth M. Sando and Shih-I Chu*

Model-Potential Methods, *G. Laughlin and G. A. Victor*

Z-Expansion Methods, *M. Cohen*

Schwinger Variational Methods, *Deborah Kay Watson*

Fine-Structure Transitions in Proton-Ion Collisions, *R. H. G. Reid*

Electron Impact Excitation, *R. J. W. Henry and A. E. Kingston*

Recent Advances in the Numerical Calculation of Ionization Amplitudes, *Christopher Bottcher*

The Numerical Solution of the Equations of Molecular Scattering, *A. C. Allison*

High Energy Charge Transfer, *B. H. Bransden and D. P. Dewangan*

Relativistic Random-Phase Approximation, *W. R. Johnson*

Relativistic Sturmian and Finite Basis Set Methods in Atomic Physics, *G. W. F. Drake and S. P. Goldman*

Dissociation Dynamics of Polyatomic Molecules, *T. Uzer*

Photodissociation Processes in Diatoniic Molecules of Astrophysical Interest, *Kate P. Kirby and Ewine F. van Dishoeck*

The Abundances and Excitation of Interstellar Molecules, *John H. Black*

Volume 26

Comparisons of Positrons and Electron Scattering by Gases, *Walter E. Kauppila and Talbert S. Stein*

Electron Capture at Relativistic Energies, *B. L. Moiseiwitsch*

The Low-Energy, Heavy Particle Collisions—A Close-Coupling Treatment, *Mineo Kimura and Neal F. Lane*

Vibronic Phenomena in Collisions of Atomic and Molecular Species, *V. Sidis*

Associative Ionization: Experiments, Potentials, and Dynamics, *John Weiner, Françoise Masnou-Sweeuws, and Annick Giusti-Suzor*

On the β Decay of ^{187}Re: An Interface of Atomic and Nuclear Physics and Cosmochronology, *Zonghau Chen, Leonard Rosenberg, and Larry Spruch*

Progress in Low Pressure Mercury-Rare Gas

Discharge Research, *J. Maya and R. Lagushenko*

Volume 27

Negative Ions: Structure and Spectra, *David R. Bates*

Electron Polarization Phenomena in Electron–Atom Collisions, *Joachim Kessler*

Electron–Atom Scattering, *I. E. McCarthy and E. Weigold*

Electron–Atom Ionization, *I. E. McCarthy and E. Weigold*

Role of Autoionizing States in Multiphoton Ionization of Complex Atoms, *V. I. Lengyel and M. I. Haysak*

Multiphoton Ionization of Atomic Hydrogen Using Perturbation Theory, *E. Karule*

Volume 28

The Theory of Fast Ion-Atom Collisions, *J. S. Briggs and J. H. Macek*

Some Recent Developments in the Fundamental Theory of Light, *Peter W. Milonni and Surendra Singh*

Squeezed States of the Radiation Field, *Khalid Zaheer and M. Suhail Zubairy*

Cavity Quantun, Electrodynamics, *E. A. Hinds*

Volume 29

Studies of Electron Excitation of Rare-Gas Atoms into and out of Metastable Levels Using Optical and Laser Techniques, *Chun C. Lin and L. W. Anderson*

Cross Sections for Direct Multiphoton Ionionization of Atoms, *M. V. Ammosov, N. B. Delone, M. Yu. Ivanov, I. I. Bondar, and A. V. Masalov*

Collision-Induced Coherences in Optical Physics, *G. S. Agarwal*

Muon-Catalyzed Fusion, *Johann Rafelski and Helga E. Rafelski*

Cooperative Effects in Atomic Physics, *J. P. Connerade*

Multiple Electron Excitation, Ionization, and Transfer in High-Velocity Atomic and Molecular Collisions, *J. H. McGuire*

Volume 30

Differential Cross Sections for Excitation of Helium Atoms and Helium-Like Ions by Electron Impact, *Shinobu Nakazaki*

Cross-Section Measurements for Electron Impact on Excited Atomic Species, *S. Trajmar and J. C. Nickel*

The Dissociative Ionization of Simple, Molecules by Fast Ions, *Colin J. Latimer*

Theory of Collisions between Laser Cooled Atoms, *P. S. Julienne, A. M. Smith, and K. Burnett*

Light-Induced Drift, *E. R. Eliel*

Continuum Distorted Wave Methods in Ion–Atom Collisions, *Derrick S. F. Crothers and Louis J. Dubé*

Volume 31

Energies and Asymptotic Analysis for Helium Rydberg States, *G. W. F. Drake*

Spectroscopy of Trapped Ions, *R. C. Thompson*

Phase Transitions of Stored Laser-Cooled Ions, *H. Walther*

Selection of Electronic States in Atomic Beams with Lasers, *Jacques Baudon, Rudolf Düren, and Jacques Robert*

Atomic Physics and Non-Maxwellian Plasmas, *Michèle Lamoureux*

Volume 32

Photoionization of Atomic Oxygen and Atomic Nitrogen, *K. L. Bell and A. E. Kingston*

Positronium Formation by Positron Impact on Atoms at Intermediate Energies, *B. H. Bransden and C. J. Noble*

Electron–Atom Scattering Theory and Calculations, *P. G. Burke*

Terrestrial and Extraterrestrial H_3^+, *Alexander Dalgarno*

Indirect Ionization of Positive Atomic Ions, *K. Dolder*

Quantum Defect Theory and Analysis of High-Precision Helium Term Energies, *G. W. F. Drake*

Electron–Ion and Ion–Ion Recombination Processes, *M. R. Flannery*

Studies of State-Selective Electron Capture in Atomic Hydrogen by Translational Energy Spectroscopy, *H. B. Gilbody*

Relativistic Electronic Structure of Atoms and Molecules, *I. P. Grant*

The Chemistry of Stellar Environments, *D. A. Howe, J. M. C. Rawlings, and D. A. Williams*

Positron and Positronium Scattering at Low Energies, *J. W. Humberston*

How Perfect are Complete Atomic Collision Experiments?, *H. Kleinpoppen and H. Handy*

Adiabatic Expansions and Nonadiabatic Effects, *R. McCarroll and D. S. F. Crothers*

Electron Capture to the Continuum, *B. L. Moiseiwitsch*

How Opaque Is a Star? *M. J. Seaton*

Studies of Electron Attachment at Thermal Energies Using the Flowing Afterglow–Langmuir Technique, *David Smith and Patrik Španěl*

Exact and Approximate Rate Equations in Atom-Field Interactions, *S. Swain*

Atoms in Cavities and Traps, *H. Walther*

Some Recent Advances in Electron-Impact Excitation of $n = 3$ States of Atomic Hydrogen and Helium, *J. F. Williams and J. B. Wang*

Volume 33

Principles and Methods for Measurement of Electron Impact Excitation Cross Sections for Atoms and Molecules by Optical Techniques, *A. R. Filippelli, Chun C. Lin, L. W. Andersen, and J. W. McConkey*

Benchmark Measurements of Cross Sections for Electron Collisions: Analysis of Scattered Electrons, *S. Trajmar and J. W. McConkey*

Benchmark Measurements of Cross Sections for Electron Collisions: Electron Swarm Methods, *R. W Crompton*

Some Benchmark Measurements of Cross Sections for Collisions of Simple Heavy Particles, *H. B. Gilbody*

The Role of Theory in the Evaluation and Interpretation of Cross-Section Data, *Barry I. Schneider*

Analytic Representation of Cross-Section Data, *Mitio Inokuti, Mineo Kimura, M. A. Dillon, Isao Shimamura*

Electron Collisions with N_2, O_2 and O: What We Do and Do Not Know, *Yukikazu Itikawa*

Need for Cross Sections in Fusion Plasma Research, *Hugh P. Summers*

Need for Cross Sections in Plasma Chemistry, *M. Capitelli, R. Celiberto, and M. Cacciatore*

Guide for Users of Data Resources, *Jean W. Gallagher*

Guide to Bibliographies, Books, Reviews, and Compendia of Data on Atomic Collisions, *E. W. McDaniel and E. J. Mansky*

Volume 34

Atom Interferometry, *C. S. Adams, O. Carnal, and J. Mlynek*

Optical Tests of Quantum Mechanics, *R. Y Chiao, P. G. Kwiat, and A. M. Steinberg*

Classical and Quantum Chaos in Atomic Systems, *Dominique Delande and Andreas Buchleitner*

Measurements of Collisions between Laser-Cooled Atoms, *Thad Walker and Paul Feng*

The Measurement and Analysis of Electric Fields in Glow Discharge Plasmas, *J. E. Lawler and D. A. Doughty*

Polarization and Orientation Phenomena in Photoionization of Molecules, *N. A. Cherepkov*

Role of Two-Center Electron–Electron Interaction in Projectile Electron Excitation and Loss, *E. C. Montenegro, W. E. Meyerhof and J. H. McGuire*

Indirect Processes in Electron Impact Ionization of Positive Ions, *D. L. Moores and K. J. Reed*

Dissociative Recombination: Crossing and Tunneling Modes, *David R. Bates*

Volume 35

Laser Manipulation of Atoms, *K. Sengstock and W. Ertmer*

Advances in Ultracold Collisions: Experiment and Theory, *J. Weiner*

Ionization Dynamics in Strong Laser Fields, *L. F. DiMauro and P. Agostini*

Infrared Spectroscopy of Size Selected Molecular Clusters, *U. Buck*

Femtosecond Spectroscopy of Molecules and Clusters, *T. Baumer and G. Gerber*

Calculation of Electron Scattering on Hydrogenic Targets, *I. Bray and A. T. Stelbovics*

Relativistic Calculations of Transition Amplitudes in the Helium Isoelectronic Sequence, *W. R. Johnson, D. R. Plante, and J. Sapirstein*

Rotational Energy Transfer in Small Polyatomic Molecules, *H. O. Everitt and F. C. De Lucia*

Volume 36

Complete Experiments in Electron–Atom Collisions, *Nils Overgaard Andersen and Klaus Bartschat*

Stimulated Rayleigh Resonances and Recoil-Induced Effects, *J.-Y. Courtois and G. Grynberg*

Precision Laser Spectroscopy Using Acousto-Optic Modulators, *W. A. van Wijngaarden*

Highly Parallel Computational Techniques for Electron–Molecule Collisions, *Carl Winstead and Vincent McKoy*

Quantum Field Theory of Atoms and Photons, *Maciej Lewenstein and Li You*

Volume 37

Evanescent Light-Wave Atom Mirrors, Resonators, Waveguides, and Traps, *Jonathan P. Dowling and Julio Gea-Banacloche*

Optical Lattices, *P. S. Jessen and I. H. Deutsch*

Channeling Heavy Ions through Crystalline Lattices, *Herbert F. Krause and Sheldon Datz*

Evaporative Cooling of Trapped Atoms, *Wolfgang Ketterle and N. J. van Druten*

Nonclassical States of Motion in Ion Traps, *J. I. Cirac, A. S. Parkins, R. Blatt, and P. Zoller*

The Physics of Highly-Charged Heavy Ions Revealed by Storage/Cooler Rings, *P. H. Mokler and Th. Stöhlker*

Volume 38

Electronic Wavepackets, *Robert R. Jones and L. D. Noordam*

Chiral Effects in Electron Scattering by Molecules, *K. Blum and D. G. Thompson*

Optical and Magneto-Optical Spectroscopy of Point Defects in Condensed Helium, *Serguei I. Kanorsky and Antoine Weis*

Rydberg Ionization: From Field to Photon, *G. M. Lankhuijzen and L. D. Noordam*

Studies of Negative Ions in Storage Rings, *L. H. Andersen, T. Andersen, and P. Hvelplund*

Single-Molecule Spectroscopy and Quantum Optics in Solids, *W. E. Moerner, R. M. Dickson, and D. J. Norris*

Volume 39

Author and Subject Cumulative Index Volumes 1–38
Author Index
Subject Index
Appendix: Tables of Contents of Volumes 1–38 and Supplements

Volume 40

Electric Dipole Moments of Leptons, *Eugene D. Commins*

High-Precision Calculations for the Ground and Excited States of the Lithium Atom, *Frederick W. King*

Storage Ring Laser Spectroscopy, *Thomas U. Kühl*

Laser Cooling of Solids, *Carl E. Mungan and Timothy R. Gosnell*

Optical Pattern Formation, *L. A. Lugiato, M. Brambilla, and A. Gatti*

ISBN 0-12-003840-4